High-Performance Computing and Artificial Intelligence in Process Engineering

Online at: https://doi.org/10.1088/978-0-7503-6174-3

High-Performance Computing and Artificial Intelligence in Process Engineering

Edited by
Mingheng Li
Department of Chemical and Materials Engineering, California State Polytechnic University, Pomona, CA, USA

Yi Heng
School of Computer Science and Engineering, Sun Yat-sen University, Guangzhou, China
and
School of Systems Science and Engineering, Sun Yat-sen University, Guangzhou, China

IOP Publishing, Bristol, UK

ISBN 978-0-7503-6174-3 (ebook)
ISBN 978-0-7503-6172-9 (print)
ISBN 978-0-7503-6175-0 (myPrint)
ISBN 978-0-7503-6173-6 (mobi)

DOI 10.1088/978-0-7503-6174-3

Version: 20250401

IOP ebooks

British Library Cataloguing-in-Publication Data: A catalogue record for this book is available from the British Library.

Published by IOP Publishing, wholly owned by The Institute of Physics, London

IOP Publishing, No.2 The Distillery, Glassfields, Avon Street, Bristol, BS2 0GR, UK

US Office: IOP Publishing, Inc., 190 North Independence Mall West, Suite 601, Philadelphia, PA 19106, USA

Contents

Preface

The intersection of artificial intelligence (AI) and high-performance computing (HPC) marks a pivotal moment in the evolution of process engineering. As industrial sectors adapt to the demands of the twenty-first century, there is a growing recognition that traditional methods of process design, control, and optimization are reaching their limits. The increasing complexity of modern industrial systems, coupled with the deluge of data from sensors, networks, and advanced automation, calls for more powerful and intelligent tools to drive efficiency, sustainability, and innovation. In this context, AI and HPC emerge as transformative forces, capable of handling the intricate and large-scale challenges that define contemporary process engineering. This book, *High-Performance Computing and Artificial Intelligence in Process Engineering*, seeks to provide an in-depth exploration of how these two advanced fields can be applied to enhance and revolutionize the practice of process engineering. AI, with its capacity for machine learning, neural networks, deep learning, and expert systems, has the potential to unlock new levels of performance by predicting process behaviors, optimizing operations in real-time, and discovering novel solutions that human operators may overlook. On the other hand, HPC, with its ability to process vast amounts of data and run complex simulations at unprecedented speeds, empowers engineers to model sophisticated systems with a level of detail and precision that was once thought impossible. At its core, process engineering is concerned with the design, operation, control, and optimization of physical and chemical processes across industries such as chemicals, energy, water, atmosphere, and manufacturing. In these fields, there is an ever-present need to balance cost, efficiency, safety, and environmental impact while responding to market demands and regulatory constraints. By integrating AI and HPC into the process engineering toolkit, we can rethink how these challenges are addressed. The ability to model more accurate representations of processes, optimize decisions in real-time, and simulate potential outcomes before implementation gives engineers unprecedented control and insight into their systems.

This book aims to cater to both the academic and professional communities involved in process engineering, AI, and HPC. Researchers will find detailed discussions on the theoretical foundations and emerging trends in AI and HPC, while practitioners can explore practical applications and real-world case studies that illustrate the effectiveness of these technologies in industrial environments. The interdisciplinary nature of this book reflects the convergence of fields necessary to push the boundaries of what is possible in modern process engineering. We begin by introducing the foundational concepts of AI, including machine learning algorithms, neural networks, and other data-driven techniques that can be leveraged for process prediction, optimization, and control. We then move into the realm of supercomputing, exploring how high-performance computing platforms can be utilized to solve large-scale problems, run simulations that would be infeasible on conventional systems, and analyse massive datasets in ways that provide actionable insights.

Specific applications spanning various industries and fields are used to showcase the ability of AI/HPC to enhance predictive accuracy and optimize system performance.

Throughout the book, we emphasize the importance of a collaborative approach that integrates expertise from AI, process engineering, data science, and computational physics. The role of interdisciplinary teams is crucial in the successful implementation of these advanced technologies, as the challenges faced by modern industry require a blend of domain knowledge, technical expertise, and innovative thinking. Moreover, we address the challenges that come with integrating AI and supercomputing into traditional engineering workflows. From the need for skilled personnel to the computational infrastructure required for deploying AI models and HPC simulations, the road to fully realizing the potential of these technologies is not without its obstacles. Nevertheless, the benefits far outweigh the challenges. As AI and HPC continue to evolve, their impact on process engineering will only grow, leading to safer, more efficient, and more sustainable industrial systems.

In conclusion, this book provides a comprehensive roadmap for engineers and scientists who wish to embrace AI and supercomputing in their practice. It is not merely about adopting new tools, but about rethinking the approach to process engineering itself, using data-driven insights and computational power to create systems that are smarter, faster, and more adaptive. We hope that this book will serve as a valuable resource for those who seek to be at the forefront of innovation in the field, and we look forward to seeing how the ideas presented here will inspire new research, applications, and solutions for the future of process engineering. We invite readers to explore the transformative potential of these technologies and to join us in shaping a new era in process engineering, where the synergy between AI, supercomputing, and human ingenuity opens up unprecedented possibilities.

Acknowledgements

Bringing this edited book to fruition has been an immensely rewarding journey, made possible by the collective efforts and unwavering support of many individuals. We are deeply grateful to the contributing authors, whose expertise, dedication, and insightful perspectives have greatly enriched this book. Their commitment to rigorous scholarship and thoughtful engagement with the subject matter has been indispensable.

Our sincere appreciation also goes to IOP Publishing for their invaluable support and guidance throughout the editorial process, ensuring the seamless development of this book. We are especially grateful to Mia Foulkes for her exceptional assistance and coordination in bringing this publication to life.

We also acknowledge the support of the National Science Foundation (CBET-2140946 for ML) and the Key-Area Research and Development Program of Guangdong Province, China (No. 2021B0101190003 for YH), whose funding made this work possible.

Editor biographies

Mingheng Li

Mingheng Li received his PhD in Chemical Engineering from UCLA. His research focuses on process systems engineering, particularly in materials, energy, and environmental applications. He has served as an editor for the American Institute of Physics Publishing.

Yi Heng

Yi Heng received his PhD in Natural Sciences from RWTH Aachen University, Germany. His research focuses on inverse problems, high performance computing, artificial intelligence, and their applications to various areas of science and engineering. Additionally, he has served as an excutive member of editorial board for *Science Bulletin*.

List of contributors

Song Bo
University of Alberta, Edmonton, AB, Canada

Junzhi Chen
Sun Yat-sen University, Guangzhou, China

Ke Chen
Sun Yat-sen University, Guangzhou, China

Benjamin Decardi-Nelson
University of Alberta, Edmonton, AB, Canada

Junxuan Deng
Sun Yat-sen University, Guangzhou, China

Jianghang Gu
Peking University, Beijing, China

Yi Heng
Sun Yat-sen University, Guangzhou, China

Min Hong
Sun Yat-sen University, Guangzhou, China

Mingming Huang
Southern Marine Science and Engineering Guangdong Laboratory (Zhuhai), Zhuhai, China

Joseph Kwon
Texas A&M University, College Station, TX, USA

Mingheng Li
California State Polytechnic University, Pomona, CA, USA

Jinfeng Liu
University of Alberta, Edmonton, AB, Canada

Yutong Lu
Sun Yat-sen University, Guangzhou, China

Jiu Luo
Soochow University, Suzhou, China

Dongchuan Mo
Sun Yat-sen University, Guangzhou, China

Nariman Piroozan
University of Southern California, USA

Parth Shah
Texas A&M University, College Station, TX, USA

Niranjan Sitapure
Texas A&M University, College Station, TX, USA

Chen Wang
Sun Yat-sen University, Guangzhou, China

Yujia Wang
National University of Singapore, Singapore

Zhe Wu
National University of Singapore, Singapore

Guohong Xie
Sun Yat-sen University, Guangzhou, China

Qingqing Yang
Sun Yat-sen University, Guangzhou, China

Jia Yi
Sun Yat-sen University, Guangzhou, China

Chapter 1

Artificial intelligence and the future of process engineering

Nariman Piroozan

1.1 Introduction

The science of efficiently transforming raw materials into commercial productions as it is known today was born in the nineteenth century [1]. The most well-known personification of process engineering is the oil refinery, however, the true origins of the field can be traced back even further to antiquity. From fermentation to distillation to extractions, each incremental technological advance over the centuries allowed our species to do everything from enjoying a glass of wine to smelting copper. Since the eighteenth century, as materials became more complex and their operating conditions more extreme, the scientific rigor on which process engineering had to be based ever increased. New divisions within this field included the emerging fields of thermodynamics, kinetics, and reactor design. The fusion of these fields would allow us to eventually be able to design materials for aircraft, spacecraft, and high performance integrated electronics. Through the twentieth century, as manufacturing methods and systems became more complex, process safety and control systems were added eventually as well. By the dawn of the twenty-first century, it seemed as if the field of process engineering had reached its zenith. Then, everything changed.

Ever since the development of the transistor in 1947, there has been an inexorable increase in the power of computing systems [2]. As the size of transistors has decreased, currently down to the nanometer scale, their density on silicon chips has increased exponentially in the past several decades. With that quantum leap in computing power, in conjunction with vastly increased capabilities in both memory and disk space, has also come the ability to run ever more complex and sophisticated applications. One of the most revolutionary and potentially the most impactful of these fields is that of artificial intelligence (AI). AI generally involves both the theory and development of computer systems with the capability to perform computations

doi:10.1088/978-0-7503-6174-3ch1

and duties typically reserved for humans. These include varied tasks such as visual perception, language translation, speech recognition, and decision making. The process by which this is implemented is through the concept of the neural network (NN) [1]. An NN is fundamentally a model that is inspired by the human brain.

NNs are algorithms based on the functionality of the human brain that are used to model complex patterns. The difference to the human brain, however, is that NNs work with data in various forms. Several different types of NNs exist, including convolution neural networks (CNNs), deep neural networks (DNNs), recursive neural networks (RNNs), among others. Some are used to accept unstructured, non-numeric data (RNNs and CNNs) and some are designed to accept structured, numeric data (DNNs) [3].

For brevity, let us discuss NNs in the most general way, regarding how NNs work. In figure 1.1, we have a schematic which details how the network architecture is built. There are three layers at work here: the input layer, hidden layer (multiple), and the output layer.

The input layer is the point at which raw data and datasets are received and is an initialization layer where information is subsequently passed to the hidden layer [2]. While the input and output layers are self-explanatory, the hidden layer is a bit more complex. The hidden layer can be though of as an extraction point for the most relevant patterns from the input and subsequently sends them to the following layer for further analysis. The redundant information is discarded and the information most relevant for the NN is preserved at each hidden layer. What is determined to be relevant and redundant is dictated by the researcher when designing the workflow [3]. One of the key ways in which this is done is through the concept of the activation function.

The activation function within an NN calculates the output of a layer based upon pre-defined inputs and their individual weights of importance within the network. Activation functions are critical in hidden layers to solve highly complex problems and transmit large amounts of data through the AI algorithm. Many different types

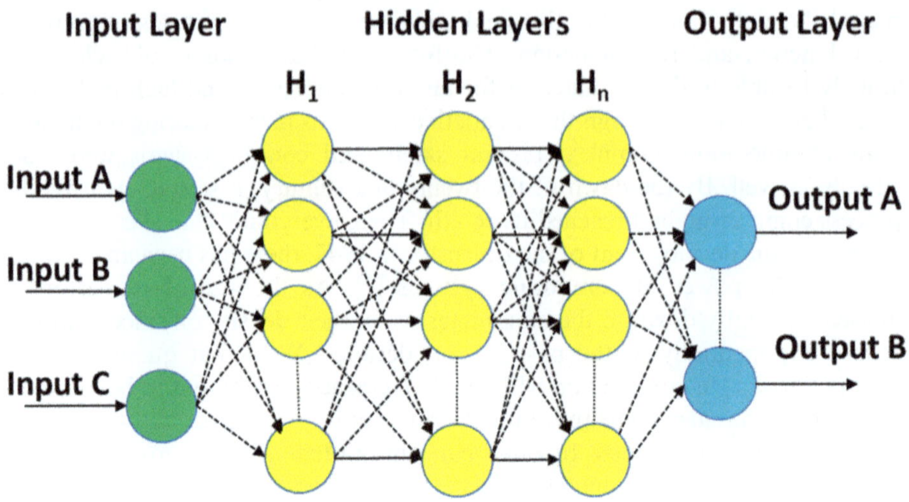

Figure 1.1. Schematic showing the design of an NN with input, hidden and output layers.

of activation functions exist, including binary, linear, and a wide array of nonlinear functions. Figure 1.2 details some of the types of activation functions that exist and what they look like. The goal of an NN is to allow for a model to behave as close to the human brain as possible, and activation functions are key at increasing the model's problem solving capabilities [4]. The data which are processed by all of the neurons in the multitude of hidden layers are then passed to the outer layer where the final series of calculations and predictions are performed. It is at this point where the gap between the current and desired output parameter is compared. If there is a discrepancy beyond the acceptable error, then the process will begin again. The flow through an NN from the input to the output layer is called the forward pass, while the flow from the output to input layer is called the backward pass. In the forward pass, the activation functions operate as a computational gateway between

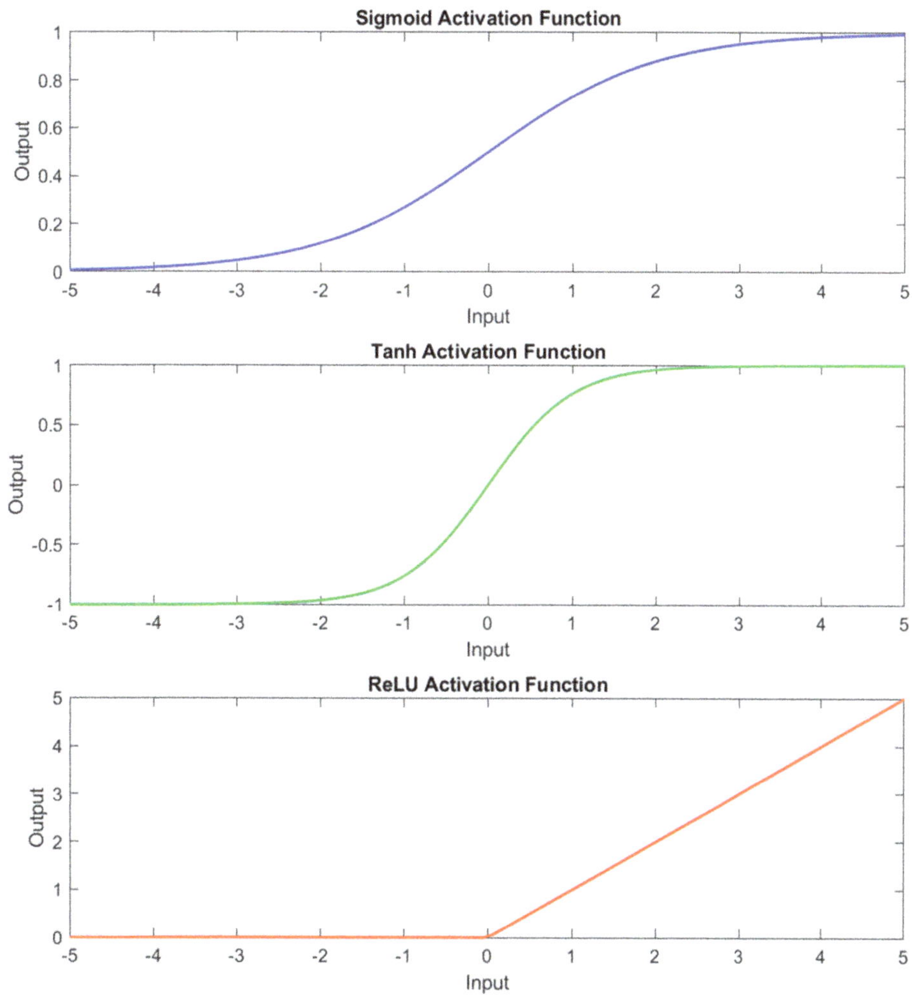

Figure 1.2. Three different activation functions represented as graphical functions and their respective equations. From left to right they are, sigmoid, tanh, and Relu.

the input data and the current node in the network [3]. The backward pass, on the other hand, adjusts the network's weights using gradients to minimize the number of passes needed to reach convergence between the current and desired output. Backward pass, subsequently, determines the level of adjustments required for the activation functions, weights and biases needed to reach convergence.

To accomplish this feat, it takes more than just machines with the computing power to handle the high amount of data involved in an NN. It also involves the software framework needed in order to support the algorithms designed for varying tasks such as climate modeling, AI-accelerated molecular dynamics, and process engineering. The most prominent of these frameworks today are TensorFlow and PyTorch, both written in Python with the latter now being used extensively for scientific applications. With respect to chemical process engineering, some of these applications would include fault diagnosis, predictive control, and process optimization.

The remainder of this chapter will describe the types of NNs in more detail and provide a discussion of the applications of NNs to chemical and process engineering [3]. This discussion will be broken down into a series of subsections relating to transport phenomena, kinetics and reactor design, thermodynamics, and process control. This review will detail the increasing use of NNs in this field and help to understand future areas for their application.

1.2 Types of neural networks

As we have discussed in the previous section, NNs, inspired by the human brain's structure and functioning, have become a cornerstone in modern AI. These networks consist of interconnected nodes which work together to process and learn from data. Over the years, researchers have developed various types of NNs. Each of these are designed to solve specific problems in any number of different domains [5]. In this section, we will explore some of the most common types of NNs and their applications.

Some of the most important types of NNs include feedforward neural networks (FNNs), convolution neural networks (CNNs), recurrent neural networks (RNNs), generative adversarial networks (GANs), and deep neural networks (DNNs).

FNNs consist of a layer of neurons wherein each layer is connected to every neuron in the next layer [6]. Information in this sort of network flows in one direction, from input to output. The associated activation functions for this sort of network include sigmoid or ReLU in order to introduce nonlinearity and enable complex mappings. Typically this class of NNs is used in applications such as image and speech recognition, classification, and regression.

CNNs are specialized for processing grid-like data such as images and consist of convolution layers, pooling layers, and fully connected layers. In this application, convolution layers apply filters to input data, capturing spatial hierarchies in the data. The pooling layers subsequently downsample the output of convolution layers, reducing the computation and controlling overfitting [7]. Applications of CNNs include image and video recognition, object detection, and image segmentation.

RNNs are designed to work with sequential data, where the order of inputs is important. They have connections that form a directed cycle, allowing information to persist. This class of NNs is typically designed for language processing, speech recognition, and time series prediction.

GANs consist of two networks, a generator and discriminator, that are trained simultaneously in a competitive setting [6]. The generator creates net data instances such as images while the discriminator evaluates the generated data. Applications of this class of NNs include data compression, anomaly detection, and generative modeling.

Finally, let us also discuss the DNN, which is arguably the most commonly used today from the standpoint of scientific applications as it pertains to AI. DNNs area class of artificial neural networks (ANNs) that are characterized by their depth, meaning they have multiple layers between the input and output layers. These networks have revolutionized the field of machine learning and are responsible for many breakthroughs in the fields of process engineering, molecular dynamics, etc. Several of these different types of NNs are of significant relevance to the application of AI in the domain of chemical and processes engineering [8]. Our specific focus as it pertains to the application of AI to chemical engineering will be focused on the four primary fields of that domain, namely:

- Transport phenomena.
- Kinetics and reactor design.
- Thermodynamics.
- Process control.

1.3 Applications in chemical and process engineering

In recent decades we have observed within the domain of chemical and process engineering a dramatic surge in the number of studies using ANNs [9]. These applications vary from molecular property prediction to process optimization and the diagnosis of faults and impurities in both fluid flow and material analysis. In order to be able to accurately and properly utilize NNs for process engineering, the use of first principles knowledge should be integrated within the network.

1.3.1 Transport phenomena

Fluid mechanics deals with the study of fluids and their properties, including their behavior under various conditions. NNs have been increasingly applied to fluid mechanics in order to enhance our understanding, modeling capabilities, and predictive accuracy in this field [10]. This section explores how NNs can be effectively applied to fluid mechanics and the benefits they would bring within the field.

NNs are used to model and predict fluid flow behavior in various scenarios, such as in pipes or around various objects within the process system. By training on datasets of flow patterns, NNs can learn complex relationships between various input variables (i.e. flow properties, boundary conditions) and output variables (i.e velocity, pressure), enabling accurate flow predictions. Another element of fluid mechanics wherein NNs can be of significant importance include the modeling of turbulent flow [9]. This regime of fluid flow is a complex phenomenon that is very challenging to model accurately using traditional methods. NNs offer a promising approach to turbulence modeling by capturing the nonlinear interactions and complex dynamics of turbulent flow through the use of large and detailed input datasets. This can lead to more accurate predictions of turbulent flow characteristics and can help to improve the design of engineering systems.

Further areas where NNs can be of importance include the design of compressors, pumps, turbines, and refrigeration systems. Through the analysis of system parameters and performance metrics, NNs can help to identify optimal operating conditions and lead to both improved efficiency and enhanced energy savings. ANNs are excellent for approximations of nonlinear functions and can be used to good effect in heat and mass transfer coefficients. Verma and Srivastava [11] successfully built an ANN model based on data available in the literature with several inputs related to the system configuration of a bubble column (Prandtl number, hole diameter, column diameter, surface diameter, among others) and one output (heat coefficient). Another example of a recent application of ANNs to the field of transport phenomena includes a case study regarding the determination of the reduced boiling point from molecular weight and acentric factor [12]. The class of NN utilized here is the feedforward ANN where the sigmoid activation function is applied.

The second field of transport phenomena, heat transfer, plays a critical role in various engineering applications. These include thermal management, energy systems, and manufacturing processes. One area where NNs can be applied to significant effect within heat transfer include the modeling of heat exchangers, electronics cooling, and building insulation. By training on experimental or simulated data, NNs can be used to learn what the relationships are between input variables such as the heat conduction coefficient or boundary conditions and the output variables such as temperature distribution and heat flux. This can be used to improve predictions for heat transfer behavior in future proposed engineering designs. Furthermore, NNs can also be used for the prediction of transient heat transfer behavior where temperatures change over time. This is useful particularly in applications where thermal management is required. This includes thermostats, heat exchangers, heat sinks and even thermal insulation. Despite these benefits, several challenges do remain within the application of NNs to heat transfer. These include the need for very large training datasets and the accuracy of NN models. Future research directions seek to correct these vulnerabilities through the development of hybrid models that combine NNs with physics-based models as well as their application to new areas within heat transfer, such as nanoscale heat transfer and phase-change heat transfer [13].

Mass transfer, the third portion of the field of transport phenomena, is a fundamental process in chemical engineering that involves the movement of molecules from one phase to another. This includes transitions such as gas to liquid or liquid to solid. NNs have been increasingly used to analyse and predict mass transfer phenomena, improving insight and areas of optimization in a variety of industries. These areas include the design of systems such as distillation columns, absorption towers, and chemical reactors. By training on experimental or simulated data, NNs can learn the relationships between input data, such as temperature, pressure, and concentrations, and output variables such as mass transfer rate. NNs can also be used to optimize process design, such as the development of separation processes and reactive systems [13]. Through the analysis of design parameters and performance metrics, ANNs can be used to determine the optimal configurations that maximize mass transfer efficiency and minimize energy consumption.

1.3.2 Kinetics and reactor design

NNs have also been applied to the second major subdivision of process engineering—kinetics and reactor design. This is, for example, applied to catalysis in order to be able to determine the relationship between the catalyst structure and its activity. With increasingly sophisticated experimentation techniques, much more detailed and accurate real data can be extracted to fully describe and characterize these experiments for chemical reactions and catalyst development [14]. With a rapidly increased amount of data extracted from such reactions, there is a pressing need for more sophisticated tools to manage such vast quantities of experimental data. To understand and to model such data is critical and, if successful, such methods could be used to optimize the catalyst's performance.

Kinetics and reactor design is a critical aspect of chemical engineering that principally involves the study of chemical reactions and the design of reactors to optimize reaction efficiency. Doing so has typically been empirical in nature and achieved through trial and error methods. Over time, however, the design tolerances for such reactions have become so stringent that new methods of data analysis must be implemented to assist in the design of optimized reactor designs. NNs have emerged as a powerful tool in this field and offer new opportunities for analysing reaction kinetics. These applications include multi-step reactions and nonlinear kinetics. By training on experimental data, NNs can learn the relationships between reactant concentrations and temperature in order to enable accurate predictions of reaction rates and product formation. NNs applied to chemical reactor design can also be used to predict reactor performance through metrics such as conversion rates, selectivity, and yield [14]. By analysing the reactor operating conditions and feed composition, ANNs can be used to create models which can predict how changes in these variables will affect reactor performance. Table 1.1 [15] contains a detailed list of recent studies within this field and presents a summary of current applications of NNs in catalytic processes.

1.3.3 Thermodynamics

Thermodynamics, the third field of chemical engineering which we will discuss in this chapter, is a branch of physical science that deals with the relationships between heat, work, and energy. NNs have found numerous applications within thermodynamics, ranging from property prediction to process optimization. NNs are used to predict the thermodynamic properties of materials and substances, such as specific heat, enthalpy, entropy and phase behavior. If we can train AI models on experimental or simulated data, NNs can learn the complex relationships between the variables that govern these properties, enabling accurate predictions of said behavior which can subsequently be used to design more effective distillation

Table 1.1. Current applications of ANNs in different catalytic processes. (Adapted from [14]. CC BY 3.0.)

Reference	Application	Study	Type of NN	Activation function	Software stack
[19]	Catalyst design	Design of a catalyst for propane ammoxidation	FF-ANN	Sigmoid	Proprietary software
[20]	Catalyst design	Design of a catalyst for methane oxidative coupling	FF-ANN	Sigmoid	Proprietary software
[21]	Catalyst selection	Catalyst selection for the WGS reaction	FF-ANN	Sigmoid	R-neuralnet
[22]	Catalyst deactivation	Dry reformer under catalyst sintering	FF-ANN	TanH	Proprietary software
[23]	Modeling of catalytic processes	Fischer–Tropsch synthesis to lower-olefins	FF-ANN	Sigmoid	R-neuralnet
[24]	Modeling of catalytic processes	Estimation of the reaction rate of methanol dehydration	FF-ANN	TanH/linear	MATLAB

columns, reactors, and other equipment generally used within process engineering. NNs can also be used to optimize thermodynamic processes within phenomena such as phase change and chemical reactions.

Several data-driven models have been employed that can predict phase equilibrium and transport phenomena coefficients for different chemical systems. These fields have some degree of empiricism in their mathematical formulations such that they can be used by NNs to better find a way to determine a functional relationship between the model variables as opposed to directly determining these constants through trial and error methods. In other words, we can use the AI model to determine what these parameters are based on the raw data which we can feed into the NN. Through the highly sophisticated methods of experimental study available today, this approach can yield a profound change in how thermodynamics is applied in the engineering design process. For example, Poort *et al* [16] studied a method to replace the conventional equations of state for property and phase stability calculations for a binary mixture of methanol–water. ANNs with data generated through thermodynamics for engineering applications (TEA) were used to represent four kinds of flash algorithms, leading to an order of magnitude improvement in the time to solution for the predictions of both properties and phases [15].

Furthermore, ANNs have also been used to predict the thermal and physical properties of ionic liquids. These properties include density and viscosity and the primary source of these values is experimental observations and the raw data extracted from them. For example, Valderrama *et al* [17] successfully developed an NN to estimate the density of ionic liquids. Table 1.2 [15] contains a detailed list of recent studies within this field and presents a summary of current applications of NNs in thermodynamics.

Table 1.2. Current applications of ANNs in thermodynamics. (Adapted from [14]. CC BY 3.0.)

Reference	Application	Study	Type of NN	Activation function	Software stack
[25]	Phase equilibrium	Vapor–liquid flash calculations	FF-ANN	Linear/sigmoid	Keras, Python
[26]	Phase equilibrium	Prediction of azeotrope formation	FF-ANN	Sigmoid	Proprietary software
[27]	Ionic liquids	Estimation of physical properties of ionic liquids	FF-ANN	TanH	MATLAB
[28]	Molecular thermodynamics	Enhancing the high throughput force field simulation	FF-ANN	Linear/ELU	PyTorch

1.3.4 Process control

Chemical process control, the final field of study in this chapter, details the use of various techniques to monitor and manipulate the variables in a chemical process to ensure that it operates safely, efficiently, and within specified parameters. These parameters include the flow rate, temperature, phase percentage of various materials within the chemical process, and rate of reaction. Chemical process control is based on the principles of feedback control, where measurements of process variables are compared to desired values or setpoints, and adjustments are made to the process inputs in order to maintain the desired conditions. The goal of this is to achieve stable and optimal operation within the chemical process in question. Process control is crucial in the chemical industry for several reasons [14]. It helps to ensure that product quality and consistency are maintained, maximizes production efficiency, minimizes energy consumption and waste generation, and also enhances process safety by monitoring and controlling critical parameters such as temperature and pressure as well as flow rate.

Whether the control strategy is feedback control, feedforward control, cascade control, or ratio control, NNs are increasingly being used in chemical process control to help improve control strategies, optimize process performance, and enable predictive maintenance [18]. By training on historical process data, NNs can learn the patterns and trends in the data and make predictions about future process behavior. ANNs can also be used in control strategies to optimize process performance. They can be used to develop advanced control algorithms that adjust process inputs in real time to maintain desired process conditions. This can lead to improved process efficiency, reduced waste, and enhanced product quality. As one of the most common applications of ANNs within chemical process engineering, these systems are built to identify habitual process behavior and recognize atypical variations in the chemical plant that can lead to an accident. For this kind of application, DNNs are generally used to extract spatial and temporal aspects of the data. This is attributed to the fact that DNNs contain several hidden layers. However, determining the various hyperparameters of DNNs places significant time demands, which is not compatible for fast paced process applications. With this limitation, an alternative NN category is necessary. Peng *et al* [19] applied a novel method to reduce the training time, the broad learning system (BLS). This NN utilizes an incremental learning procedure and modifies the size of the network, reducing the time to training considerably. Peng [19] and his team successfully utilized this strategy in a batch fermentation process for fault detection. Table 1.3 [15] contains a detailed list of recent studies within this field and presents a summary of current applications of NNs in chemical process control.

1.3.5 Sample study for applying neural networks in transport phenomena

In order to more properly and fully understand the implications of using NNs towards a process control system, let us discuss and go through the details of a series of example problems and how the utilization of AI-based approaches may be beneficial towards potentially faster and more accurate solutions to such problems. For the first example,

Table 1.3. Current applications of ANNs in chemical process control. Adapted from [15]. CC BY 3.0.

Reference	Application	Study	Type of NN	Activation function	Software stack
[29]	Fault detection	Penicillin fermentation process	LSTM	Sigmoid	MATLAB
[30]	Soft sensors	pH control in a chemical process	RNN	TanH	No description
[31]	Hybrid model in an MPC	Two consecutive CSTR reactors	RNN	TanH	PyTorch
[32]	Reaction chemistry model	Reaction process in a CSTR	FF-ANN	TanH	MATLAB

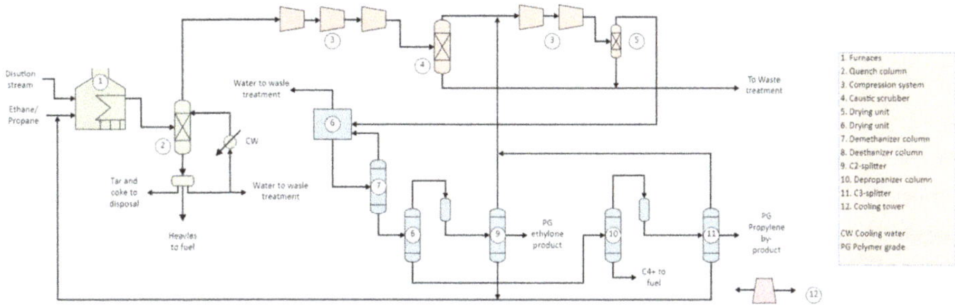

Figure 1.3. Schematic of the ethane cracker process flow.

let us consider an ethylene plant. Ethylene [20] is a critical raw material used in the chemical industry in the production of a series of key products. These include plastics, antifreeze, and detergents. The primary manner in which ethylene is produced is through steam cracking, a highly energy-intensive and complex process [21].

Raw hydrocarbon feedstocks such as ethane, propane, naphtha, and butane can be used to produce ethylene through the process of steam cracking. In this process, the feedstock is heated in a furnace at temperatures ranging from 700 °C to 900 °C in the presence of steam. Such a high temperature environment breaks down the long-chain hydrocarbons into lighter, valuable by-products with ethylene being the primary product. Some of the other by-products include propylene, methane, acetylene, and butadiene. Traditionally, the use of NNs is either to accelerate the process or to be used to improve the quality of outputs [22]. To this end, we must further break down the steam-cracking process to identify how NNs can best be of use. Figure 1.3 details the specifics of the portion of an ethylene plant that is most relevant to this discussion. Indeed, here, we see the steam-cracking process for producing ethylene from an ethane–propane mixture. We can divide this process into three primary portions. The first is cracking and quenching, followed by compression and drying, and finally separation. Initially, an ethane–propane mixture is fed into furnaces where it is cracked under high-stress conditions, producing ethylene, propylene, and other subsequent by-products. This is the specific step most relevant to our discussion in this portion of the chapter. To prevent further reactions and the formation of undesirable by-products, the furnace outlet stream is then fed into a water-based quench.

Regarding the quality of outputs, we are referring to the yield and energy consumption of the steam-cracking process. Some of the key factors which influence this include the feedstock composition, cracking temperature, the steam-to-carbon ratio, residence time, and catalyst activity. Let us break down what is relevant regarding each of these factors. For the first factor, feedstock composition, the different hydrocarbons can yield varying amounts of ethylene as product; knowing which to use, based on available data, is important to consider. Second, regarding the cracking temperature, the higher the temperature the more effective the reaction and therefore the higher yield of product. However, this is also associated with a much higher energy consumption thereby much higher operating costs for the plant. Furthermore, higher residence times can improve product yield as can increased use of catalysts. The key here is how long and how much of each to use in order to optimize the product yield and minimize operating costs. By definition, such a system is highly nonlinear and can benefit from the utilization of NNs.

An example of the use of AI for feedstock composition was the study carried out by Pyl *et al* [23], who developed an ANN to determine the detailed molecular composition of naphtha typically used in cracking processes. Further work by Niaei *et al* [24] and later Sedighi *et al* [25] utilized ANNs to model reactor effluent compositions for a given utilized feedstock. Of course all of these methods depend on the quality of the input data provided to the NN. Ideally, we would want experimental data to be available to provide the most accurate understanding of the system being studied. However, in some instances where such data are not accessible, computational data may suffice.

NNs can be applied across various other facets of ethylene production in order to help in optimizing operations in ways which are conducive towards improving yield and cost and in ways that traditional methods cannot achieve. Some of the primary areas in which NNs are employed are described in this section. Regarding the prediction and optimization of ethylene yield, NNs can be utilized in novel ways [26]. Due to the fact that the cracking process itself is highly nonlinear, small changes in temperature, feedstock composition, or reaction conditions can cause significant shifts in yield. NNs excel at handling such nonlinear relationships and can predict the ethylene yield based on a large set of data. Let us dig deeper into how this can be accomplished.

The mathematical aspects of ANNs were described previously in this chapter. All inputs to the NN are weighted by their respective weights and then summed. A constant bias term is added to this weighted sum, where the activation function introduces nonlinearity into the network. Commonly used activation functions that are utilized include the sigmoid, hyperbolic tangent, rectified linear unit (ReLU), and softmax functions [27]. The ANNs [28] which are traditionally used for this particular application are trained via back-propagation algorithms, which update the network layer weights by passing down the error from one layer to the next, starting at the output layer. A gradient descent optimization approached can also be utilized to minimize the objective function. Some of the frequently utilized error metrics in the objective function are the root mean square deviation (RMSD), mean absolute error (MAE), and mean absolute percentage error (MAPE) [29]. Several iterations through the complete training set are typically required in order to optimize the weights. These iterations are referred to as epochs. Within one epoch, the training set is further split into batches and the network weights are updated once per batch. Small batch sizes can result in faster training for certain problem sizes, but this must be tuned to any given problem with a given number of NN layers. This basic structure of an NN can then be utilized for the problem of optimizing the ethylene production process via the utilization of open source frameworks such as PyTorch or TensorFlow.

Having set up the foundation of the problem, the background of how NNs can aid in our purpose, and the background on the type of NN to be used, it is now time to detail what type of control system we can use to take advantage of the benefits which an ANN can provide [30]. This is shown in figure 1.4 as the offset-free model predictive control (MPC) system, and will be detailed here. Offset-free MPC [31] is a type of advanced control strategy that aims to predict the future behavior of a system and optimize control actions while ensuring that the system performance meets the desired specifications. One of the major enhancements in offset-free MPC compared to traditional MPC is its ability to handle model mismatches, disturbances, and external factors that could cause a deviation from the desired setpoints. In particular, offset-free MPC [32] is designed to eliminate or minimize any steady-state offsets (or biases) that might arise due to disturbances or imperfect models, ensuring that the controlled system reaches the desired target even in the presence of such issues.

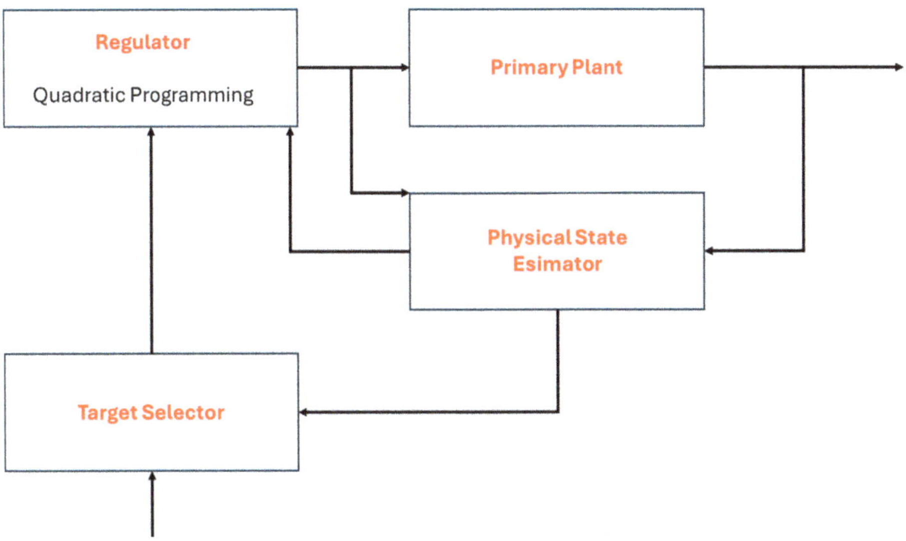

Figure 1.4. Offset-free MPC.

Traditional MPC systems take advantage of an optimal control strategy that involves solving an optimization problem at each time step. The problem minimizes a cost function over a finite time horizon, subject to system dynamics, input constraints, and state constraints [31]. It utilizes the dynamic model of the system to predict future system states, and optimizes control inputs to minimize deviations from desired states or setpoints over the horizon. The control input is then updated by solving this optimization problem in real time, typically using an optimization solver.

The offset, as part of the offset-free MPC, refers to the steady-state difference between the actual system output and the desired output (setpoints). In traditional MPC, even if the model is imperfect or there are external disturbances, the controller might not fully reject stead-state errors (offsets). This may lead to a constant deviation from the setpoints. Offset-free MPC is designed to eliminate or mitigate steady-state offsets. Some of the types of disturbance observers used in offset-free MPCs include proportional-derivation disturbance observers, the Kalman filter, or the inversion-based disturbance estimator. This system, when integrated with the NN, can help to both reduce the effects of offset and improve performance of the process control regulator.

1.4 Conclusions

The application of NNs to chemical process engineering has revolutionized the field by offering advanced modeling, prediction, and optimization capabilities. NNs are used in various aspects of chemical process engineering, including process modeling, control, optimization, fault detection, and diagnosis. In process modeling, NNs are used to capture complex relationships in process data, enabling accurate predictions of process behavior such as product quality, yields, and energy consumption. They are also used in control strategies to develop advanced algorithms that adjust process inputs in real time to maintain desired conditions, leading to improved efficiency, reduced waste, and enhanced product quality. NNs are also valuable in fault detection and diagnosis, where they analyse process data to detect abnormalities and identify potential faults. This allows operators to take corrective actions and prevent costly downtime. Overall, the application of NNs in chemical process engineering offers numerous benefits, including improved process efficiency and product quality, enhanced safety through real-time monitoring and control, reduced downtime and maintenance costs, and optimized resource utilization and energy consumption. However, there are challenges to using NNs in chemical process engineering, including the need for large, high-quality datasets, the complexity of NN models, and the interoperability of NN predictions. Future research directions include the development of more advanced NN architectures, the integration of NNs with other control strategies, and the application of NNs to new areas within chemical engineering, such as process intensification and sustainability. In conclusion, NNs offer significant potential in improving the efficiency, safety, and sustainability of chemical processes. By leveraging the power of NNs, researchers and engineers can continue to advance the field of chemical process engineering and drive innovation in the industry.

Bibliography

[1] Alves R M B and Nascimento C A O 2004 *Neural Network Based Approach Applied to for Modeling and Optimization Industrial Isoprene Unit Production* (Austin, TX: American Institute of Chemical Engineers)

[2] Bishop C M and Nasrabadi N M 2006 *Pattern Recognition and Machine Learning* **vol 4** (Berlin: Springer)

[3] Dick S and Fernandez-Serra M 2020 Machine learning accurate exchange and correlation functionals of the electronic density *Nat. Commun.* **11** 3509

[4] Gemperline P J, Long J R and Gregoriou V G 1991 Nonlinear multivariate calibration using principal components regression and artificial neural networks *Anal. Chem.* **63** 2313–23

[5] Goodfellow I 2016 *Deep Learning* (Cambridge, MA: MIT Press)

[6] Haykin S 2001 *Redes Neurais: Princípios e Prática* (Lisbon: Bookman)

[7] Himmelblau D M 2008 Accounts of experiences in the application of artificial neural networks in chemical engineering *Ind. Eng. Chem. Res.* **47** 5782–96

[8] Hirschfeld L, Swanson K, Yang K, Barzilay R and Coley C W 2020 Uncertainty quantification using neural networks for molecular property prediction *J. Chem. Inform. Model.* **60** 3770–80

[9] Khezri V, Yasari E, Panahi M and Khosravi A 2020 Hybrid artificial neural network–genetic algorithm-based technique to optimize a steady-state gas-to-liquids plant *Ind. Eng. Chem. Res.* **59** 8674–87

[10] Hornik K, Stinchcombe M and White H 1989 Multilayer feedforward networks are universal approximators *Neural Netw.* **2** 359–66

[11] Vashishtha M 2011 Application of artificial neural networks in prediction of vapour liquid equilibrium data *25th European Conference on Modelling and Simulation (Krakow, 7–10 June)* (Caserta: ECMS) pp 142–45

[12] Hemmati-Sarapardeh A, Ameli F, Varamesh A, Shamshirband S, Mohammadi A H and Dabir B 2018 Toward generalized models for estimating molecular weights and acentric factors of pure chemical compounds *Int. J. Hydrogen Energy* **43** 2699–717

[13] LeCun Y, Bengio Y and Hinton G 2015 Deep learning *Nature* **521** 436–44

[14] Schmidhuber J 2015 Deep learning in neural networks: an overview *Neural Netw.* **61** 85–117

[15] Cavalcanti F M, Kozonoe C E, Pacheco K A and Alves R M B D 2021 Application of artificial neural networks to chemical and process engineering *Deep Learning Applications* (London: InTech)

[16] Poort J P, Ramdin M, van Kranendonk J and Vlugt T J H 2019 Solving vapor–liquid flash problems using artificial neural networks *Fluid Phase Equilib.* **490** 39–47

[17] Valderrama J O, Reategui A and Rojas R E 2009 Density of ionic liquids using group contribution and artificial neural networks *Ind. Eng. Chem. Res.* **48** 3254–59

[18] Venkatasubramanian V 2019 The promise of artificial intelligence in chemical engineering: is it here, finally? *AIChE J.* **65** 466–78

[19] Peng C and ChunHao D 2022 Monitoring multi-domain batch process state based on fuzzy broad learning system *Expert Syst. Appl.* **187** 115851

[20] Yuan B, Zhang Y, Du W, Wang M and Qian F 2019 Assessment of energy saving potential of an industrial ethylene cracking furnace using advanced exergy analysis *Appl. Energy* **254** 113583

[21] Al Jitan S, Alkhoori S A and Yousef L F 2018 Phenolic acids from plants: extraction and application to human health *Studies in Natural Products Chemistry* **vol 58** (Amsterdam: Elsevier) pp 389–417

[22] Ciric A, Krajnc B, Heath D and Ogrinc N 2020 Response surface methodology and artificial neural network approach for the optimization of ultrasound-assisted extraction of polyphenols from garlic *Food Chem. Toxicol.* **135** 110976

[23] Pyl S P, Van Geem K M, Reyniers M-F and Marin G B 2010 Molecular reconstruction of complex hydrocarbon mixtures: an application of principal component analysis *AIChE J.* **56** 3174–88

[24] Niaei A, Towfighi J, Khataee A R and Rostamizadeh K 2007 The use of ANN and the mathematical model for prediction of the main product yields in the thermal cracking of naphtha *Petrol. Sci. Technol.* **25** 967–82

[25] Sedighi M, Keyvanloo K and Towfighi J 2011 Modeling of thermal cracking of heavy liquid hydrocarbon: application of kinetic modeling, artificial neural network, and neuro-fuzzy models *Ind. Eng. Chem. Res.* **50** 1536–47

[26] Fonseca A P, Stuart G, Oliveira J V and Lima E 1999 Using hybrid neural models to describe supercritical fluid extraction processes *Braz. J. Chem. Eng.* **16** 267–78

[27] Arora S, Du S, Hu W, Li Z and Wang R 2019 Fine-grained analysis of optimization and generalization for overparameterized two-layer neural networks *Int. Conf. on Machine Learning* pp 322–32

[28] Demmig-Adams B, Stewart J J, López-Pozo M, Polutchko S K and Adams W W III 2020 Zeaxanthin, a molecule for photoprotection in many different environments *Molecules* **25** 5825

[29] Brunner G 1984 Mass transfer from solid material in gas extraction *Ber. Bunsenges. Phys. Chem.* **88** 887–91

[30] Alper E 1983 Introduction to liquid-liquid extraction with chemical reaction *Mass Transfer with Chemical Reaction in Multiphase Systems; Volume I: Two-Phase Systems; Volume II: Three-Phase Systems* (Berlin: Springer) pp 577–611

[31] Pannocchia G and Rawlings J B 2003 Disturbance models for offset-free model-predictive control *AIChE J.* **49** 426–37

[32] Pannocchia G and Bemporad A 2007 Combined design of disturbance model and observer for offset-free model predictive control *IEEE Trans. Autom. Control* **52** 1048–53

IOP Publishing

Chapter 2

Machine learning in optimal control and process modeling

Yujia Wang and Zhe Wu

Recent technological advances in the process manufacturing industries have led to increasingly complex processes, systems, and products, which pose challenges in design, analysis, manufacturing, and management. Process systems engineering (PSE) serves as a strategic framework for addressing the complexities and demands that arise from the latest technological innovations in the process industry. Machine learning (ML) has shown great potential in the PSE domain due to its ability to analyse data and make predictions for complex systems [1–9]. Therefore, the integration of ML with PSE has shown tremendous potential, offering intelligent and efficient solutions to the challenges associated with industrial manufacturing processes.

This chapter focuses on the development of reinforcement learning (RL) schemes for optimal control of nonlinear chemical processes. Specifically, we start with a brief introduction of three popular ML methods—supervised learning, unsupervised learning, and reinforcement learning—in PSE, and provide a review of RL algorithms. We will focus on RL and discuss safety and process modeling issues in the context of RL. Chemical process examples will be used to show the applications of the proposed RL algorithms.

2.1 Introduction to machine learning

ML is a branch of artificial intelligence (AI) that encompasses computer science and statistics to endow machines with autonomous learning capabilities. By progressively enhancing performance through data-driven knowledge acquisition, ML obviates the need for explicit rule-based programming, addressing challenges in complex, uncertain, or dynamic tasks. The operational paradigm of ML can be analogized to the human learning process. Initially, it required historical data, which

doi:10.1088/978-0-7503-6174-3ch2

can comprise labeled input–output pairs or unlabeled data, serving as the foundation for ML. Subsequently, the training process is carried out, in which various ML algorithms are employed to enable machines to discover patterns and regularities from the data. The outcome of this training is a model, representing an abstract representation of the pattern within the data. There are three main types of algorithms used to train ML models: supervised learning, unsupervised learning, and reinforcement learning.

2.1.1 Supervised learning

Supervised learning is a method in ML that trains a model using labeled data, allowing the model to predict the labels of new data, as indicated in figure 2.1. In supervised learning, the process relies on known input–output pairs, where an input variable corresponds to an output variable. An algorithm is then employed to learn the mapping function from the given inputs to the corresponding outputs. This learning method is called supervised learning because we utilize labeled data to 'supervise' the model's training process. In short, supervised learning is a way to learn complex (nonlinear) input–output relationships from labeled data. Although supervised learning can learn complex models, it generally requires a large amount of labeled training data. Additionally, the quality and representativeness of training data may have a significant impact on the accuracy of the model. Supervised learning methods have been widely used in the PSE field for data-driven modeling of complex systems that cannot be modeled well by first-principles knowledge.

2.1.2 Unsupervised learning

Unlike supervised learning, unsupervised learning models operate on unlabeled data, autonomously discovering patterns and insights without any explicit guidance or instruction, as shown in figure 2.2. This method entails training models on raw and unlabeled data, commonly utilized to identify patterns and trends in the original dataset or to cluster similar data into specific groups. Typically used in the initial exploration stage to gain a better understanding of the dataset, unsupervised ML is well suited for unveiling hidden trends and relationships within the data. The primary goals of unsupervised learning include dimensional reduction, feature extraction, and

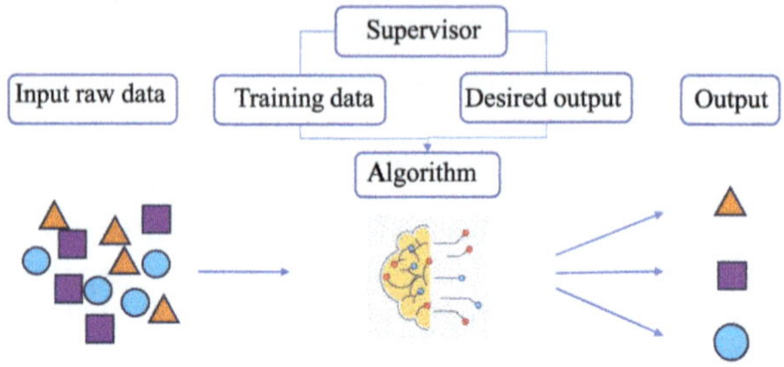

Figure 2.1. Supervised learning.

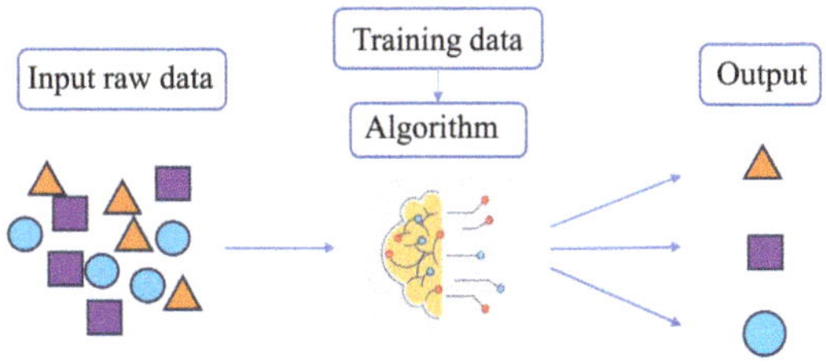

Figure 2.2. Unsupervised learning.

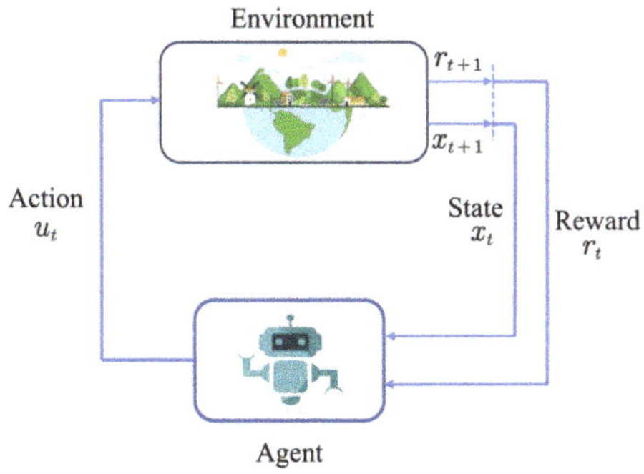

Figure 2.3. Reinforcement learning.

clustering. However, due to the absence of extensive human supervision, careful attention must be paid to the explainability of unsupervised ML. Unsupervised learning plays a crucial role in monitoring and identifying abnormal situations in industrial processes, such as cyberattacks and process faults, based on collected data.

2.1.3 Reinforcement learning

In RL, the learning machine does not receive input–output pairs; instead, it uses a scalar reward signal through trial and error to find optimal outputs for each input. This approach is suitable for problems without predefined optimal mappings, but with input–output pairs that can be evaluated. In this sense, RL can be considered an integration of supervised and unsupervised learning, with the reward signal acting as a form of supervision. Additionally, RL aims to achieve a balance between the use of existing knowledge and the exploration of undiscovered space. In figure 2.3, the agent receives stimulus or state information (x_t) at time t, determining its action (u_t). This action changes the state of the environment (x_{t+1}), and the resulting evaluation determines the reward (r_{t+1}). This reward, with the new state, guides the agent's next action, reinforcing positive behavior and discouraging negative actions. In control engineering, reward implies a reduction in control costs, while punishment increases costs. Therefore, RL algorithms can be used to learn an optimal behavior that ensures the survival and growth of the agent.

2.2 Reinforcement learning for optimal control

RL draws inspiration from biology, with its fundamental concept rooted in experiential learning and the application of reward and punishment principles borrowed from living organisms, including humans and animals. In this process, the agent (or controller) executes actions, evaluates them based on reward and cost functions, and receives rewards (positive reinforcement) or incurs penalties (negative reinforcement). This heuristic approach involves the agent's pursuit of maximizing future rewards, where, in the field of control engineering, maximizing rewards is equivalent to minimizing costs. Through this iterative process, the agent strives to formulate an optimal strategy. As an ML approach, RL has gained significant attention in various process industries. It learns optimal behavior in an environment by performing actions and observing the results of actions, and provides an efficient tool for leveraging data to achieve optimal control [10]. While most existing research on the implementation of RL in process control has focused on improving its performance [11–15], the increasing importance of RL safety is driven by the possibility of hazardous outcomes and catastrophic consequences resulting from the inappropriate use of RL technologies in process operations. Specifically, since data collection and learning in RL are carried out through trial and error and do not respect safety constraints during learning, real-world systems may encounter unsafe scenarios that cause significant damage to the learning process. Therefore, how to safely learn optimal control policies using RL algorithms remains a critical challenge. In this section, we provide a review of RL in the context of control fields, followed by an illustration of the application of RL for optimal control while considering safety.

2.2.1 Review of reinforcement learning

RL is a branch of ML that aims to empower computer systems to make decisions and take actions within a given environment, learning appropriate decisions through continuous interaction with the environment [8, 10]. It is considered a learning framework for goal-oriented problems through interactive learning. Regardless of the specific problem, it can be simplified as the exchange of signals between the agent and the environment, including decisions (actions) made by the agent, decision making based on information (state), and goals (rewards) of the agent.

Environment

The environment is the dynamic system from which we generate data. Examples of environments include robots, linear and nonlinear dynamical systems, and games such as Atari and Go. This environment takes actions as input and generates variables, i.e. states, based on its own rules. To effectively address challenges in RL, one step involves creating a mathematical model of the environment. Markov decision processes (MDPs) serve as a widely utilized framework for modeling environments, because nearly all RL problems can be represented using MDPs. They can effectively represent and simplify RL problems, transforming the resolution of RL problems into finding the optimal solution of the Markov decision model. Additionally, in control systems, the environments represent dynamical systems. Here, the state transition laws are deterministic and are often derived from the dynamics of the underlying physical system to be modeled.

Agent

An agent refers to an RL system that can interact with the environment and take actions to complete tasks. It mainly consists of three parts: the model, policy, and value function.

- The model is the agent's interpretation of the environment, showing the operational mechanism of the environment from the perspective of an agent, with the expectation that the model can simulate the interaction mechanism between the environment and the agent. Given a state and behavior, the model can predict the next state and immediate return. Specifically, based on the observed state x at each time step, an action u is selected from the available set of actions, and the environment transitions to a new state x' under the influence of u. The decision maker, upon observing the new state x', then makes a new decision, takes action u', and this process is repeated iteratively. According to the deterministic or stochastic policy π, i.e. $u_t = \pi(x_t)$ or $P(u_t|x_t) = \pi(u_t|x_t)$, the system undergoes a transformation from the current state x_t to the next state $x_{t+1} = x'$ according to the state transition function $x_{t+1} = F(x_t, u_t)$ or probability $P(x_{t+1} = x'|x_t = x, u_t = u)$ and, at the same time, receives an immediate reward, denoted by $r_{t+1} = r(x_t, u_t, x_{t+1})$. In general, when we refer to a known model, it means acquiring state transition models and rewards. The model is specific to the agent and serves as an approximation of the actual operational mechanism of the environment. Constructing a model is not a necessary component for building an agent in some RL algorithms, as agents do not attempt to construct a model, such as in Monte Carlo (MC) and temporal difference (TD) methods.
- A policy is a mechanism that determines the behavior of an agent by mapping states to actions. It defines the agent's behavior, specifying the various possible ways in which the agent can act in each state. There are two types of policies: deterministic policies and stochastic policies. A deterministic policy outputs an action based on a specific state, such as $\pi(x) = u$. In contrast, a stochastic policy outputs the probability of each action based on the state, $\pi(u|x) = P(u|x)$, where the probability values range from 0 to 1, forming a probability distribution. The purpose of the policy is to describe the behavior of an agent at a particular time.
- The value function represents the performance of an agent in a given state or the desirability of taking a specific action in a given state. The notion of desirability is quantified using the future reward, and the reward is contingent on the strategy used. Let $0 < \gamma \leqslant 1$ denote the discount factor. The goal of RL is to find a policy π to maximize/minimize the accumulated reward,

namely the total return denoted by $G_t = \sum_{k=0}^{\infty} \gamma^k r_{t+k+1}$, starting from an initial state. The state-value function, $V_\pi(x)$, represents the return obtained by starting from state x under a policy π. This value is used to assess the desirability of a state, guiding the agent in selecting actions that transition it to states with higher value functions. Mathematically, it is expressed as $V_\pi(x) = G_t|_{x_t = x} = \sum_{k=0}^{\infty} \gamma^k r_{t+k+1}|_{x_t = x}$. Another value function for a policy π is known as the action-value function, denoted as $Q_\pi(x, u)$. It represents the reward obtained by taking an action u in a given state x while executing the policy π, which is described as $Q_\pi(x, u) = G_t|_{x_t = x, u_t = u} = \sum_{k=0}^{\infty} \gamma^k r_{t+k+1}|_{x_t = x, u_t = u}$.

Bellman equation

The Bellman equation, also known as the dynamic programming equation, is used to solve problems with certain optimality properties. It decomposes the problem into sub-problems and derives the solution to the original problem from these subproblem solutions. This equation is foundational in RL algorithms, as nearly all methods, including dynamic programming (DP), MC, TD, and others, rely on the Bellman equation to identify optimal policies. The Bellman equation expresses the relationship between the current value function (or action-value function) and its successor value function (or action-value function), as well as the relationship between the value function and the action-value function. According to Bellman's principle of optimality, the optimal state-value function $V_\pi(x)$ and the optimal action-value function $Q_\pi(x, u)$ can be rewritten as follows [16]:

$$V^*(x) = \max_u \{r + \gamma V^*(x')\} \tag{2.1}$$

$$Q^*(x, u) = r + \gamma \max_{u'} Q^*(x', u'). \tag{2.2}$$

Let π^* denote the optimal policy and it is determined by

$$\pi^*(x) = \arg\max_u Q^*(x, u) = \arg\max_u \{r + \gamma V^*(x')\}. \tag{2.3}$$

Methods of reinforcement learning

In reinforcement learning, there are three main types of algorithms to find optimal policies: DP, MC, and TD [10, 16]. Specifically, DP refers to a set of algorithms with the ability to find optimal policies assuming that a perfect model is available. The fundamental concept of DP is grounded in Bellman equations. It involves breaking down the problem into several sub-problems, solving these sub-problems initially, and subsequently deriving the solution to the original problem from the solutions of these sub-problems. The two most popular methods in DP are policy iteration and value iteration. Unlike DP, MC methods do not require a model of the environment (i.e. they are model-free methods). MC methods determine the optimal policy by initially estimating the average returns for different policies through the sampling of many sequences of states, actions, and rewards under a given policy. TD methods stand out as one of the most widely used RL algorithms today due to their simplicity and relatively low computational cost. TD methods do not necessitate a model of the system, as they learn the dynamics directly from interactions. Additionally, TD methods can update immediately after the state and reward are received. Different RL methods can be selected based on various application scenarios. For instance, in the case where the environment is known, model-based approaches such as DP can be employed. In the case of unknown environments, model-free methods such as MC and TD methods can be chosen.

2.2.2 Class of nonlinear systems and problem formulation

In this section, RL is used to solve the optimal control problem, where the environment model is described as follows:

$$\dot{x} = f(x) + g(x)u, \quad x(t_0) = x_0, \tag{2.4}$$

where $x \in \mathbb{R}^n$ is the measurable state vector, and $g(x)$ and $f(x)$ are matrix and vector functions of dimensions $n \times m$, and $n \times 1$, respectively. The control input vector is represented by $u \in \mathcal{U}_c \subset \mathbb{R}^m$, and is constrained by the upper bound u_{max} and the lower bound u_{min}. Note that $x(t)$ is a function of time in the nonlinear dynamic system of equation (2.4). To simplify the notation, we denote $x(t)$ by x.

Safety is an important issue in RL as the system may encounter unsafe states that correspond to unsafe operating conditions in real-world problems during the learning stage. Suppose that there exists a set of states within the state-space that corresponds to unsafe operating conditions, which can be described as an open and bounded set $\mathcal{D}$, excluding the steady-state. We define a set of safe states, denoted as $\mathcal{X}_s$, that satisfies the condition $\mathcal{X}_s \cap \mathcal{D} = \varnothing$. Ensuring system safety requires that the state of equation (2.4) always remains within the safe operating region, and never enters the set of unsafe states. The definitions of closed-loop stability and safety are presented below.

Definition 1. (Closed-loop stability) Consider a class of nonlinear systems represented by equation (2.4). The control policy $u = \varphi(x) \in \mathcal{U}_c$ is said to ensure the closed-loop stability of the system if, for any initial state x_0 within a certain operating region, the closed-loop state $x(t)$, $\forall t \geqslant 0$, remains bounded within the operating region and ultimately converges to a small neighborhood around the steady-state under $u = \varphi(x) \in \mathcal{U}_c$.

Definition 2. (Safety) Consider a class of nonlinear systems represented by equation (2.4) defined on a state-space that includes an unsafe set $\mathcal{D}$ and a safe set $\mathcal{X}_s$. The control policy $u = \varphi(x) \in \mathcal{U}_c$ is said to ensure system safety if, for any initial state x_0 in the region $\mathcal{X}_s$, the evolution of the system states over time under $u = \varphi(x) \in \mathcal{U}_c$ is maintained within the safe set $\mathcal{X}_s$ and avoids entering $\mathcal{D}$, $\forall t \geqslant 0$.

The objective is to develop a safe optimal control strategy for the nonlinear system described by equation (2.4) to ensure that the state is confined to the safe region at all times and converges to the origin in an optimal manner while avoiding moving into the unsafe region. To achieve this goal, a performance function is designed to optimize the control policy. The formulation of the safe optimal control problem is presented below:

$$\begin{aligned} \min_u J(x(t), u(t)) &= \int_t^{\infty} \mathcal{R}(x(\tau), u(\tau))\mathrm{d}\tau \\ \text{s. t. } \dot{x} &= f(x) + g(x)u \\ x(t) &\in \mathcal{X}_s, \quad \forall t \geqslant 0 \\ u(t) &\in \mathcal{U}_c, \quad \forall t \geqslant 0, \end{aligned} \tag{2.5}$$

where $\mathcal{R}(x, u)$ denotes the utility function, and is typically formulated as $\mathcal{R}(x, u) = x^T Qx + u^T Ru$ in traditional optimal control problems without accounting for safety issues, where Q and R are symmetric positive-definite matrices. However, this standard form of $\mathcal{R}(x, u)$ is not applicable to address the safe optimal control problem. To overcome this challenge, some recent works [17] use barrier functions $B(x)$ to map the unconstrained variables $x \in (-\infty, \infty)$ to the constrained variables $x_c \in (\underline{x}, \bar{x})$. Subsequently, by designing the utility function with x replacing x_c, the variable x_c is guaranteed to remain within the bounds of $(\underline{x}, \bar{x})$ (e.g. the safe region $\mathcal{X}_s$) provided that the resulting optimal controller u achieves a finite value function. Additionally, another approach to improve safety in optimal control is to augment the performance index with a barrier function $B'(x)$ [18]. The utility function for the safe optimal control problem of equation (2.5) can be constructed in the form of $\mathcal{R}(x, u) = x^T Qx + u^T Ru + B'(x)$, where $B'(x)$ approaches infinity near the boundary of $\mathcal{D}$. By minimizing the performance function, the safe optimal controller can ensure system safety and achieve a finite value function. However, the derivation of provable safety and stability results for the above design of safe optimal controllers remains an open question, since it is challenging to obtain a control policy that ensures system safety and closed-loop stability simultaneously throughout the learning process.

The following assumptions and definitions are first presented and will be used in the development of the safe optimal controller of equation (2.5).

Assumption 1. It is assumed that the system model described by equation (2.4) is known with $f(0) = 0$. Additionally, the nonlinear functions $f(x)$ and $g(x)$ are sufficiently smooth functions defined on a compact set $\mathcal{X} \subset \mathbb{R}^n$ that includes the origin.

Assumption 2. The nonlinear system of equation (2.4) is assumed to be controllable in the set $\mathcal{X}$, i.e. there is a continuous control policy $u = \varphi(x) \in \mathcal{U}_c$ that renders the origin asymptotically stable for all initial conditions $x_0 \in \mathcal{X}$.

Definition 3. (Admissible control policy) Given a nonlinear system described by equation (2.4) subject to a control policy $\varphi(x) \in \mathcal{U}_c$ for all $x \in \mathcal{X}$, with its associated value function:

$$V(x(t)) = \int_t^{\infty} \mathcal{R}(x(\tau), \varphi(x(\tau)))\mathrm{d}\tau, \tag{2.6}$$

we say that the control policy $\varphi(x)$ is admissible with respect to equation (2.6) on $\mathcal{X}$, denoted by $\varphi(x) \in \mathcal{U}_{ca}(\mathcal{X})$, provided that the following conditions are met: 1. $\varphi(x)$ is continuous over the set $\mathcal{X}$; 2. $\varphi(0)$ is equal to zero; 3. the asymptotic stability of the system described in equation (2.4) at the steady-state can be achieved under $\varphi(x)$; and 4. the value function V is finite.

Assumption 3. There exists an admissible policy with respect to the value function of equation (2.6) for any state in the set $\mathcal{X}$, for the nonlinear system of equation (2.4).

2.2.3 Stabilization and safety via CLBF

The control Lyapunov function (CLF) is an extension of the Lyapunov function for systems with manipulated inputs [19]. The definition of CLF is given as follows.

Definition 4. (Control Lyapunov function) A continuously differentiable, proper, and positive-definite function $V_l(x)$ is a CLF for equation (2.4) if

$$L_f V_l(x) < 0, \forall x \in \{z \in \mathbb{R}^n \backslash \{0\} \mid L_g V_l(z) = 0\}, \tag{2.7}$$

and it satisfies the small control property stated in [20]. Specifically, there exists $\delta > 0$ for any $\varepsilon > 0$, such that for all $x \in \mathcal{B}_\delta(0)$, there exists a control input u satisfying $|u| < \varepsilon$ that induces a negative derivative of the Lyapunov candidate V_l along the system trajectories, i.e. $\dot{V}_l(x) = L_f V_l(x) + L_g V_l(x)u < 0$.

Let $\mathcal{X}$ represent the set consisting of the origin and the state x that satisfies $\dot{V}_l(x) < 0$ under the control policy $\varphi(x)$. Within $\mathcal{X}$, let the set $\mathcal{X}_b$ represent all the states that meet the inequality $V_l(x) \leqslant b$ with a positive constant b (i.e. a level set of V_l). For any initial state x_0 starting within $\mathcal{X}_b$, the state of the system will remain within this invariant set under the control policy for all $t \geqslant t_0$, due to the fact that $\mathcal{X}_b$ is a forward invariant subset inside $\mathcal{X}$. A candidate control law $\varphi_s(x)$ that considers input constraints is the following saturated Sontag [21]:

$$k(x) = \begin{cases} -\dfrac{L_f V_l + \sqrt{L_f V_l^2 + \left(L_g V_l L_g V_l^T\right)^4}}{L_g V_l L_g V_l^T} L_g V_l^T, & L_g V_l \neq 0 \\ 0, & L_g V_l = 0, \end{cases} \tag{2.8a}$$

$$\varphi_s(x) = \begin{cases} u_{\max} & k(x) > u_{\max} \\ k(x) & u_{\min} \leqslant k(x) \leqslant u_{\max} \\ u_{\min} & k(x) < u_{\min}, \end{cases} \tag{2.8b}$$

where $k(x)$ is the control law prior to considering the saturation of the control action, and $\varphi_s(x)$ is the saturated control input that takes into account the input constraint.

The control barrier function (CBF) is a useful tool for designing a control law that ensures the safety of the closed-loop system by preventing the states from entering the unsafe region [22]. The definition of CBF is given as follows.

Definition 5. (Control barrier function) Let $\mathcal{D}$ represent a region of unsafe system states. A function $\mathcal{B}(x)$ is considered a CBF, provided that it is a $\mathcal{C}^1$ function and meets the following conditions [22]:

$$\mathcal{X}_{\mathcal{B}} := \{x \in \mathbb{R}^n \mid \mathcal{B}(x) \leqslant 0\} \neq \varnothing \tag{2.9a}$$

$$\mathcal{B}(x) > 0, \quad \forall x \in \mathcal{D} \tag{2.9b}$$

$$L_f\mathcal{B}(x) \leqslant 0, \quad \forall x \in \left\{z \in \mathbb{R}^n \backslash \mathcal{D} \mid L_g\mathcal{B}(z) = 0\right\}. \tag{2.9c}$$

Barrier functions can be designed in various ways, such as the reciprocal barrier function and zeroing barrier function. They have been integrated with CLFs through weighted sum or quadratic programming combination for stabilization of nonlinear systems while incorporating safety concerns (e.g. [18, 23]). However, closed-loop stability and safety cannot be guaranteed through a simple integration of CBFs and CLFs (e.g. $V_l(x) + \mathcal{B}(x)$) since a negative value of the sum of $\dot{V}_l$ and $\dot{\mathcal{B}}$ does not necessarily guarantee the simultaneous stability and safety of the closed-loop system. Therefore, to obtain provable stability and safety guarantees for the system described in equation (2.4) with input constraints, Romdlony and Jayawardhana [24] developed a new function, termed CLBF, by integrating a CLF and a CBF. Then, by replacing V_l with a CLBF, the Sontag controller of equation (2.8) guarantees that the state of the system remains within the level set of the CLBF function, thus remaining outside the unsafe regions. Our previous work [21] proposed a constrained CLBF that addresses input constraints and the presence of saddle points in the design of CLBF, and improved the closed-loop performance of equation (2.4) under model predictive control using CLBFs. Herein, we provide the following definition for the constrained CLBF.

Definition 6. (Constrained CLBF) A lower-bounded and proper $\mathcal{C}^1$ function $V(x)$: $\mathbb{R}^n \to \mathbb{R}$ is a constrained CLBF for the nonlinear system described in equation (2.4) associated with an unsafe region $\mathcal{D}$ in state-space, if a minimum of $V(x)$ is achieved at the origin and there exists a positive constant ρ_c, and a class $\mathcal{K}$ function $\gamma(\cdot)$ such that the following equations hold:

$$\mathcal{X}_{\rho_c} := \left\{x \in \phi_{uc} \mid V(x) \leqslant \rho_c\right\} \neq \varnothing \tag{2.10a}$$

$$L_f V(x) < 0, \quad \forall x \in \left\{z \in \phi_{uc} \backslash \left(\mathcal{D} \cup \{0\} \cup \mathcal{X}_e\right) \mid L_g V(z) = 0\right\} \tag{2.10b}$$

$$\left| \frac{\partial V(x)}{\partial x} \right| \leqslant \gamma(|x|) \tag{2.10c}$$

$$V(x) > \rho_c, \quad \forall x \in \mathcal{D} \subset \phi_{uc}, \tag{2.10d}$$

where $\phi_{uc} := \left\{x \in \mathbb{R}^n \mid \dot{V} = L_f V + L_g Vu < 0\right\} \cup \{0\} \cup \mathcal{X}_e$. The set $\mathcal{X}_e$ represents the set of all the saddle points with $\partial V(x)/\partial x = 0$ within the set ϕ_{uc} excluding those in $\mathcal{D} \cup 0$, i.e.$\mathcal{X}_e := \{x \in \phi_{uc} \backslash (\mathcal{D} \cup 0) \mid \partial V(x)/\partial x = 0\}$.

One approach to designing the safe controller $\varphi(x) \in \mathcal{U}_{ca}(\mathcal{X})$ is to use CLBFs. Specifically, this design strategy adopts the saturated Sontag controller of equation (2.8) using $V(x)$ in place of $V_l(x)$. It has been proven by Wu *et al* [21] that the CLBF-based controller ensures that the closed-loop state $x(t)$ of equation (2.4) remains within the safe operating region $\forall t \geqslant 0$. To handle saddle points, discontinuous control actions can be applied to drive the system to escape saddle points [21].

2.2.4 Control Lyapunov-barrier function-based RL

In this section, a novel safe reinforcement learning (SRL) design using CLBFs is developed. The design of a performance function such that, given an admissible control policy, its associated value function satisfies the properties of the CLBF is first presented. Then we discuss the incorporation of CLBFs into the design of safe RL for optimal control with safety guarantees.

CLBF-based value function

Based on the construction method of a constrained CLBF in [21], we propose a new method for solving the safe optimal control problem of equation (2.5), where the utility function $\mathcal{R}(x, u)$ is designed as $\mathcal{R}(x, u) = p(x) + u^T Ru$ such that the value function V under an admissible control policy $u = \varphi(x) \in \mathcal{U}_{ca}(\mathcal{X})$ satisfies the properties of constrained CLBFs in equation (2.10a):

$$V(x(t)) = \int_t^\infty (p(x(\tau)) + \varphi^T(x(\tau))R\varphi(x(\tau)))\mathrm{d}\tau, \tag{2.11}$$

where $p(x) = l(x) + \mu b(x) + v$, and $l(x)$ is a state penalty function designed to be positive-definite and is commonly expressed in quadratic form as $l(x) = x^T Qx$. The function $b(x)$ is a continuously differentiable barrier function designed to penalize the value function when the state of the system is close to the unsafe region. The constant v is designed to ensure that the minimum of the function $p(x)$ is zero at the origin such that a finite value of V of equation (2.11) can be achieved under an admissible control policy. A systematic way to design v is given in equation (2.16b).

Since the control policy $\varphi(x)$ is a function of x, $\mathcal{R}(x, \varphi(x))$ can be rewritten as $\mathcal{R}(x, \varphi(x)) = r(x) := l(x) + \varphi^T(x)R\varphi(x) + \mu b(x) + v = q(x) + \mu b(x) + v$, where $q(x)$ is defined as follows: $q(x) = l(x) + \varphi^T(x)R\varphi(x)$. To satisfy the requirements of CLBFs, the functions $q(x)$ and $b(x)$ are constructed as follows:

$$b(x) > \eta_1, \forall x \in \mathcal{D} \tag{2.12}$$

$$b(x) \geqslant \eta_2, \forall x \in \mathcal{H} \tag{2.13}$$

$$b(x) = \eta_2, \forall x \in \mathbb{R}^n \backslash \mathcal{H}, \tag{2.14}$$

$$c_1|x|^2 \leqslant q(x) \leqslant c_2|x|^2, \ \forall x \in \mathbb{R}^n, c_2 > c_1 > 0, \tag{2.15}$$

where $\eta_1 > \eta_2$, and the connected and compact set $\mathcal{H}$ is inside the region ϕ_{uc} that satisfies $\mathcal{D} \subset \mathcal{H} \subset \phi_{uc}$, $0 \notin \mathcal{H}$. The weight coefficients in the value function of equation (2.11) are designed to satisfy the following equations:

$$\mu > \frac{c_2 c_3 - c_1 c_4}{\eta_1 - \eta_2} \tag{2.16a}$$

$$v = -\mu\eta_2 \tag{2.16b}$$

$$c_3 := \max_{x \in \partial\mathcal{H}} |x|^2 \tag{2.16c}$$

$$c_4 := \min_{x \in \partial\mathcal{D}} |x|^2. \tag{2.16d}$$

Additionally, $V(x)$ of equation (2.11) is required to satisfy (figure 2.4)

$$L_f V(x) < 0, \quad \forall x \in \{z \in \phi_{uc} \backslash (\mathcal{D} \cup \{0\} \cup \mathcal{X}_e) \mid L_g V(z) = 0\} \tag{2.17}$$

$$V(x) < \infty, \forall x \in \{z \in \phi_{uc} \backslash \mathcal{D}\} \tag{2.18}$$

$$\left| \frac{\partial V(x)}{\partial x} \right| \leqslant \gamma(|x|). \tag{2.19}$$

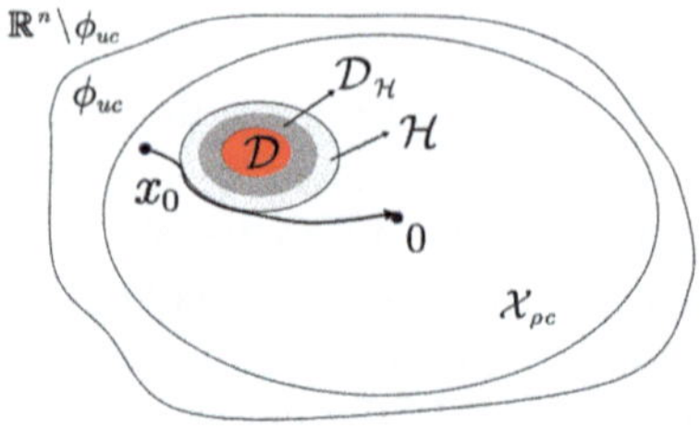

Figure 2.4. A schematic of all the sets ϕ_{uc}, $\mathcal{X}_{\rho_c}$, $\mathcal{H}$, $\mathcal{D}_\mathcal{H}$ and $\mathcal{D}$. (Reproduced with permission of Wiley from [25]. Copyright 2023 American Institute of Chemical Engineers.)

Reinforcement learning for solving the optimal control policy
We denote the optimal value function as $V^*(x)$, which is defined as follows:

$$V^*(x(t)) = \min_{u \in \mathcal{U}_{ca}} \int_t^{\infty} (p(x(\tau)) + u^T(x(\tau))Ru(x(\tau)))\mathrm{d}\tau. \tag{2.20}$$

Consider the system described by equation (2.4) with the value function of equation (2.6). We define the following Hamiltonian of the optimal control problem in equation (2.5):

$$H(x, u, \nabla V) = p(x) + u^T Ru + \nabla V(f(x) + g(x)u). \tag{2.21}$$

Substituting the optimal value function V^* into equation (2.21), we have

$$H(x, u, \nabla V^*) = p(x) + u^T Ru + \nabla V^*(f(x) + g(x)u). \tag{2.22}$$

Using the stationarity condition [26], the following optimal control policy is derived:

$$\begin{aligned} u^* &= \operatorname*{argmin}_{u \in \mathcal{U}_{ca}} H(x, u, \nabla V^*) \\ &= -\frac{1}{2} R^{-1} L_g V^{*T}, \end{aligned} \tag{2.23}$$

where $L_g V^* = \nabla V^* g(x)$. The Hamilton–Jacobi–Bellman (HJB) equation for the system of equation (2.4) is given as follows:

$$H(x, u^*, \nabla V^*) = p(x) + u^{*T} Ru^* + \nabla V^*(f(x) + g(x)u^*) = 0. \tag{2.24}$$

Once the optimal value V^* is solved from equation (2.24), we can determine the optimal control input u^* using equation (2.23). However, it should be noted that solving the HJB partial differential equation is generally challenging. As a result, the policy iteration (PI) method [27], as shown in algorithm 1, is employed to solve the solution of equation (2.24). Specifically, each iteration includes two stages: the policy evaluation step computes the value function under the current control policy, and the policy improvement step updates the control policy. To initiate the policy iteration algorithm in algorithm 1, an initial admissible control policy $u^{(0)}$ must be provided. Subsequently, we obtain $V^{(0)}$ by substituting $u^{(0)}$ into equation (2.25). The control policy $u^{(1)}$ is updated using $V^{(0)}$. This process is repeated until $|V^{(i+1)} - V^{(i)}|$ is not greater than the predetermined threshold ε_v. Note that in the first step of algorithm 1, a careful selection of an admissible control policy u^0 is required to stabilize the system of equation (2.4) and yield a finite value function V^0. A detailed proof of convergence and optimality can be found in [28, 29] and is omitted here.

Algorithm 1. Policy iteration

Step 1 (Initialization): Design an initial admissible control policy $u^{(0)}$, and an arbitrarily small positive constant ε_v.
Step 2 (Policy evaluation): Compute the value function $V^{(i)}$, where $i = 0, 1, 2, \ldots$, by solving the following equation:

$$p(x) + u^{(i)T} R u^{(i)} + L_f V^{(i)} + L_g V^{(i)} u^{(i)} = 0. \tag{2.25}$$

Step 3 (Policy improvement): Update the control policy $u^{(i+1)}$ using

$$u^{(i+1)} = -\frac{1}{2} R^{-1} L_g V^{(i)T}. \tag{2.26}$$

Step 4: If $|V^{(i+1)} - V^{(i)}| > \varepsilon_v$, return to step 2; otherwise, terminate the iteration.

Although the convergence of $V^{(i)}$ to the optimal value function and the stability of the closed-loop system of equation (2.4) with $u^{(i)}$ can be achieved by the PI algorithm, solving the GHJB equation of equation (2.25), which is a simpler linear PDE compared to the HJB of equation (2.24), is still challenging [29, 30]. To address this issue, neural networks (NNs) that provide effective tools to model complex nonlinear functions can be used to approximate the value function $V^{(i)}$. Specifically, we construct an NN, denoted as $V_{\text{nn}}(x)$, which possesses the properties of CLBFs to approximate the CLBF-based value function $V(x)$.

The value function $V(x)$ can be written as the sum of two components, i.e. $V(x) = V_l(x) + V_b(x)$, using equation (2.11), where $V_l(x) = \int_t^{\infty} \left(l(x) + u^T R u \right) \mathrm{d}\tau$ and $V_b(x) = \int_t^{\infty} \left(\mu b_{\text{nn}}(x) + v \right) \mathrm{d}\tau$. Thus, the CLBF-NN $V_{\text{nn}}(x)$ is developed with two parts, i.e. an FNN V_{lnn} that is constructed to approximate the function V_l (termed the Lyapunov neural network (LNN)), and a barrier neural network (BNN) V_{bnn} that is constructed using the architecture of FNN to approximate the function V_b. The input of the CLBF-NN $V_{\text{nn}}^{(i)}(x)$ is the state vector x, and its output is the predicted solution $V^{(i)}(x)$ to equation (2.25). As the true value of $V^{(i)}(x)$ is unknown, there are no labeled outputs available to train the CLBF-NN $V_{\text{nn}}^{(i)}(x)$. It is observed that the predicted output of CLBF-NN $V_{\text{nn}}^{(i)}(x)$ should be the solution to equation (2.25). We design the following loss function for training $V_{\text{nn}}^{(i)}(x)$:

$$\mathcal{L}_i = \frac{1}{N_v} \sum_{k=1}^{N_v} e_{v_i}(x_{t_k}, u_{t_k}^{(i)})^2, \tag{2.27}$$

with $e_{v_i}\left(x, u^{(i)}\right) = p(x) + u^{(i)\mathrm{T}} R u^{(i)} + L_f V_{\text{nn}}^{(i)} + L_g V_{\text{nn}}^{(i)} u^{(i)}$, where N_v represents the number of training samples. Here, x_{t_k} and $u_{t_k}^{(i)}$ denote the state vector x and the manipulated input $u^{(i)}$ of the nonlinear system described in equation (2.4) at the t_kth time step. Note that in practice, the continuous-time system of equation (2.4) is solved using numerical methods (e.g. the explicit Euler method) with a sufficiently small-time step, and it is assumed that the state measurement is available at every time step for data collection purposes. The algorithm for SRL is presented in algorithm 2, where W_i, $i = 0, 1, \ldots$, represents the weight of $V_{\text{nn}}^{(i)}$. Specifically, the algorithm involves the initialization of a CLBF-NN $V_{\text{nn}}^{(0)}(x, W_0)$ that satisfies the conditions specified in equations (2.10a)–(2.10d). Then, the initial admissible control policy $u^{(1)}$ can be designed by using equation (2.28) with $V_{\text{nn}}^{(0)}$. Subsequently, the exact solution $V^{(i)}$ to equation (2.25) is approximated using the constructed CLBF-NN $V_{\text{nn}}^{(i)}$, which is trained by minimizing the loss function of equation (2.27). It should be noted that $V_{\text{nn}}^{(i)}$ may not satisfy the properties of CLBF after each learning iteration. To address this challenge, when $V_{\text{nn}}^{(i)}$ is not a CLBF, we redesign the learning rate as $lr_j \leftarrow lr_{j-1}/2$, $j = 1, 2, \ldots$ and retrain the NN $V_{\text{nn}}^{(i)}$. In this case, W_j is the jth weight of the NN $V_{\text{nn}}^{(i)}$, and lr_j denotes the jth learning rate for training $V_{\text{nn}}^{(i)}$. Finally, the control policy is improved using equation (2.28) with the updated CLBF-NN V_{nn}. The convergence of this SRL algorithm can be found in [25].

Algorithm 2. SRL algorithm

Initialization with $V_{\text{nn}}^{(0)}(x, W_0)$ under which the control policy $u^{(1)}(x)$ (designed using Sontag's formula) is safe and admissible, and other parameters such as $i = 1$, epoch number N_e, an error tolerance ε_v and learning rate lr_0

while $|V_{\text{nn}}^{(i+1)} - V_{\text{nn}}^{(i)}| \geqslant \varepsilon_v$ **do**

for $t_k \leftarrow t_0$ **to** t_{N_v} **do**

Apply the control policy $u^{(i)} = \varphi(x_{t_k}) \in \mathcal{U}_c$ to the system of equation (2.4). Obtain the next state $x_{t_{k+1}}$, and save the data $(x_{t_k}, \ u_{t_k}^{(i)})$

end for

Set $j = 1$

while $V_{\text{nn}}^{(i)}$ is not a CLBF on the grid points in S **do**

$W_j \leftarrow W_i$

$lr_j \leftarrow lr_{j-1}/2$

for Epoch $\leftarrow$ 0 **to** N_e **do**

Train $V_{\text{nn}}^{(i)}$ by minimizing the cost function designed as equation (2.27)

end for

$j = j + 1$

end while

Update control policy $u^{(i+1)}$ using the following equation:

$$u^{(i+1)} = -\frac{L_f V_{\text{nn}}^{(i)} + \sqrt{L_f V_{\text{nn}}^{(i)2} + L_g V_{\text{nn}}^{(i)} R^{-1} L_g V_{\text{nn}}^{(i)T} p(x)}}{L_g V_{\text{nn}}^{(i)} L_g V_{\text{nn}}^{(i)T}} L_g V_{\text{nn}}^{(i)T} \tag{2.28}$$

end while

2.2.5 Application to a chemical process example

In this section, to assess the efficacy of the proposed control scheme, the simulation study of a well-mixed, non-isothermal continuous-stirred tank reactor (CSTR) is used:

$$\begin{aligned}\frac{\mathrm{d}C_A}{\mathrm{d}t} &= \frac{F}{V_L}(C_{A0} - C_A) - k_0 e^{\frac{-E}{RT}} C_A \\ \frac{\mathrm{d}T}{\mathrm{d}t} &= \frac{F}{V_L}(T_0 - T) - \frac{\Delta H k_0}{\rho_L C_p} e^{\frac{-E}{RT}} C_A + \frac{Q}{\rho_L C_p V_L},\end{aligned} \tag{2.29}$$

where the reactant concentration is represented by C_A. The reactor temperature and the heat supply/removal rate are denoted by T and Q, respectively. The volume of liquid in the reactor is denoted by V_L. In the feed stream to the reactor, the volumetric flow rate is represented by F, the temperature is T_0, and the reactant concentration is C_{A0}. Throughout the process, the liquid density is constant and is denoted as ρ_L, and the heat capacity is represented by C_p. The pre-exponential factor, activation energy, and enthalpy of the reaction are denoted by k_0, E, and ΔH, respectively. The parameter values for the aforementioned system can be obtained in [21] and are omitted here.

Let (C_{As}, T_s) denote the equilibrium point of equation (2.29). The aim is to optimize the performance of the system described by equation (2.29) by minimizing the following performance function while maintaining the state within the safe operating region and driving the state of the system to the equilibrium point $(0.57\ \text{kmol m}^{-3},\ 395.3\ \text{K})$:

$$V(x(t)) = \int_t^{\infty} (x^{\mathrm{T}}(\tau) Q x(\tau) + u^{\mathrm{T}}(x(\tau)) R u(x(\tau)) + \mu b(x(\tau)) + v) \mathrm{d}\tau, \tag{2.30}$$

with $Q = [9.35, 0.41; 0.41, 0.02]$ and $R = [1, 0; 0, 1]$. The manipulated inputs, namely the inlet concentration of the species A, $\Delta C_{A0} = C_{A0} - C_{A0_s}$, and the heat input rate $\Delta Q = Q - Q_s$, are subject to physical constraints. Specifically, the manipulated inputs $|\Delta Q|$ and $|\Delta C_{A0}|$ are saturated by $0.0167\ \text{kJ min}^{-1}$ and $1\ \text{kmol m}^{-3}$, respectively. We utilize deviation variables to rewrite the input and state vectors as $u = [\Delta C_{A0}, \Delta Q]^T$ and $x = [C_A - C_{As}, T - T_s]^T$, respectively.

To show the efficacy of the proposed SRL-based optimal control scheme in ensuring the optimality, stability, and safety of the closed-loop system, we consider the same bounded unsafe region embedded in the stability region as shown in [21]. Specifically, we design the unsafe region as $\mathcal{D} := \left\{ x \in \mathbb{R}^n \mid F(x) = \frac{(x_1 + 0.22)^2}{1} + \frac{(x_2 - 0.46)^2}{1 \times 10^4} < 2 \times 10^{-4} \right\}$, and the region $\mathcal{H}$ is designed as $\mathcal{H} := \{x \in \mathbb{R}^n \mid F(x) < 2.5 \times 10^{-4}\}$ such that $\mathcal{D}$ is a subset of $\mathcal{H}$, and $\mathcal{H}$ is a subset of ϕ_{uc}. The initial LNN and the trained LNN are shown in figures 2.5 and 2.6, from which it is demonstrated that the value function after training has been significantly reduced compared to the initial value function. The set ϕ_{uc} is represented by the blue region where the time derivative of the CLBF-NN V_{nn} is non-positive. When $|V_{\text{nn}}^{(i+1)} - V_{\text{nn}}^{(i)}| < 0.001$ is met, the approximate

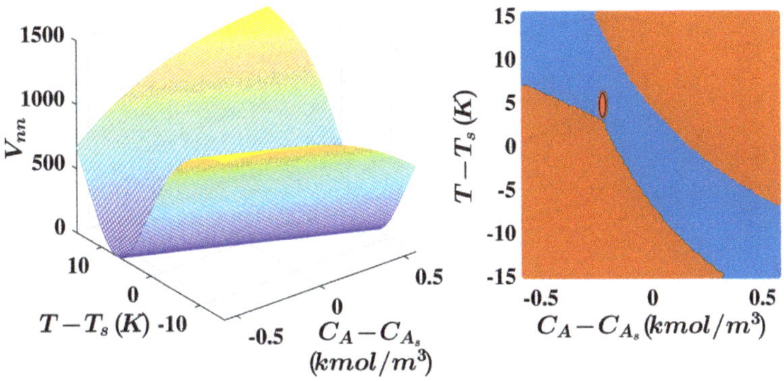

Figure 2.5. Predicted output of the initial CLBF-NN used to design the initial admissible control policy $u^{(0)}$ (left panel), and the set ϕ_{uc} within which $\dot{V}_{\text{nn}}$ is rendered negative under the controller $u^{(0)}$ (denoted by the blue region), and the unsafe region $\mathcal{D}$ represented by the orange elliptical region (right panel). (Reproduced with permission of Wiley from [25]. Copyright 2023 American Institute of Chemical Engineers.)

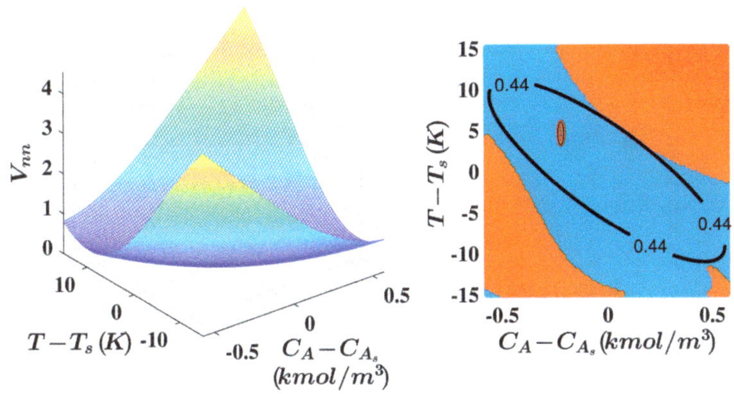

Figure 2.6. Predicted output of the CLBF-NN after 11 iterations (left panel), and the set ϕ_{uc} (blue region) and the unsafe region $\mathcal{D}$ (red elliptical region) for the CSTR under the control policy $u^{(13)}$ with the updated CLBF-NN (right panel). (Reproduced with permission of Wiley from [25]. Copyright 2023 American Institute of Chemical Engineers.)

optimal control policy is obtained and it is shown in figure 2.6. The largest level set contained in the stable region is 0.44. The approximated optimal control can ensure that the closed-loop state trajectory stays within the largest level set and never enters the unsafe region, which is the orange elliptical area in the picture for any initial state in this set.

The trajectories of the state variables over time arising from the different initial states are shown in figure 2.7. The state trajectories are shown to remain outside the region $\mathcal{D}$ that is unsafe in operation and ultimately be bounded within the terminal set under the safe RL algorithm. The control inputs that are constrained by (2 kmol m^{-3}, 0.167 K) for the initial condition (-0.26 kmol m^{-3}, 7 K) are shown

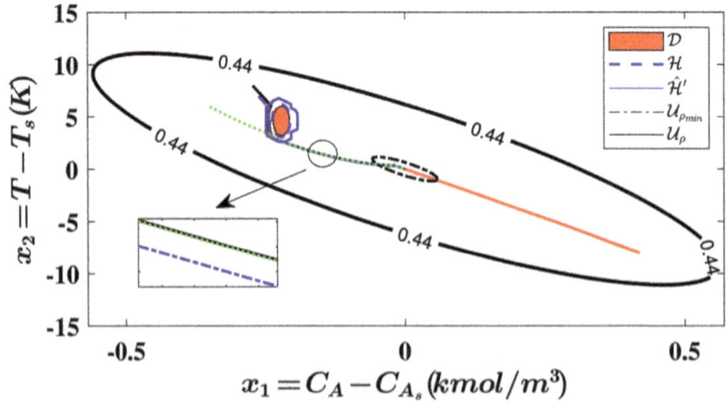

Figure 2.7. Closed-loop state trajectories under the safe optimal control policy, where the red, black, blue, and green lines represent the closed-loop state trajectories from the initial conditions (0.42 kmol m^{-3}, -8 K), (-0.27 kmol m^{-3}, 8 K), (-0.26 kmol m^{-3}, 7 K), and (-0.35 kmol m^{-3}, 6 K), respectively. (Reproduced with permission of Wiley from [25]. Copyright 2023 American Institute of Chemical Engineers.)

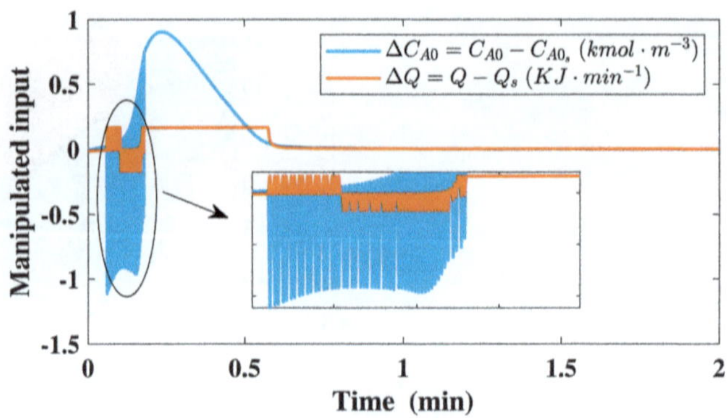

Figure 2.8. Manipulated input profiles for the initial condition (-0.26 kmol m^{-3}, 7 K) under the safe optimal control policy. (Reproduced with permission of Wiley from [25]. Copyright 2023 American Institute of Chemical Engineers.)

in figure 2.8. In particular, the state trajectory displays oscillatory behavior when approaching the boundary of $\hat{\mathcal{H}}'$, as control actions start to oscillate heavily from the initial stage due to the underlying dynamics of the closed-loop system. However, it is noticed that the oscillations are reduced during the training iterations, since SRL counteracts this unfavorable behavior by minimizing its performance function.

2.3 Supervised learning in modeling nonlinear systems

Parts of this section have been reproduced from [31] with permission from Elsevier.

In the previous section, we developed RL-based optimal control for nonlinear systems under the assumption that an accurate first-principles model is available.

However, in practical applications, this assumption is often challenging to meet due to the complex dynamics of industrial processes such as chemical, bioengineering, and energy systems. These processes exhibit intricate dynamics involving coupling, nonlinearity, and multivariable interactions, presenting a persistent research challenge. Therefore, supervised learning has garnered increased attention in recent years, particularly for modeling nonlinear systems [30–37]. In this subsection, we will introduce two supervised learning-based modeling methods: feedforward neural networks (FNNs) and recurrent neural networks (RNNs).

2.3.1 Data-driven modeling using feedforward neural network

This section presents the construction of an FNN with the control-affine form as demonstrated in our previous work [31] to predict the drift function and the control input gain function of equation (2.4). Specifically, the process of generating training data from the open-loop simulation is first presented. Subsequently, an FNN is developed, which has a control-affine structure that is the same as the nonlinear system of equation (2.4).

Data generation

The training data for NNs can be collected from various sources, such as real-world systems, experiments, and computer simulations. In this chapter, we introduce the data generation method using open-loop simulations of the system of equation (2.4) under various initial conditions x_s within a certain region in the state-space and control inputs $u_s \in \mathcal{U}_c$, following the method in [38–41]. Note that unlike real-world chemical processes that often operate around steady-states, we have more freedom to build a representative dataset using computer simulation by choosing different initial conditions and control actions. Specifically, we conduct N open-loop simulation runs and obtain N sets of data samples. Each open-loop simulation is performed for a fixed duration, for example, for one sampling period t_s. During each simulation, the control input u_s is executed using a sample-and-hold scheme (i.e. $u_s(t) = u_s(t_p)$, for $\forall\, t \in [t_p, t_p + t_s)$, where t_s is one sampling period). Using the explicit forward difference technique with a sufficiently small integration time step, equation (2.4) can be solved in simulation. The time derivative of the state, denoted as $\dot{x}_s$, can be approximated using some numerical methods (e.g. forward difference method) based on two consecutive sampled states. We denote each state-action tuple as $\{(x_s, u_s)\}$, and its corresponding transition as $\dot{x}_s$. Open-loop simulations produce N data samples denoted as $\{(x_{si}, u_{si}, \dot{x}_{si})\}_{i=1}^{N}$. Thus, the training dataset for FNN is obtained, where the FNN input data are $\{(x_{si}, u_{si})\}_{i=1}^{N}$ and the labeled output data are $\{(\dot{x}_{si})\}_{i=1}^{N}$.

FNN architecture and training process

After generating the training dataset, we will develop an FNN model to approximate the nonlinear system described by equation (2.4). The output of a general class of FNN models can be represented as follows:

$$Y = \omega_\ell \alpha_{\ell-1}(\omega_{\ell-1}\alpha_{\ell-2}(\cdots \alpha_1(\omega_1 X))), \tag{2.31}$$

where X is the input vector of the FNN, and Y denotes the output vector, respectively. In this work, the FNN takes the manipulated input u and the state x as its inputs, and $\dot{\hat{x}}$ as its output for modeling the nonlinear system described by equation (2.4). α_i (for $i = 1, 2, \ldots, \ell$) is the activation function of the ith layer of the FNN, where ℓ represents the total number of layers in the FNN. Additionally, $\omega_i \in \mathbb{R}^{d_i \times d_{i-1}}$ represents the weight matrix that connects the ith and $(i-1)$th layers of the FNN, with d_i and d_{i-1} denoting the dimensions of the ith and $(i-1)$th layer, respectively.

We develop an architecture of FNN that is of the same control-affine form of equation (2.4). Unlike the conventional fully connected structure of NNs that has been adopted in many process modeling works, (e.g. [32, 36]), the proposed FNN model, as depicted in figure 2.9, can not only predict the future states (or the time derivative of state $\dot{\hat{x}}$), but also estimate the values of $f(\cdot)$ and $g(\cdot)$ when provided with the current state measurement. Therefore, the novel design of FNN will significantly benefit the design of model-based controllers, e.g. Sontag law, and optimal control policies that require information of system dynamics. Note that the target values of $f(x)$ and $g(x)$ are unknown since they cannot be measured in general. Therefore, the training process of the proposed FNN model is still carried out using the training samples $\{(x_{si}, u_{si}, \dot{x}_{si})\}_{i=1}^{N}$. The structure of the proposed control-affine FNN, denoted as $\mathcal{N}$, is shown in figure 2.9, where the mapping from FNN input x to the unknown function $f(x)$ in the last hidden layer is denoted as $\mathcal{N}_f$, and the one that predicts $g(x)$ given the current state x is denoted as $\mathcal{N}_g$. To differentiate the notations of prediction from those of the true values in equation (2.4), the symbols $\hat{f}(x)$ and $\hat{g}(x)$ are used to represent the predicted values derived from the FNN model depicted in figure 2.9.

Let $\dot{\hat{x}}$ denote the output of the FNN. Following the construction of the FNN in figure 2.9, the predicted output is represented as

$$\dot{\hat{x}} = H_{\text{nn}}(x, u) := \hat{f}(x) + \hat{g}(x)u. \tag{2.32}$$

The approximation error between the labeled output $\dot{x} = H(x, u) := f(x) + g(x)u$ (i.e. the nonlinear system of equation (2.4) without disturbances) and its prediction $\dot{\hat{x}}$ can be written as

$$e_x = \dot{x} - \dot{\hat{x}}. \tag{2.33}$$

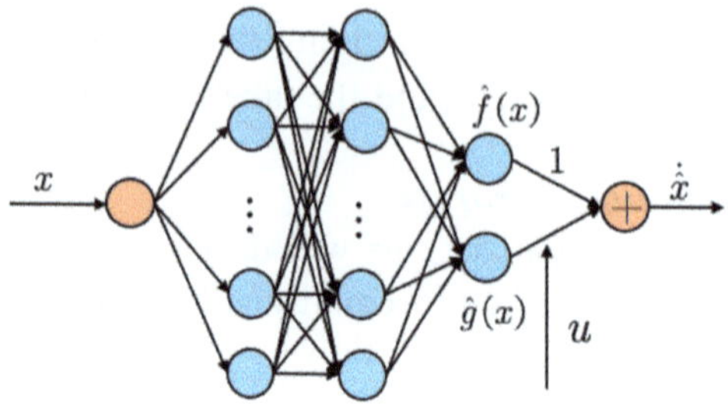

Figure 2.9. Structure of a control-affine FNN.

The following loss function is designed to train the FNN model:

$$L = \frac{1}{N}\sum_{i=1}^{N}|\dot{x}_{si} - \hat{\dot{x}}_{si}|^2 = \frac{1}{N}\sum_{i=1}^{N}|e_{xi}|^2. \tag{2.34}$$

To optimize the weights of the FNN and minimize the loss function, an optimizer, such as Adam or stochastic gradient descent (SGD), can be used for the training process. Since the approximation error indicates the ability of an NN to predict new samples based on the patterns and rules it has learned from the training dataset, it is important to theoretically study the approximation performance of the proposed FNN model $\mathcal{N}$ that is optimized by minimizing the loss function of equation (2.34). The detailed proofs can be found in [42] and are omitted here.

2.3.2 Application of NN model-based RL

After constructing the control-affine NN model, this section utilizes the continuous-stirred tank reactor described by equation (2.29) to demonstrate the effectiveness of the NN model-based RL algorithm in dealing with the optimal control problem with unknown process dynamics, and safety constraints.

In this simulation, the bound of the manipulated inputs is defined as $|u_1| \leqslant 4$ kmol m^{-3} and $|u_2| \leqslant 0.167$ kJ min^{-1}. The bounded unsafe region is designed to be the same as in the simulation example in section 2.2.5. The objective is to achieve stability for the CSTR system by maintaining it at its steady-state. The parameters of the cost function are designed as follows: $Q = [9.35, 0.41; 0.41, 0.02]$ and $R = [1/100, 0; 0, 1/500]$. The RL algorithm has been demonstrated in section 2.2.4. In this simulation, the system model is assumed to be unknown, and an FNN with control-affine architecture consisting of two hidden layers with 128 and 64 neurons is developed following the method in section 2.3.1. To show the efficacy of the control policy derived from the learned optimal value function and NN model, closed-loop simulations are performed. We choose a few different initial states within the stability region, and show the corresponding closed-loop simulation results in figure 2.10. In this figure, the orange region in the upper right corner

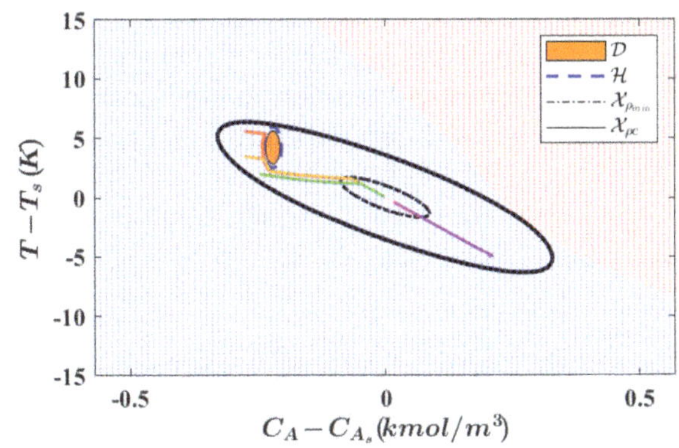

Figure 2.10. Closed-loop state trajectories under the approximate optimal controller with the four initial conditions of (-0.31 kmol m^{-3}, 5.25 K), (-0.28 kmol m^{-3}, 3.5 K), (-0.28 kmol m^{-3}, 5.6 K), and (0.2 kmol m^{-3}, -5 K), respectively.

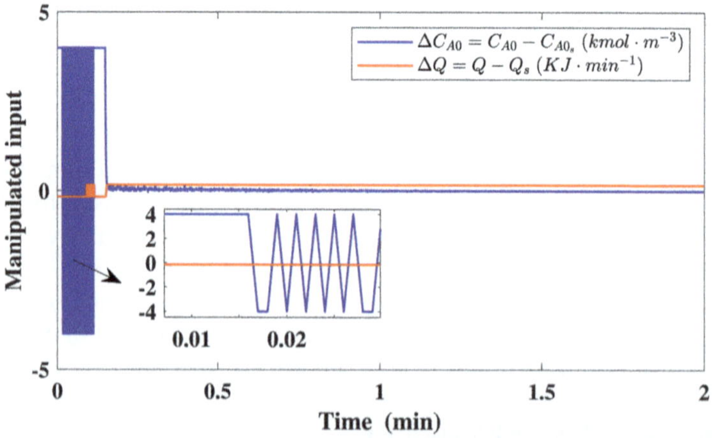

Figure 2.11. Manipulated inputs for the initial condition (−0.31 kmol m^{-3}, 5.25 K) under the safe optimal control policy.

indicates the set of states where $\mathrm{d}\hat{V}/\mathrm{d}t > 0$, while the remaining area represents the stability region where $\mathrm{d}\hat{V}/\mathrm{d}t \leqslant 0$. The set $\mathcal{X}_{\rho c}$ is the largest invariant set in the stability region. It is found that in the presence of disturbances, all trajectories starting from different initial conditions converge to the region $\mathcal{X}_{\rho_{\min}}$ while avoiding the unsafe region $\mathcal{D}$. Figure 2.11 shows the control inputs for the initial condition (−0.31 kmol m^{-3}, 5.25 K). The simulation results of the chemical reactor example obtained using the proposed machine-learning-based RL scheme demonstrate that the FNN-based optimal controller achieves the desired control performance even if the system dynamics are complex and unknown.

2.3.3 Application of NN model-based MPC

RNNs are an efficient approach to modeling nonlinear dynamical systems using time-series data. RNN architectures include feedback loops that introduce past information stored in hidden neurons at earlier time steps to the current network. Thus, the RNN internal states preserve past network information and can be considered the memory of an RNN. It allows RNN models to capture process dynamic behavior in a way conceptually similar to nonlinear dynamical systems that can be described by state-space ordinary differential equations. A schematic of an RNN is shown in figure 2.12. It is observed that the state information derived from earlier inputs is fed back into the network, which exhibits a dynamic behavior. In this section, we consider a one-hidden-layer RNN with hidden states $\mathbf{x}_i \in \mathbf{R}^{d_h}$ computed as follows:

$$\mathbf{x}_{i,t} = \sigma_h(U\mathbf{x}_{i,t-1} + W\mathbf{u}_{i,t}), \tag{2.35}$$

where σ_h is the element-wise nonlinear activation function (e.g. ReLU). $U \in \mathbf{R}^{d_h \times d_h}$ and $W \in \mathbf{R}^{d_h \times d_x}$ are weight matrices connected to the hidden states and input vector, respectively. The output layer $\mathbf{o}_{i,t}$ is computed as follows:

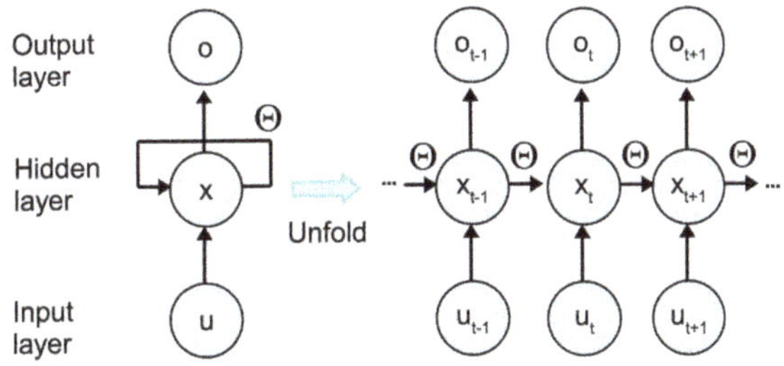

Figure 2.12. An RNN and its unfolded structure, where Θ is the weight matrix, x is the state vector, u is the input vector and o is the output vector (for the nonlinear system in the form of equation (2.4), the output vector is equal to the state vector).

$$\mathbf{o}_{i,t} = \sigma_y(V\mathbf{x}_{i,t}), \tag{2.36}$$

where $V \in \mathbf{R}^{d_y \times d_h}$ is the weight matrix, and σ_y is the element-wise activation function in the output layer (typically linear unit for regression problems).

In addition to RL, NNs that model nonlinear chemical processes have been widely used for model predictive control that optimizes control actions over a finite, receding horizon by solving an optimization problem [43, 44]. We use a chemical process example to illustrate the application of MPC using RNN models to stabilize the system at its steady-state. The formulation of the RNN-based Lyapunov-based MPC (LMPC) is first given as follows:

$$\mathcal{J} = \min_{u \in S(\Delta)} \int_{t_k}^{t_{k+N}} L(\tilde{x}(t), u(t))\mathrm{d}t \tag{2.37a}$$

$$\text{s. t.} \quad \dot{\tilde{x}}(t) = F_{\text{nn}}(\tilde{x}(t), u(t)) \tag{2.37b}$$

$$u(t) \in \mathcal{U}_c, \quad \forall \ t \in [t_k, t_{k+N}) \tag{2.37c}$$

$$\tilde{x}(t_k) = x(t_k) \tag{2.37d}$$

$$\dot{\hat{V}}(x(t_k), u) \leqslant \dot{\hat{V}}(x(t_k), \Phi_{\text{nn}}(x(t_k)), \quad \text{if } x(t_k) \in \Omega_{\hat{\rho}} \backslash \Omega_{\rho_{\text{nn}}} \tag{2.37e}$$

$$\hat{V}(\tilde{x}(t)) \leqslant \rho_{\text{nn}}, \quad \forall \ t \in (t_k, t_{k+N}), \quad \text{if } x(t_k) \in \Omega_{\rho_{\text{nn}}}, \tag{2.37f}$$

where $\tilde{x}$ is the predicted state trajectory, $S(\Delta)$ is the set of piecewise constant functions with period Δ, and N is the number of sampling periods in the prediction horizon. F_{nn} represents the RNN model of equations (2.35)–(2.36) that is developed for the nonlinear system of equation (2.4). $\dot{\hat{V}}(x, u)$ is used to represent $\frac{\partial \hat{V}(x)}{\partial x}(F_{\text{nn}}(x, u))$. The optimal input trajectory computed by the LMPC is denoted by $u^*(t)$, which is calculated over the entire prediction horizon $t \in [t_k, t_{k+N})$. The control action computed for the first sampling period of the prediction horizon $u^*(t_k)$ is sent by the LMPC to be applied over the first sampling period and the LMPC is resolved at the next sampling time. In the optimization problem of equation (2.37a), the objective function of equation (2.37a) is the integral of $L(\tilde{x}(t), u(t))$ over the prediction horizon. The constraint of equation (2.37b) is the RNN model that is used to predict the states of the closed-loop system. Equation (2.37c) defines the input constraints applied over the entire prediction horizon. Equation (2.37d) defines the initial condition $\tilde{x}(t_k)$ of equation (2.37b), which is the state measurement at $t = t_k$. The constraint of equation (2.37e) forces the closed-loop state to move towards the origin if $x(t_k) \in \Omega_{\hat{\rho}} \backslash \Omega_{\rho_{nn}}$. However, if $x(t_k)$ enters $\Omega_{\rho_{nn}}$, the states predicted by the RNN model of equation (2.37b) will be maintained in $\Omega_{\rho_{nn}}$ for the entire prediction horizon.

The chemical process example we considered is a well-mixed, non-isothermal continuous-stirred tank reactor (CSTR) that is similar to the one in section 2.2.5, but with an irreversible second-order exothermic reaction, which can be described by the following material and energy balance equations:

$$\frac{dC_A}{dt} = \frac{F}{V}(C_{A0} - C_A) - k_0 e^{\frac{-E}{RT}} C_A^2 \tag{2.38a}$$

$$\frac{dT}{dt} = \frac{F}{V}(T_0 - T) + \frac{-\Delta H}{\rho_L C_p} k_0 e^{\frac{-E}{RT}} C_A^2 + \frac{Q}{\rho_L C_p V}. \tag{2.38b}$$

The process parameter values can be found in [32] and are omitted here. The CSTR is initially operated at the unstable steady-state $(C_{As}, \ T_s) = (1.95 \text{ kmol m}^{-3}, 402 \text{ K})$. The manipulated inputs are the inlet concentration of species A and the heat input rate, which are represented by the deviation variables $\Delta C_{A0} = C_{A0} - C_{A0_s}, \Delta Q = Q - Q_s$, respectively. The manipulated inputs are bounded as follows: $|\Delta C_{A0}| \leqslant 3.5 \text{ kmol m}^{-3}$ and $|\Delta Q| \leqslant 5 \times 10^5 \text{ kJ hr}^{-1}$. Therefore, the states and the inputs of the closed-loop system are $x^T = [C_A - C_{As} \ \ T - T_s]$ and $u^T = [\Delta C_{A0} \ \ \Delta Q]$, respectively, such that the equilibrium point of the system is at the origin of the state-space, (i.e. $(x_s^*, u_s^*) = (0, 0)$). The Lyapunov function $V(x) = x^T P x$ is designed with a positive-definite matrix $P = [1060, 22; 22, 0.52]$. The control objective of this simulation is to operate the CSTR at the unstable equilibrium point $(C_{As}, \ T_s)$ by manipulating the heat input rate ΔQ and the inlet concentration ΔC_{A0} under the LMPC using RNN models.

The RNN model is designed to have two hidden recurrent layers consisting of 96 and 64 recurrent units, respectively. The closed-loop stability region $\Omega_{\hat{\rho}}$ for the CSTR system described by the RNN model is characterized as the largest level set of $\hat{V}$ in $\hat{\phi}_u$ and also a subset of Ω_ρ (i.e. $\Omega_{\hat{\rho}} \subset \Omega_\rho$) with $\hat{\rho} = 368$. Additionally, $\rho_{\min} = 2$ is determined through extensive simulations for $u \in U_c$. The LMPC cost function of equation (2.37a) is designed to be $L(x, u) = |x|_{Q_1}^2 + |u|_{Q_2}^2$, where $Q_1 = [500 \ \ 0; \ 0 \ 0.5]$ and $Q_2 = [1 \ \ 0; \ 0 \ \ 8 \times 10^{-11}]$, such that the minimum value of L is achieved at the origin. To illustrate the effectiveness of the proposed LMPC of equation (2.37a) using RNN models, we also compare it with the LMPC using a linear state-space model and the first-principles model of equation (2.38a), respectively. The linear state-space model for the CSTR system of equation (2.38a) is identified with the following form:

$$\dot{x} = A_s x + B_s u, \tag{2.39}$$

where x and u are the state vector and the manipulated input vector, and A_s and B_s are coefficient matrices for the state-space model. Following the system identification method in [45], the numerical algorithms for subspace state-space system identification are utilized to obtain A_s and B_s as follows: $A_s = 100 \times [-0.154, -0.003; 5.19, 0.138]$ and $B_s = [4.03, 0; 1.23, 0.004]$.

We perform simulation results under the LMPC using the RNN model and the first-principles model of equation (2.38a), respectively. It should be noted that the ML approach is used when only data are available. The first-principles model in the following simulations substitutes for the role of the experimental/industrial process. In other words, the MPC using the first-principles model only serves as a benchmark to determine the best performance that any data-driven modeling method can achieve. Simulation results are shown in figures 2.13 and 2.14. Specifically, figure 2.13 shows the comparison of state trajectories for the closed-loop system under the LMPC using an RNN model, the state-space model of equation (2.39) and the first-principles model of equation (2.38a), respectively. It is demonstrated that in all cases, the state of the closed-loop system of equation (2.38a) is maintained within $\Omega_{\hat{\rho}}$ for all times and driven to $\Omega_{\rho_{\min}}$ under LMPC for an initial condition $x_0 = (-1, 63.6)$. However, through the comparison of state profiles under the LMPC using three different models in figure 2.14, it is shown that the state trajectory under the RNN model stays closer to the one under the actual nonlinear model of equation (2.38a), and thus, takes less time to settle to the steady-state compared to the LMPC using the state-space model. It is also noted that although the LMPC using the state-space model performs well for some initial conditions close to the origin, it shows oscillation for initial conditions near the boundary of the closed-loop stability region $\Omega_{\hat{\rho}}$ because the linear state-space model of equation (2.39) is not able to capture the nonlinearities of the CSTR in this region. Therefore, the LMPC using the RNN model outperforms the one using the state-space model in terms of faster convergence speed and improved closed-loop stability.

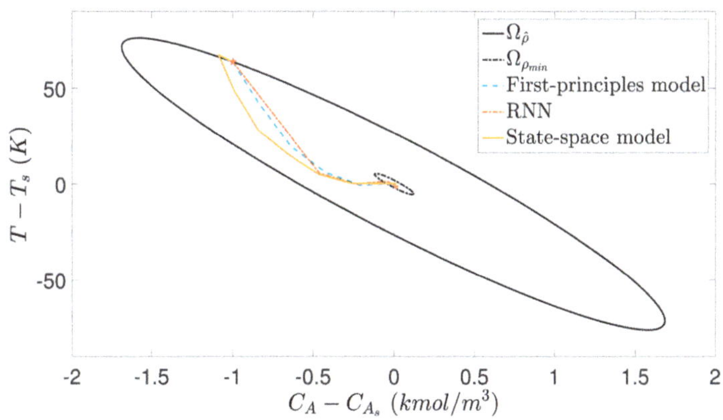

Figure 2.13. The state-space profiles for the closed-loop CSTR under the LMPC using the following models: the first-principles model (blue trajectory), the RNN model (red trajectory), and the linear state-space model (yellow trajectory) for an initial condition (–1, 63.6). (Reproduced with permission of Wiley from [36]. Copyright 2019 American Institute of Chemical Engineers.)

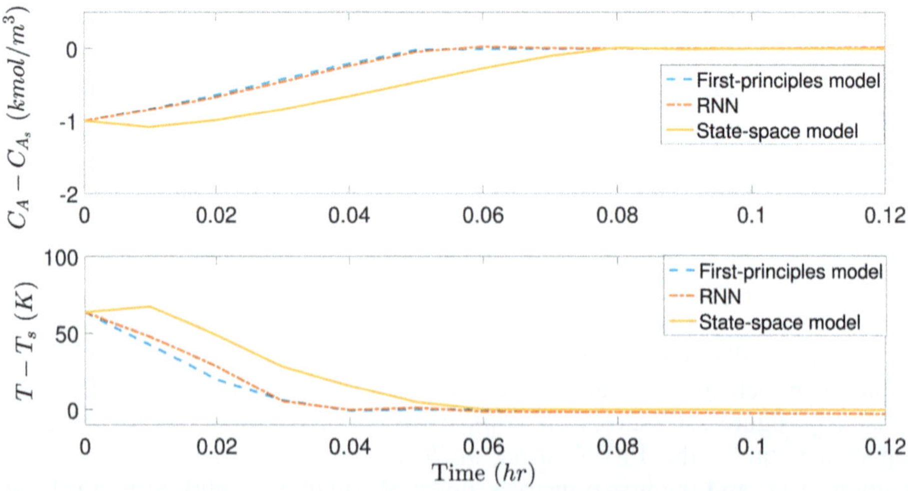

Figure 2.14. The state profiles ($x_1 = C_A - C_{As}$ and $x_2 = T - T_s$) for the initial condition (–1, 63.6) under the LMPC using the following models: the first-principles model (blue trajectory), the RNN model (red trajectory), and the linear state-space model (yellow trajectory). (Reproduced with permission of Wiley from [36]. Copyright 2019 American Institute of Chemical Engineers.)

2.4 Conclusion

This chapter discussed the applications of ML in process systems engineering. Specifically, we focused on RL and discussed its application in the optimization and control of nonlinear chemical processes. Subsequently, we introduced the use of supervised learning in modeling the nonlinear dynamics of chemical processes, and showed the development of NN-based RL and NN-based MPC, with applications to chemical reactor examples. It was demonstrated that the applications of ML in the PSE domain offer more flexible and efficient solutions for data analysis, design, and control. As technology continues to evolve, the exploration of ML in PSE awaits further advancement to promote innovation and development in engineering practice.

Bibliography

[1] Bangi M S F and Kwon J S-I 2023 Deep hybrid model-based predictive control with guarantees on domain of applicability *AIChE J.* **69** e18012

[2] Choi H-K and Kwon J S-I 2019 Multiscale modeling and control of kappa number and porosity in a batch-type pulp digester *AIChE J.* **65** e16589

[3] Daoutidis P *et al* 2023 Machine learning in process systems engineering: challenges and opportunities *Comput. Chem. Eng.* **181** 108523

[4] Kwon J S-I, Nayhouse M, Christofides P D and Orkoulas G 2014 Modeling and control of crystal shape in continuous protein crystallization *Chem. Eng. Sci.* **107** 47–57

[5] Lee J H, Shin J and Realff M J 2018 Machine learning: overview of the recent progresses and implications for the process systems engineering field *Comput. Chem. Eng.* **114** 111–21

[6] Qin S J and Chiang L H 2019 Advances and opportunities in machine learning for process data analytics *Comput. Chem. Eng.* **126** 465–73
[7] Shang C and You F 2019 Data analytics and machine learning for smart process manufacturing: recent advances and perspectives in the big data era *Engineering* **5** 1010–6
[8] Shin J, Badgwell T A, Liu K H and Lee J H 2019 Reinforcement learning-overview of recent progress and implications for process control *Comput. Chem. Eng.* **127** 282–94
[9] Wu Z and Christofides P D 2021 *Process Operational Safety and Cybersecurity* (Berlin: Springer)
[10] Nian R, Liu J and Huang B 2020 A review on reinforcement learning: introduction and applications in industrial process control *Comput. Chem. Eng.* **139** 106886
[11] Bangi M S F and Kwon J S-I 2021 Deep reinforcement learning control of hydraulic fracturing *Comput. Chem. Eng.* **154** 107489
[12] Chen C, Xie L, Jiang Y, Xie K and Xie S 2023 Robust output regulation and reinforcement learning-based output tracking design for unknown linear discrete-time systems *IEEE Trans. Autom. Control* **68** 2391–8
[13] Singh Sidhu H, Siddhamshetty P and Kwon J S 2018 Approximate dynamic programming based control of proppant concentration in hydraulic fracturing *Mathematics* **6** 132
[14] Wang T, Wang Y, Yang X and Yang J 2023a Further results on optimal tracking control for nonlinear systems with nonzero equilibrium via adaptive dynamic programming *IEEE Trans. Neural Netw. Learn. Syst.* **34** 1900–10
[15] Wang Y and Wu Z 2024b Physics-informed reinforcement learning for optimal control of nonlinear systems *AIChE J.* **70** e18542
[16] Sutton R S and Barto A G 2018 *Reinforcement Learning: An Introduction* (Cambridge, MA: MIT Press)
[17] Xu J, Wang J, Rao J, Zhong Y and Wang H 2022 Adaptive dynamic programming for optimal control of discrete-time nonlinear system with state constraints based on control barrier function *Int. J. Robust Nonlinear Control* **32** 3408–24
[18] Marvi Z and Kiumarsi B 2021 Safe reinforcement learning: a control barrier function optimization approach *Int. J. Robust Nonlinear Control* **31** 1923–40
[19] Lin Y and Sontag E D 1991 A universal formula for stabilization with bounded controls *Syst. Control Lett.* **16** 393–7
[20] Sontag E D 1989 A 'universal' construction of Artstein's theorem on nonlinear stabilization *Syst. Control Lett.* **13** 117–23
[21] Wu Z, Albalawi F, Zhang Z, Zhang J, Durand H and Christofides P D 2019a Control Lyapunov-barrier function-based model predictive control of nonlinear systems *Automatica* **109** 108508
[22] Wieland P and Allgöwer F 2007 Constructive safety using control barrier functions *IFAC Proc.* **40** 462–7
[23] Zhang X, Peng Y, Luo B, Pan W, Xu X and Xie H 2021 Model-based safe reinforcement learning with time-varying state and control constraints: an application to intelligent vehicles arXiv: 2112.11217
[24] Romdlony M Z and Jayawardhana B 2016 Stabilization with guaranteed safety using control Lyapunov-barrier function *Automatica* **66** 39–47
[25] Wang Y and Wu Z 2024a Control Lyapunov-barrier function-based safe reinforcement learning for nonlinear optimal control *AIChE J.* **70** e18306
[26] Lewis F L, Vrabie D and Syrmos V L 2012 *Optimal Control* (New York: Wiley)

[27] Howard R A 1960 *Dynamic Programming and Markov Processes* (New York: Wiley)
[28] Abu-Khalaf M and Lewis F L 2005 Nearly optimal control laws for nonlinear systems with saturating actuators using a neural network HJB approach *Automatica* **41** 779–91
[29] Beard R W, Saridis G N and Wen J T 1997 Galerkin approximations of the generalized Hamilton–Jacobi–Bellman equation *Automatica* **33** 2159–77
[30] Saridis G N and Lee C S G 1979 An approximation theory of optimal control for trainable manipulators *IEEE Trans. Syst. Man Cybern.* **9** 152–9
[31] Wang Y, Kadakia Y, Wu Z and Christofides P D 2024 An overview of control methods for process operational safety and cybersecurity *Method of Process Systems in Energy Systems: Current System Part I* **vol 8** (Amsterdam: Elsevier) p 1
[32] Wu Z, Tran A, Rincon D and Christofides P D 2019b Machine learning-based predictive control of nonlinear processes. Part I: theory *AIChE J.* **65** e16729
[33] Sitapure N and Kwon J S-I 2023 CrystalGPT: enhancing system-to-system transferability in crystallization prediction and control using time-series-transformers *Comput. Chem. Eng.* **177** 108339
[34] Son S H, Choi H-K and Kwon J S-I 2021 Application of offset-free Koopman-based model predictive control to a batch pulp digester *AIChE J.* **67** e17301
[35] Wang Y, Wang T, Yang X and Yang J 2023b Gradient descent-Barzilai Borwein-based neural network tracking control for nonlinear systems with unknown dynamics *IEEE Trans. Neural Netw. Learn. Syst.* **34** 305–15
[36] Wu Z, Tran A, Rincon D and Christofides P D 2019c Machine-learning-based predictive control of nonlinear processes. Part II: computational implementation *AIChE J.* **65** e16734
[37] Zheng Y, Wang X and Wu Z 2022 Machine learning modeling and predictive control of the batch crystallization process *Ind. Eng. Chem. Res.* **61** 5578–92
[38] Bangi M S F, Kao K and Kwon J S-I 2022 Physics-informed neural networks for hybrid modeling of lab-scale batch fermentation for β-carotene production using *Saccharomyces cerevisiae Chem. Eng. Res. Des.* **179** 415–23
[39] Bhadriraju B, Kwon J S-I and Khan F 2023 An adaptive data-driven approach for two-timescale dynamics prediction and remaining useful life estimation of Li-ion batteries *Comput. Chem. Eng.* **175** 108275
[40] Bonassi F, Farina M, Xie J and Scattolini R 2022 On recurrent neural networks for learning-based control: recent results and ideas for future developments *J. Process Control* **114** 92–104
[41] Sitapure N and Kwon J S-I 2024 Machine learning meets process control: unveiling the potential of LSTMc *AIChE J.* **70** e18356
[42] Wang Y and Wu Z 2025 Machine learning model-based optimal tracking control of nonlinear affine systems with safety constraints *Int. J. Robust Nonlinear Control* **35** 511–35
[43] Hwang G, Sitapure N, Moon J, Lee H, Hwang S and Kwon J S-I 2022 Model predictive control of lithium-ion batteries: development of optimal charging profile for reduced intracycle capacity fade using an enhanced single particle model (SPM) with first-principled chemical/mechanical degradation mechanisms *Chem. Eng. J.* **435** 134768
[44] Narasingam A and Kwon J S-I 2019 Koopman Lyapunov-based model predictive control of nonlinear chemical process systems *AIChE J.* **65** e16743
[45] Kheradmandi M and Mhaskar P 2018 Data driven economic model predictive control *Mathematics* **6** 51

IOP Publishing

High-Performance Computing and Artificial Intelligence in Process Engineering

Mingheng Li and Yi Heng

Chapter 3

Graph-based control invariant set approximation and its applications

Song Bo, Benjamin Decardi-Nelson and Jinfeng Liu

3.1 Introduction

Motivated by sustainability and profitability goals, modern chemical process systems have undergone continuous evolution, becoming increasingly complex with diverse operating constraints [1]. Characterizing the domain within which a system can feasibly operate while adhering to all constraints is crucial for designing controllers. This task, however, presents challenges, particularly when dealing with nonlinear dynamics, a common occurrence in chemical process systems. In the presence of poorly defined feasible operating regions, there is a risk of significantly reduced control performance or even instability [2]. The term control invariant set (CIS) is employed in control theory to describe such feasible operating regions [3]. The concept of CIS is closely interconnected with notions such as reachable set [4], null controllable set [5], and viability kernel [6].

Characterizing the CIS of a system poses challenges due to the necessity of considering all potential trajectories. The majority of efforts have focused on approximating CISs of linear systems [7–11]. Many of these studies rely on dynamic programming [12], offering convergence to the largest CIS. However, for general nonlinear systems, only limited results exist [13–15]. In [2, 5, 16], an algorithm was developed to estimate the null controllable regions of nonlinear systems by enlarging an initial estimate of a control invariant set. Another algorithm, presented in [13], employs the difference of two convex functions to estimate convex robust control invariant sets for nonlinear systems. In [4], a grid-based algorithm solving a time-dependent Hamilton–Jacobi formulation was used to estimate reachable sets of continuous systems with uncertainties. Notably, these algorithms face challenges in determining the largest or the maximum CIS, and all exhibit limited scalability to the dimensionality of the system.

doi:10.1088/978-0-7503-6174-3ch3

Motivated by the successful application of graph theory in analysing large-scale nonlinear networks [17, 18], this chapter introduces graph theoretical algorithms for approximating CISs of general nonlinear systems. Graph theory has previously proven effective in approximating invariant sets of autonomous systems [17–20], where system trajectories are represented as directed graphs and analysed using graph-based methods, offering enhanced scalability concerning system dimensions. This chapter will extend these algorithms to address controlled systems, specifically targeting the approximation of potentially the maximum robust CIS (RCIS) of general nonlinear systems in the presence of unknown disturbances. These algorithms provide both inner and outer approximations with high precision. When integrated with the subdivision technique, the presented algorithms exhibit significantly improved computational efficiency compared to existing grid-based methods.

In the subsequent sections of this chapter, we will delve into the applications of CIS in the realms of economic model predictive control (MPC) and in the development of safe reinforcement learning (RL) agents. In the first application, we will explore how CIS can be used in MPC, particularly in the context of zone tracking, ensuring stability and later in the design of economic MPC to ensure both stability and economic optimality. MPC stands as the established solution for advanced process control within the process industry [21]. Economic MPC (EMPC) represents a novel class of MPC algorithms that directly integrate economic optimization considerations. Traditional MPC commonly follows a reference trajectory, often an optimal steady-state operating point [21]. However, this approach may constrain the controller's flexibility and compromise process performance during transient operations. EMPC, on the other hand, directly incorporates a general economic objective in its dynamic optimizations [22, 23], earning recognition as a promising next-generation advanced process control algorithm.

Zone tracking is often associated with addressing economic performance in system operations [24, 25]. It is a natural objective that emerges in many real-world problems. For instance, in agricultural practices, maintaining the soil moisture in a specific range proves to be more feasible than aiming for a precise level. The literature has documented various applications of MPC with zone tracking, such as diabetes treatment [26, 27], regulation of building heating systems [28], control of coal-fired boiler-turbine generating systems [29], and management of irrigation systems [30, 31]. MPC with zone tracking (ZMPC) not only addresses these objectives but also offers a natural framework for simultaneously handling economic optimization and operational safety or tracking goals [32, 33]. Building upon this concept, ZMPC was integrated with a secondary economic objective, prioritizing the zone-tracking objective over the economic one [24, 25]. This approach aligns with the principles of Lypuanov-based economic MPC (LEMPC) [34] and can be viewed as a generalized version of LEMPC. The forthcoming sections will present a design for zone economic MPC (ZEMPC), demonstrating how a CIS can be leveraged to ensure stability. Notably, this design stands out by eliminating the need for an optimal steady-state, a requirement present in other EMPC approaches. While this chapter will focus on ZEMPC, CIS may also be incorporated into many other MPC designs. For example, in recent years, there have emerged many data-driven MPCs such as [35–39]. The approach discussed in this chapter may also be applicable to these data-driven MPCs for enhanced stability.

A second application of CIS or the maximum RCIS finds its utility in shaping the design of safe RL agents. RL, as a class of optimal control algorithms, empowers machines to learn an optimal policy or closed-loop control law by iteratively maximizing future rewards through interactions with the environment or system [40]. However, the standard RL approach lacks the incorporation of state constraints in its design, and it cannot guarantee closed-loop stability, thereby limiting its applicability in real-world scenarios [41]. To overcome these limitations, the field has witnessed the development of safe RL algorithms explicitly addressing state constraints during training to ensure the closed-loop stability of the learned policy. There are different approaches to safe RL [41–43]. One common strategy involves formulating the problem as a constrained Markov decision process (CMDP), where a cost function defined in terms of safety state constraints guides the optimization of policies [44, 45]. This transformation shifts the focus from solely maximizing returns to balancing return maximization with the minimization of constraint violations. Another approach integrates safety constraints into the exploration process by modifying the value function, encouraging RL agent to favor safer actions during exploration [46, 47]. Despite these efforts, guaranteeing state constraint satisfaction and closed-loop stability remains a challenge. An alternative strategy combines MPC with RL, treating MPC as parameterized value or policy neural networks [48, 49]. While this approach allows explicit handing of safety state constraints in MPC, it involves recursively solving MPC optimization problems, introducing a notable computational burden.

In the realm of control theory, the critical role of CISs in ensuring the stability of control systems is widely recognized [50]. These sets define the states for which a feedback control law is consistently available to keep the system within the set [3]. Incorporating CIS in RL design provides a promising way to ensure stability guarantee by constraining the agent's interactions to controllable states. Researchers have explored two main approaches in this context. The first assumes that the CIS is unknown and is concurrently approximated along with the policy. This approach seeks to constrain the policy based on the estimated CIS to achieve safety, though the safety guarantee may be compromised if the CIS approximation is inaccurate. This approach indeed aligns with the CMDP idea [51]. The second approach assumes knowledge of CIS and involves filtering or projecting risky actions into safe ones, often incorporating a standalone safety filter after the RL policy [52–55]. In this chapter, we will show how the maximum RCIS can act as a designated state space for the RL agent to explore safely to ensure state constraint satisfaction and stability guarantee.

This chapter is organized as follows. Section 3.2 presents the preliminaries including the notation used and the class of systems considered in this chapter. Section 3.3 presents the graph-based algorithms for CIS approximation taking into account model uncertainty. Section 3.4 shows how CIS can be used in ZEMPC design to ensure stability. Section 3.5 introduces safe RL and shows how the maximum RCIS can be used in the offline training and online implementation of RL to impose state constraint and ensure closed-loop stability. Section 3.6 concludes the chapter.

3.2 Preliminaries

Parts of this section have been reproduced with permission from [57]. Copyright 2021 Elsevier.

3.2.1 Notation

Throughout this chapter, $\mathbb{Z}$ denotes the set of integers $\{\dots, -2, -1, 0, 1, 2, \dots\}$. $\mathbb{Z}_+$ denotes the set of non-negative integers $\{0, 1, 2, \dots\}$. $\mathbb{Z}_M^N$ represents the set of integers from M to N $\{M, M+1, \cdots, N\}$. $\{z_k\}_{k\in\mathbb{Z}_+}$ denotes an ordered set of numbers according to $k \in \mathbb{Z}_+$ $\{z_0, z_1, z_2, \dots\}$. $\mathbb{B}$ denotes the unit ball in $\mathbb{R}^n$ with respect to the infinity norm. The operator $|\cdot|$ denotes the Euclidean norm of a scalar or vector and the operator $||\cdot||_p$ denotes the p-norm of a scalar or a vector. A directed graph is denoted as $G = (\mathbb{V}, \mathbb{E})$ with $\mathbb{V}$ denoting the set of vertices of the graph and $\mathbb{E}$ denoting the set of ordered pairs of vertices known as edges. A function $f: \mathbb{X} \to \mathbb{X}$ is said to be homomorphic in $\mathbb{X}$ if it is continuous with continuous inverse in $\mathbb{X}$. The operator diam$(\cdot)$ denotes the diameter of a set and is defined as: $\text{diam}(\mathbb{C}_i) = \sup\{|x - y|: x, y \in \mathbb{C}_i\}$. The operators $\max\{\cdot\}$ and $\min\{\cdot\}$ find the maximum and minimum of elements in a vector variable, respectively. The operator $a \odot b$ represents element-wise multiplication between vectors a and b.

3.2.2 System description

In this chapter, our focus is on a class of discrete-time nonlinear systems described by the following model:

$$x_{k+1} = f(x_k, u_k, w_k), \tag{3.1}$$

where $x \in \mathbb{R}^n$ represent the system state, $u \in \mathbb{R}^m$ is the control input, and $w \in \mathbb{R}^n$ denotes the unknown disturbance input. The variable $k \in \mathbb{Z}_+$ denotes the discrete sampling time. It is assumed that the state, control and disturbance are subject to the following constraints:

$$x \in \mathbb{X} \subseteq \mathbb{R}^n, \quad u \in \mathbb{U} \subseteq \mathbb{R}^m \text{ and } w \in \mathbb{W} \subseteq \mathbb{R}^n, \tag{3.2}$$

where $\mathbb{X}$, $\mathbb{U}$ and $\mathbb{W}$ are compact sets and $\mathbb{U}$ and $\mathbb{W}$ have the origin in their interiors. It is also assumed that the function $f: \mathbb{X} \times \mathbb{U} \times \mathbb{W} \to \mathbb{X}$ is sufficiently smooth and for each $x \in \mathbb{X}$, $u \in \mathbb{U}$ and $w \in \mathbb{W}$, and $f(x, u, w)$ is uniquely defined.

For system (3.1), we are interested in determining the maximum RCIS contained in $\mathbb{X}$. A set $R \subseteq \mathbb{X}$ is said to be an RCIS for system (3.1) if for every $x \in R$, there exists a control input $u \in \mathbb{U}$ such that $f(x, u, w) \in R$ for all $w \in \mathbb{W}$ [3].

3.3 CIS approximation

Parts of this section have been reproduced with permission from [57]. Copyright 2021 Elsevier.

In section 3.3.1, we will introduce some important concepts and results in invariant set approximation for nonlinear autonomous systems. Based upon the concepts and results in section 3.3.1, we will then extend the results to systems with control and disturbance inputs as described by (3.1).

3.3.1 Set invariance conditions for autonomous systems

For now, we consider discrete-time autonomous systems as shown below:

$$x_{k+1} = \hat{f}(x_k), \tag{3.3}$$

where $\hat{f} : \mathbb{X} \to \mathbb{X}$ is a homomorphism on a compact domain $\mathbb{X} \subseteq \mathbb{R}^n$. A symbolic image of (3.3) is a directed graph approximation of its dynamics. To obtain such a symbolic image, the state space $\mathbb{X}$ is discretized to a finite covering, $\mathbb{C} = \{\mathbb{C}_1, \cdots, \mathbb{C}_l\}$, which is a collection of closed sets, known as cells, $\mathbb{C}_i$, $i = 1, \cdots, l$, such that

$$\mathbb{X} \subseteq \cup_{\mathbb{C}_i \in \mathbb{C}} \mathbb{C}_i, \tag{3.4a}$$

$$\mathbb{C}_i \cap \mathbb{C}_j = \varnothing, \quad \forall \mathbb{C}_i, \mathbb{C}_j \in \mathbb{C} \text{ with } i \neq j. \tag{3.4b}$$

The diameter of the covering $\mathbb{C}$ is defined as $\text{diam}(\mathbb{C}) := \max_{\mathbb{C}_i \in \mathbb{C}} \text{diam}(\mathbb{C}_i)$.

Definition 3.1. (Symbolic image [56]) Let G be a directed graph with l vertices where each vertex is a cell $\mathbb{C}_i$ in a finite covering $\mathbb{C}$ of $\mathbb{X}$ of system (3.3). The vertices $\mathbb{C}_i$ and $\mathbb{C}_j$ are connected by a directed edge $\mathbb{C}_i \to \mathbb{C}_j$ if

$$\mathbb{C}_j \cap \hat{f}(\mathbb{C}_i) \neq \varnothing,$$

where $\hat{f}(\mathbb{C}_i) := \{y \mid y = \hat{f}(x), x \in \mathbb{C}_i\}$. The graph G is called a symbolic image of (3.3) with respect to the covering $\mathbb{C}$.

The image of a cell $\hat{f}(\mathbb{C}_i)$ can be approximated by sampling sufficiently large points in the cell and then finding the images of the points. The image of the cell is approximated as the union of the images of the sampled points in the cell. Once the symbolic image graph of system (3.3) is constructed, it can be investigated to obtain an outer approximation of the largest forward invariant set of system (3.3). To proceed, we recall some more relevant definitions.

Definition 3.2. (Admissible path [56]) A sequence $\{z_k\}_{k \in \mathbb{Z}_+}$ with each element z_k taking a value from the set of vertices of G is called an admissible path if for each $k \in \mathbb{Z}_+$, the graph G contains the edge $z_k \to z_{k+1}$.

Definition 3.3. (Out-degree of a vertex [56]) The out-degree of a vertex in a directed graph is the number of edges (including self-edges) going out of the vertex.

If a vertex $\mathbb{C}_i$ of the symbolic image of system (3.3) has zero out-degree, then its image $\hat{f}(\mathbb{C}_i)$ does not intersect with $\mathbb{C}_i$ or any other vertices of the symbolic image.

That is, $\hat{f}(\mathbb{C}_i) \cap \mathbb{X} = \varnothing$. This implies that $\hat{f}(\mathbb{C}_i)$ lies outside $\mathbb{X}$ and any trajectory from $\mathbb{C}_i$ will exit the state constraint $\mathbb{X}$.

Definition 3.4. (Strongly connected graph [56]) A directed graph $G = (\mathbb{V}, \mathbb{E})$ is said to be strongly connected if there is an admissible path in both directions between each pair of vertices of the graph.

Definition 3.5. (Non-leaving cells [56]) The set of vertices of the directed graph G with infinite admissible paths passing through them, denoted as $\mathbb{I}^+(G)$, is the union of vertices of the largest strongly connected component subgraph of the directed graph G and any vertex of G that is not in the largest strongly connected components but has a path to a vertex in the largest strongly connected component subgraph.

Theorem 3.1 below describes how to obtain an outer approximation of the largest forward invariant set of system (3.3) [56].

Theorem 3.1. Let $G = (\mathbb{V}, \mathbb{E})$ with vertices $\mathbb{V}$ and a set of ordered pairs of vertices $\mathbb{E}$ be a symbolic image of system (3.3) with respect to a finite covering $\mathbb{C}$ of $\mathbb{X}$. Then

i. The vertices of the largest strongly connected component subgraph $G_s = (\mathbb{V}_s, \mathbb{E}_s)$ of G have infinite admissible paths passing through them.
ii. Any element of $\mathbb{V}$ but not $\mathbb{V}_s$ with a path to at least one vertex of G_s also has an infinite admissible path passing through it.
iii. The union of the elements of (*i*) and (*ii*), $\mathbb{I}^+(G)$, is a closed neighborhood of the largest forward invariant set $\hat{R}$ of (3.3) in $\mathbb{X}$. That is,

$$\hat{R} \subseteq \mathbb{I}^+(G). \tag{3.5}$$

For a more detailed discussion on theorem 3.1, readers are referred to [56, 57].

3.3.2 Robust control invariance conditions and approximation

In this section, we extend the results for autonomous systems introduced in the previous section to constrained dynamical systems with control and disturbance inputs as in (3.1). For this purpose, we need to consider the set-valued map of f defined as follows:

$$F(x, w) := f(x, \mathbb{U}, w) = \{f(x, u, w)\}\bigcup_{u\in\mathbb{U}}, \tag{3.6}$$

where F is a inclusion map that sends each (x, w) pair to a point in the set of feasible next states when u takes values in $\mathbb{U}$. If we denote the next state as x^+, this also implies that

$$x^+ \in F(x, w). \tag{3.7}$$

This inclusion will be used to construct the symbolic image of (3.1) for RCIS approximation. Its function is similar to $\hat{f}(\mathbb{C}_i)$ in the autonomous case presented in the previous section. This inclusion map can be approximated by sampling sufficient x, w and u in their admissible sets and evaluating the corresponding f.

Due to the presence of the disturbance w, the construction of the symbolic image of (3.1) for RCIS approximation is not a straightforward extension of the procedures used in the autonomous case. For an RCIS, it should be ensured that for all possible w realization, there is a u value that can maintain the state x within the RCIS. This implies that we need to consider the worst-case scenario for w's influence on the system dynamics. In the worst-case scenario, the control input seeks to enlarge the robust control invariant set while the disturbance seeks to make it smaller.

To address this, we construct individual graphs for each w realization using the inclusion map (3.7) and analyse them using theorem 3.1. The intersection of the resulting sets for each w outer approximates the largest robust control invariant set. Let us denote $G_w = (\mathbb{V}_w, \mathbb{E}_w)$ as the symbolic image of $F(\mathbb{X}, w)$ with respect to the finite covering $\mathbb{C}$ of $\mathbb{X}$. For each graph G_w, we can find its set of vertices with infinite admissible paths $\mathbb{I}^+(G_w)$. Let us denote the intersection of these sets as $\mathbb{K}$. That is,

$$\mathbb{K} := \bigcap\nolimits_{w \in \mathbb{W}} \mathbb{I}^+(G_w). \tag{3.8}$$

The set $\mathbb{K}$ represents the cells with infinite admissible paths passing through them irrespective of the disturbance realization, which provides an outer approximation of the largest RCIS of system (3.1). Theorem 3.2 below summarizes this result.

Theorem 3.2. If $\mathbb{K}$ is obtained according to (3.8), then the set $\mathbb{K}$ is a closed neighbourhood of the largest robust control invariant set $R(\mathbb{X})$ of system (3.1) with constraint sets (3.2); that is,

$$R(\mathbb{X}) \subseteq \mathbb{K}. \tag{3.9}$$

A proof of theorem 3.2 can be found in [57]. It should be pointed out that the computational complexity of evaluating the set $\mathbb{K}$ based on sampling could be high and increases quickly with the increase of dimensions of x, u, or w. In the remainder of this section, we present an algorithm for computing the maximum RCIS of system (3.1) incorporating subdivision technique [58] to significantly improve the computational efficiency.

The algorithm adopts an iterative procedure and starts with an initial covering $\mathbb{C}_{d_0}$ of the state space $\mathbb{X}$ with diameter d_0. The initial covering of the system could be rough and it is refined iteratively. At each iteration, it contains three main steps: subdivision, graph construction and cell selection. When at the pth iteration, the calculations in the three steps are as follows:

- *Subdivision*: In this step, a finer covering $\hat{\mathbb{C}}_{d_p}$ is generated by dividing each of the cells into two along one of the dimensions. If $\mathbb{C}_{d_{p-1}}$ is the covering at the previous iteration, then $d_p < d_{p-1}$. The two coverings $\hat{\mathbb{C}}_{d_p}$ and $\mathbb{C}_{d_{p-1}}$ cover the same set such that $\bigcup_{\mathbb{C}_i \in \hat{\mathbb{C}}_{d_p}} \mathbb{C}_i = \bigcup_{\mathbb{C}_j \in \mathbb{C}_{d_{p-1}}} \mathbb{C}_j$. The set covered by the two coverings does not change. The new covering has cells with smaller diameters. In each iteration of the algorithm, the dimension along which the cells are divided is rotated. In algorithm 3.1, the function subdivide() is used to perform the subdivision operations.

- *Graph construction*: once the subdivision is done, graphs should be constructed based on the updated subdivision. Note that the covering $\hat{\mathbb{C}}_{d_p}$ provides a discretized representation of the state space. When constructing the graphs, sampling of w and u should be performed. For a specific cell (quantized x value), for each w realization, all the potential u values (samples) should be considered and f should be evaluated to find the next state. For each w realization, a graph can be constructed accordingly. Therefore, for the considered realizations of w, a collection of graphs can be constructed. Let us denote the collection as $\mathbb{G}_p = \{G_w^p = (\mathbb{V}_w^p, \mathbb{E}_w^p), \ \ \forall w \in \mathbb{W}\}$, where G_w^p denotes the graph corresponding to one w realization with the vertices being the cells in the covering $\hat{\mathbb{C}}_{d_p}$ ($\mathbb{V}_w^p = \hat{\mathbb{C}}_{d_p}$) and the edges generated by propagating the states one step ahead $\left(\mathbb{E}_w^p = \left\{\left(\mathbb{C}_i, \mathbb{C}_j\right) \in \hat{\mathbb{C}}_{d_p} \times \hat{\mathbb{C}}_{d_p} \mid F\left(\mathbb{C}_i, w\right) \cap \mathbb{C}_j \neq \varnothing\right\}\right)$. In algorithm 3.1, the function graph() is used to perform the graph construction computing.

  The graph construction step can be computationally expensive. There are a few approaches to reduce the computational complexity or to increase the computation efficiency. First, the structure of system (3.1) may be taken into account to reduce the computational complexity. For example, if the system is affine in control and disturbance inputs, feedback linearization can be used to cancel out some or all the disturbance inputs as demonstrated in [57]. Second, the evaluation of f for the cells can be parallelized and implemented in GPUs to significantly improve the computational efficiency [59]. Moreover, adaptive cell subdivision can be used to significantly reduce the number of cells in graph construction [59].
- *Cell selection*: this step involves the selection of the set of cells that have infinite paths passing through them irrespective of the realization of w using the conditions stated in theorem 3.2. The covering $\hat{\mathbb{C}}_{d_p}$ is updated to obtain $\mathbb{C}_{d_p}$. That is,

  $$\mathbb{C}_{d_p} = \{\mathbb{C}_i \in \hat{\mathbb{C}}_{d_p} \mid \mathbb{C}_i \in \bigcap\nolimits_{G \in \mathbb{G}_k} \mathbb{I}^+(G)\}.$$

  The cells that are not selected are discarded while the selected ones go on to the next iteration. This is represented in algorithm 3.1 as the function select().

The complete algorithm is summarized in algorithm 3.1. The algorithm is initialized using the state constraint $\mathbb{X}$, which ensures that the state constraints are satisfied when searching for the RCIS. In the above discussion, the implementation of the state propagation (3.7) is realized through sampling of the disturbance and control inputs. An alternative is to treat the control and disturbance sets as intervals and use interval arithmetic for the computations. Notice that the computational load of the algorithm depends heavily on the number of cells generated at each iteration. This grows exponentially as the algorithm progresses. Similar to other algorithms for numerically computing invariant sets, there is a trade off between computational load and accuracy. Let us denote by R_k the collection of closed sets after the kth iteration of algorithm 3.1. That is, $R_p = \cup_{\mathbb{C}_i \in \mathbb{C}_{d_p}} \mathbb{C}_i$. Theorem 3.3 summarizes the convergence property of algorithm 3.1.

Algorithm 3.1. Computing the outer approximations of maximum RCIS.

Input: System (3.1), constraint sets (3.2) and maximum number of iterations N
Output: largest robust control invariant set

$\mathbb{C}_{d_0} \leftarrow \mathbb{X}$ `// Initialization`
for $p \leftarrow 1, 2, 3, \cdots, N$ **do**
    if $\mathbb{C}_{d_p} = \emptyset$ **then**
        $\mathbb{C}_{d_N} \leftarrow \mathbb{C}_{d_p}$
        break
    if $\mathbb{C}_{d_p} = \mathbb{C}_{d_{p-1}}$ **then**
        $\mathbb{C}_{d_N} \leftarrow \mathbb{C}_{d_p}$
        break
    $\hat{\mathbb{C}}_{d_p} \leftarrow \mathit{subdivide}(\mathbb{C}_{d_p})$
    $\mathbb{G}_p \leftarrow \mathit{graph}(\hat{\mathbb{C}}_{d_p})$
    $\mathbb{C}_{d_{p+1}} \leftarrow \mathit{select}(\mathbb{G}_p)$
return $\mathbb{C}_{d_N}$

Theorem 3.3. Let $R(\mathbb{X})$ be the largest robust control invariant set of system (3.1) with constraint sets (3.2). Consider the sequence $\{R_p\}_{p \in \mathbb{Z}_+}$ generated by algorithm 3.1. Then $R(\mathbb{X}) = R_\infty$.

Algorithm 3.1 provides outer approximations of the largest RCIS contained in $\mathbb{X}$. In many cases, inner approximations of the RCIS are preferred. Algorithm 3.2 provides an algorithm to obtain inner approximations of the maximum RCIS. This is achieved by modifying system (3.1) to the following form:

$$x_{k+1} = f(x_k, u_k, w_k) + \varepsilon\mathbb{B}, \tag{3.10}$$

where $\varepsilon > 0$ is a small positive constant and $\mathbb{B}$ denotes the unit ball in $\mathbb{R}^n$ with respect to the infinity norm. Note that in algorithm 3.2, the stopping criterion in line

6 plays an important role in seeking an inner approximation of the control invariant set. Further, if the system is affine in w, then the modification in (3.10) is equivalent to increasing the disturbance set W by $\varepsilon\mathbb{B}$ ($W_\varepsilon = W + \varepsilon\mathbb{B}$).

Theorem 3.4 below summarizes the convergence properties of algorithm 3.2 to an inner approximation of the largest RCIS.

Algorithm 3.2. Computing the inner approximation of maximum RCIS.

Input: system (3.10), constraint sets (3.2) and ε

Output: Inner approximation of largest robust control invariant set

1 $\mathbb{C}_{d_0} \leftarrow X$ `// Initialization`

2 **for** $k \leftarrow 1, 2, 3, \cdots$ **do**

3   **if** $\mathbb{C}_{d_k} = \emptyset$ **then**

4         $\mathbb{C}_{d_N} \leftarrow \mathbb{C}_{d_k}$

5   **break**

6   **if** $\mathbb{C}_{d_{k-1}} \subseteq \mathbb{C}_{d_k} + \varepsilon\mathbb{B}$ **then**

7         $\mathbb{C}_{d_N} \leftarrow \mathbb{C}_{d_k}$

8   **break**

9     $\hat{\mathbb{C}}_{d_k} \leftarrow subdivide(\mathbb{C}_{d_k})$

10     $\mathbb{G}_k \leftarrow graph(\hat{\mathbb{C}}_{d_k})$

11     $\mathbb{C}_{d_{k+1}} \leftarrow select(\mathbb{G}_k)$

12 **return** $\mathbb{C}_{d_N}$

Theorem 3.4. Consider system (3.10) and constraint sets (3.2). Let R_k^ε denote set generated by algorithm 3.2 in the pth iteration ($R_p^\varepsilon = \cup_{\mathbb{C}_i \in \mathbb{C}_{d_p}} \mathbb{C}_i$). Then, there exists $p' \in \mathbb{Z}_+$ and $\varepsilon > 0$ such that the following condition,

$$R_p^\varepsilon \subseteq R_{p+1}^\varepsilon + \varepsilon\mathbb{B}, \tag{3.11}$$

holds for all $p \geqslant p'$. For any $p \geqslant p'$, the set R_{p+1}^ε is robust control invariant and is an inner approximation of the maximum RCIS.

Proofs of theorems 3.3 and 3.4 can be found in [57].

3.3.3 Application to a chemical process

In this section, we apply the algorithms introduced in the previous section to a chemical process. Let us consider a well-mixed continuously stirred tank reactor (CSTR) where an elementary first-order reaction of the form $A \rightarrow B$ takes place. It is assumed that the reaction is exothermic and a cooling jacket is equipped with the reactor to remove excessive heat. Based on mass and energy balances, the following modeling equations are obtained to describe the dynamics of the CSTR:

$$\frac{\mathrm{d}c_A}{\mathrm{d}t} = \frac{q}{V}(c_{Af} - c_A) - k_0 \exp\left(-\frac{E}{RT}\right)c_A \tag{3.12a}$$

$$\frac{\mathrm{d}T}{\mathrm{d}t} = \frac{q}{V}(T_f - T) + \frac{-\Delta H}{\rho c_p}k_0 \exp\left(-\frac{E}{RT}\right)c_A + \frac{UA}{V\rho c_p}(T_c - T), \tag{3.12b}$$

where c_A (mol L^{-1}) and T (K) respectively indicate the concentration of the reactant A and the temperature of the reaction mixture, T_c (K) is the coolant temperature, q (L min^{-1}) represents the volumetric flow rate of the inlet and outlet flows, c_{Af} (mol L^{-1}) is the concentration of reactant A in the inlet flow and T_f (K) is the temperature of flow, V (L) is the volume of the reaction mixture, k_0 (min^{-1}) is the pre-exponential factor of the reaction rate, E (J) denotes the activation energy, R (J K^{-1}) is the universal gas constant, ρ (g L^{-1}) is the density of the mixture, c_p (J (g $\cdot$ K)$^{-1}$) denotes the specific heat capacity of the mixture, ΔH (J mol^{-1}) is the heat of reaction and UA (J (min·K)$^{-1}$) is the heat transfer coefficient. The values of the parameters used in the simulations are presented in table 3.1.

The model (3.12) is discretized with a step size $h = 0.1$ min to obtain a discrete-time nonlinear system in the form of (3.1). For the CSTR, the state is $x = [c_A, \quad T]^T$ and the control input is $u = T_c$. It is considered that the state and control input are constrained to be in the following sets: $0.0 \leqslant x_1 \leqslant 1.0$, $345.0 \leqslant x_2 \leqslant 355.0$, and $285.0 \leqslant u \leqslant 315.0$. Disturbances on c_{Af} (w_1) and T_f (w_2) are considered. They are assumed to be bounded and additive to the two variables. The bounds on the two disturbances are as follows: $|w| \leqslant [0.1, 2.0]^T$.

The discretized CSTR model and the constraints on x and u are considered as inputs to algorithm 3.2. The obtained maximum CIS (without considering w) and the maximum RCIS (taking into account w) of the CSTR are shown in figure 3.1. To obtain the CIS and RCIS, in algorithm 3.2, ϵ is tuned in a way such that $\epsilon\mathbb{B}$ is

Table 3.1. Parameter values.

Parameter	Unit	Value
q	L min^{-1}	100.0
V	L	100.0
c_{Af}	mol L^{-1}	1.0
T_f	K	350.0
E/R	K	8750.0
k_0	min^{-1}	7.2×10^{10}
$-\Delta H$	J mol^{-1}	5.0×10^4
UA	J (min·K)$^{-1}$	5.0×10^4
c_p	J (g $\cdot$ K)$^{-1}$	0.239
ρ	g L^{-1}	1000.0

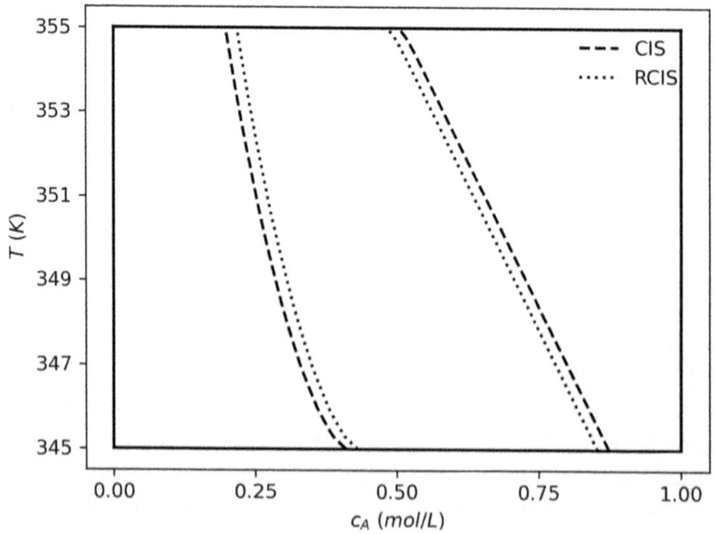

Figure 3.1. The maximum CIS (dashed line) and RCIS (dotted line) of the CSTR obtained by algorithm 3.2 within the constraint set $\mathbb{X}$ (solid line).

sufficiently small. The criterion in line 6 of algorithm 3.2 is to check whether there is any change in the shape of the found CIS. For each cell, five state samples were sampled within the cell; for each sampled state, five manipulated input values uniformly distributed across the constraint set were considered; when computing the RCIS, only values of w at the two extremes were considered.

It can be seen that when disturbance presents, the CIS becomes smaller. Extensive simulations were performed to verify the obtained CIS sets through taking random state samples inside CISs and then trying various u values to see whether the next step state is within the CISs. The test results verified that the found CIS and RCIS are indeed control invariant.

3.4 CIS in economic MPC through zone tracking

Parts of this section have been reproduced with permission from [60]. Copyright 2023 Elsevier.

In this section, we show how a CIS can be used in the design of economic MPC through zone tracking (ZEMPC) to ensure stability while optimizing the economic objective. We first present an MPC with zone-tracking (ZMPC) design and then discuss how an economic objective can be considered in the ZMPC and how a CIS can be taken advantage to ensure the closed-loop stability.

3.4.1 MPC with zone tracking

In MPC with zone tracking, the control objective is to drive the state of system (3.1) into a target tracking set (zone) $\mathbb{X}_t \subseteq \mathbb{X}$ and keep the state inside $\mathbb{X}_t$ thereafter in the presence of uncertainties.

Equation (3.13) shows the design (a dynamic optimization) of a ZMPC at time instant k:

$$\min_{u_0,\cdots,u_{N-1}} \sum_{i=0}^{N-1} \ell_z(\tilde{x}_i) \tag{3.13a}$$

$$\text{s. t. } \tilde{x}_{i+1} = f(\tilde{x}_i, u_i, 0), \quad i \in \mathbb{Z}_0^{N-1} \tag{3.13b}$$

$$\tilde{x}_0 = x_k \tag{3.13c}$$

$$\tilde{x}_i \in \mathbb{X}, \quad i \in \mathbb{Z}_0^{N-1} \tag{3.13d}$$

$$u_i \in \mathbb{U}, \quad i \in \mathbb{Z}_0^{N-1} \tag{3.13e}$$

$$\tilde{x}_N \in \tilde{\mathbb{X}}_f, \tag{3.13f}$$

where (3.13b) is the nominal process model with $w_i \equiv 0$ for all i, (3.13c) is the initial condition, (3.13d) and (3.13e) are the state and input constraints, respectively, (3.13f) denotes the terminal constraint where $\tilde{\mathbb{X}}_f$ denotes the terminal set, N is a positive constant that represents the prediction horizon of the controller, and $\tilde{x}_i$ denotes the predicted states. In (3.13a), $\ell_z(\cdot)$ represents the zone-tracking objective and is defined as follows:

$$\ell_z(z) = \min_{z_z} c_1||z - z_z||_1 + c_2||z - z_z||_2^2 \tag{3.14a}$$

$$\text{s. t. } z_z \in \tilde{\mathbb{X}}_t, \tag{3.14b}$$

where c_1 and c_2 are non-negative weighting factors, $\tilde{\mathbb{X}}_t \subset \mathbb{X}_t \subseteq \mathbb{X}$ denotes a modified target zone used within the ZMPC, and z_z is a slack variable taking values from the modified target zone $\tilde{\mathbb{X}}_t$ as in (3.14b). The zone-tracking objective is a weighted summation of the 1-norm and squared 2-norm of the minimum difference between the actual state and the slack variable, which is a representation of the distance between the system state and the modified target set $\tilde{\mathbb{X}}_t$. When the system state is outside the modified target set, ℓ_z is positive. When the system state converges to the modified target set, ℓ_z equals zero.

Taking the definition of (3.14) into consideration, the objective of controller (3.13) is to find the optimal input sequence for N future steps such that the system state is driven towards and kept inside the modified target zone $\tilde{\mathbb{X}}_t$ while constraints (3.13b)–(3.13e) are satisfied. In addition, the terminal state $\tilde{x}_N$ is forced to converge to $\tilde{\mathbb{X}}_f$. The terminal set $\tilde{\mathbb{X}}_f$ should be a CIS within the largest CIS $\tilde{\mathbb{X}}_t^M$ found in the modified target set $\tilde{\mathbb{X}}_t$. That is, $\tilde{\mathbb{X}}_f \subseteq \tilde{\mathbb{X}}_t^M \subseteq \tilde{\mathbb{X}}_t \subset \mathbb{X}_t$. The modified target set is a subset of the original target set such that $\tilde{\mathbb{X}}_t \subset \mathbb{X}_t$. The original target set is reduced to the modified target set to take into account the impact of uncertainty.

The ZMPC is implemented in the receding horizon fashion. At time $k \in \mathbb{Z}_+$, the optimal control input trajectory $u^*(i|k), \ i \in \mathbb{Z}_0^{N-1}$ for N future steps is obtained by

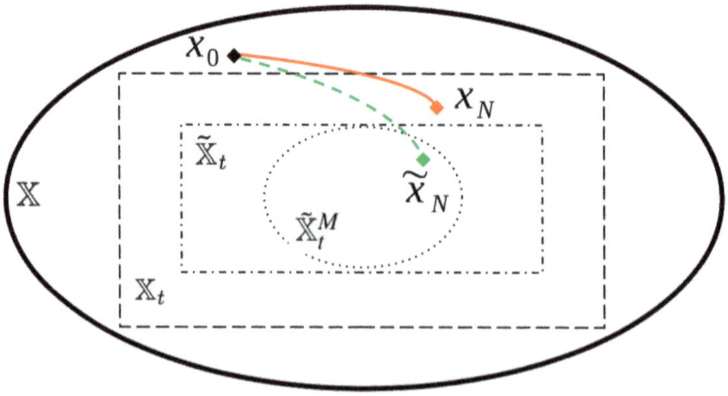

Figure 3.2. Sample nominal system state trajectory (green dashed line) and the actual state trajectory (red solid line) of the closed-loop system under the ZMPC.

solving the optimization problem (3.13). Only the first step control input for $i = 0$ is applied to system (3.1) at time k. At $k + 1$, the optimization problem (3.13) is evaluated again with updated initial condition (state information at $k + 1$). Figure 3.2 illustrates how the ZMPC works. In the figure, the solid black line indicates the state constraint set $\mathbb{X}$, the dashed rectangle represents the actual control target set $\mathbb{X}_t$, the dash-dotted rectangle is the modified target zone $\tilde{\mathbb{X}}_t$, and the dotted ellipsoid represents the maximum CIS within the modified target zone $\tilde{\mathbb{X}}_t^M$.

For the ZMPC designed based on the nominal system, if the initial condition x_0 is not within the target set ($x_0 \in \mathbb{X}$ and $x_0 \notin \mathbb{X}_t$), the ZMPC drives the predicted nominal system state $\tilde{x}$ towards the terminal set $\tilde{\mathbb{X}}_f$ within the modified target set $\tilde{\mathbb{X}}_t$. It has been shown in [24] that as long as $\tilde{\mathbb{X}}_f \subseteq \tilde{\mathbb{X}}_t^M$, the nominal system will converge to the maximum CIS $\tilde{\mathbb{X}}_t^M$ thus to $\tilde{\mathbb{X}}_t$. The modified target set $\tilde{\mathbb{X}}_t$ is a subset of the actual target set $\mathbb{X}_t$. The purpose of the modified target set is to introduce a buffer zone to tolerant process uncertainties such that if the predicted terminal state $\tilde{x}_N$ is driven into $\tilde{\mathbb{X}}_t$, the actual system state x_N is guaranteed to be within the actual target set $\mathbb{X}_t$ [60].

3.4.2 Economic MPC through zone tracking

Typically, the solution space of the ZMPC design (3.13) has extra degrees of freedom that can be taken advantage of to address a secondary control objective such as economic optimization. The ZMPC design (3.13) that handles an additional economic objective is presented as follows:

$$\begin{aligned} \min_{u_0,\cdots,u_{N-1}} \quad & \sum_{i=0}^{N-1} \ell_z(\tilde{x}_i) + \ell_e(\tilde{x}_i, u_i) \\ \text{s. t.} \quad & (3.13b) - (3.13f). \end{aligned} \tag{3.15}$$

With weighting factors being large enough on the zone control objective, namely c_1, c_2 in (3.14a), the controller will prioritize the zone-tracking objective over the economic objective such that the economic objective will be optimized only after the system enters the target zone [61]. Thus the stability properties of the controller (3.15) remain unchanged from that of (3.13).

3.4.3 Application to the CSTR

In this section, we use the CSTR described in section 3.3.3 as the benchmark process to demonstrate the effectiveness of the ZEMPC. For the CSTR, the control input $u = T_c$ is determined by the ZEMPC.

The zone control objective is to hold the reactor temperature T between 348.0 K to 352.0 K. This implies the target zone is

$$\mathbb{X}_t = \{x\text{: } [0.0,\ 348.0]^T \leqslant x \leqslant [1.0,\ 352.0]^T\}.$$

To estimate the modified target zone $\tilde{\mathbb{X}}_t$, we can perform sensitivity analysis to determine how sensitive the system state to the uncertainty at extreme points. Specifically, a few steps were performed. First, the largest CIS $\mathbb{X}_t^M$ within the target zone $\mathbb{X}_t$ is found using the graph-based algorithm discussed earlier; second, the sensitivities of x to w at points with extreme values of $x \in \mathbb{X}_t$, $u \in \mathbb{U}$, $w \in \mathbb{W}$ are calculated; third, the maximum sensitivity value is determined and the target zone is shrunk by this value. For this specific example, since the target zone is only on the temperature, the shrinking is only performed on the temperature. It was found that maximum sensitivity value is about 0.5. The modified target zone $\tilde{\mathbb{X}}_t$ is then defined as follows:

$$\tilde{\mathbb{X}}_t = \{x\text{: } [0.0,\ 348.5]^T \leqslant x \leqslant [1.0,\ 351.5]^T\}.$$

Once the modified target zone is determined, the maximum CIS within the modified target zone $\tilde{\mathbb{X}}_t^M$ is found and used as the terminal set in the ZEMPC. The economic objective is to minimize the concentration of the reactant inside the reactor, i.e. $\ell_e = c_A$. For the simulations, the prediction horizon $N = 5$. The value of the disturbance w_k at any time step is generated randomly within the bounds for the disturbance.

Figure 3.3 shows the state trajectories of the CSTR under the control of the ZEMPC design (3.15) and the ZEMPC without considering model uncertainty. In the design where the process uncertainty is not considered, the original tracking zone $\mathbb{X}_t$ is tracked and the maximum CIS within $\mathbb{X}_t$, $\mathbb{X}_t^M$, is used as the terminal set. It can be seen that ZEMPC (3.15) can maintain the state of the CSTR within the target zone once it enters the target zone; the other ZEMPC cannot maintain the state of the CSTR within the target zone. The zone control objective is violated often.

This example shows the effectiveness of the ZEMPC design (3.15) and in the design the control invariant terminal set $\tilde{\mathbb{X}}_f$ plays an important role.

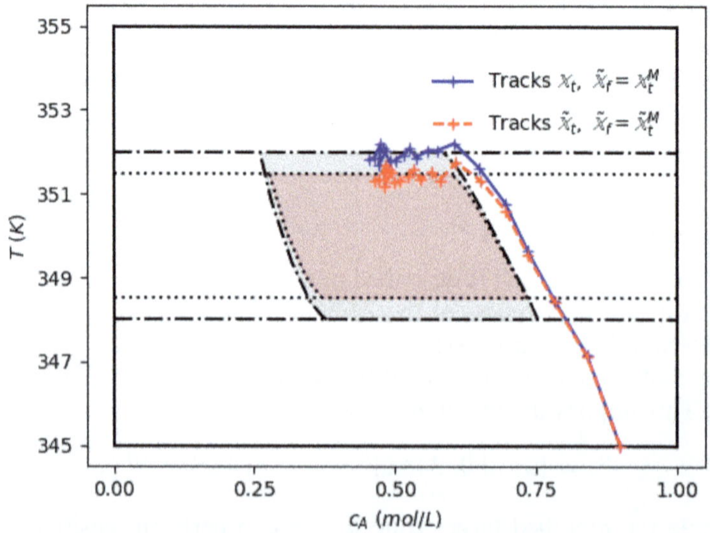

Figure 3.3. State trajectories of the CSTR under the ZEMPC (3.15) which tracks a modified target zone $\tilde{\mathbb{X}}_t$ and uses the maximum CIS with the modified target zone $\tilde{\mathbb{X}}_t^M$ as the terminal set (red dashed line), and a ZEMPC design which tracks the original target zone $\mathbb{X}_t$ and uses the maximum CIS with the target zone $\mathbb{X}_t^M$ as the terminal set (blue solid line).

3.5 RCIS in RL

In this section, we first give a brief introduction to safe RL and then present an application of the maximum RCIS in safe RL design. The safe RL is also applied to the CSTR described in section 3.3.3 to show its performance.

3.5.1 Safe RL

Parts of this section have been reproduced with permission from [62]. Copyright 2023 Elsevier.

RL represents a class of algorithms in which an agent learns a closed-loop control law or policy $u = \pi(x)$ by interacting with system (3.1). In the learning process, at each sampling time k, the RL agent prescribes a control input u_k; the system applies the control input u_k and evolves from the current state x_k to the next state x_{k+1}; the system sends the new state x_{k+1} and the associated reward r_{k+1} back to the RL agent. The reward r_{k+1} provides feedback on the 'goodness' of the control input u_k. In RL's learning process, $(x_k, u_k, r_{k+1}, x_{k+1})$ forms a basic data tuple.

Generally, the standard RL problem can be formulated as follows [40]:

$$\pi^* = \underset{\pi}{\operatorname{argmax}}\, \mathbb{E}[G], \tag{3.16}$$

where G denotes the predicted accumulated reward from now into the future. The optimal policy π^* is the one that maximizes the expected return. In (3.16), constraints on u and x are not considered. While the constraint on u ($u \in \mathbb{U}$) may be implemented by only prescribing or sampling control input in the constraint set, the constraint on x is not trivial to impose. Moreover, the standard RL cannot guarantee the closed-loop stability.

In recent years, CISs, in particular, RCISs have been adopted in RL designs to incorporate state constraints and to ensure closed-loop stability [62]. By incorporating the knowledge of an RCIS of system (3.1), a safe RL optimization can be obtained:

$$\pi^* = \underset{\pi}{\text{argmax}}\, \mathbb{E}[G] \tag{3.17a}$$

$$\text{s. t. } x_k \in R \subseteq \mathbb{X} \tag{3.17b}$$

$$u_k \in \mathbb{U}, \tag{3.17c}$$

where constraint (3.17b) takes into account the state constraint with R denoting an RCIS of system (3.1). Preferably, R is the maximum RCIS of system (3.1). In this section, it is assumed that the RCIS is expressed in the form of a polytope bounded by a finite number of hyperplanes as follows:

$$R = \{x \in \mathbb{X}:\ Ax - b \leqslant 0\}, \tag{3.18}$$

where the matrix $A \in \mathbb{R}^{c \times n}$ and the vector $b \in \mathbb{R}^c$ define the hyperplanes bounding the polytope, comprising a total of c affine constraints.

To ensure that the state constraint is met and the closed-loop stability is guaranteed, the offline training of the safe RL in (3.17) and its online implementation both need to be adjusted. In the offline training, the RCIS is used to guide the RL to achieve a near-optimal policy. While the RCIS is used in offline training, the offline trained policy does not guarantee the closed-loop stability. In order to ensure the closed-loop stability, a safety supervisor is employed during the implementation of the RL. Figure 3.4 illustrates the approach. In figure 3.4, x and $\hat{x}$ denote the actual state of the system and the predicted state, respectively.

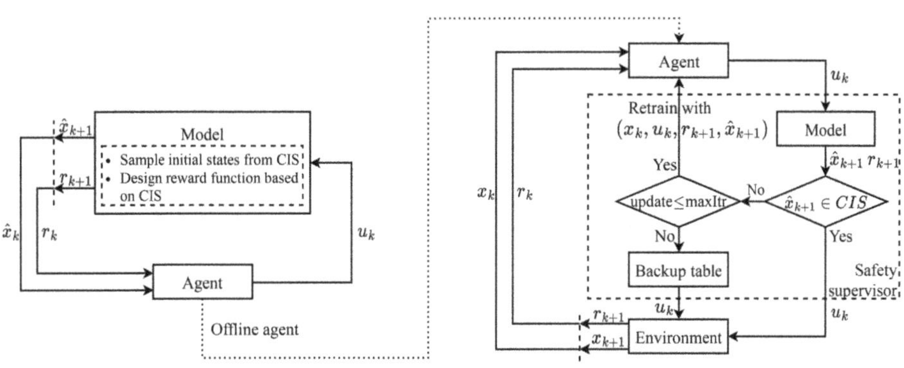

Figure 3.4. Flow diagram of the offline training and online implementation of the safe RL.

It is considered that an RCIS of system (3.1), R, is known. Preferably, the RCIS is the maximum one within the constraint set $\mathbb{X}$, which gives the RL the maximum feasible operating region to explore during the training. R is used in the design of the reward to guide the RL to maintain the system state within R as follows:

$$r_{k+1}(x_k, u_k) = \begin{cases} r_a, & \text{if } \hat{x}_{k+1} \in R \\ r_b, & \text{otherwise} \end{cases}, \tag{3.19}$$

where $r_a, r_b \in \mathbb{R}$ denote the rewards associated with the current control input u_k prescribed by the current RL based on the state x_k, and $\hat{x}_{k+1}$ indicates the predicted next step state based on the model starting from x_k and applying u_k. In the design of the rewards, r_a should be greater than r_b ($r_a > r_b$) to guide the RL to learn a control law that can maintain the system state within R. The details of offline training and online implementation are explained below.

Offline training

Parts of this section have been reproduced with permission from [62]. Copyright 2023 Elsevier.

The nominal model of system (3.1) ($w \equiv 0$) is used in the offline training. At time instant k, the RL prescribes the control action u_k based on the available current state information $\hat{x}_k$; u_k is then applied to the nominal system model to get a prediction of the next state $\hat{x}_{k+1}$; if $\hat{x}_{k+1}$ is within R ($\hat{x}_{k+1} \in R$), the RL will receive a higher reward r_a; if $\hat{x}_{k+1}$ is outside of R ($\hat{x}_{k+1} \notin R$), indicating u_k resulting in an unstable operation, the RL receives a relatively lower reward (or penalty) r_b.

In RL offline training, an RL needs to sample the initial state of the system randomly many times. In safe RL, it is proposed to sample the initial state $\hat{x}_0$ of the system within the RCIS R. If R is the maximum RCIS, and if the system starts from an initial state outside of R, the RL is not able to stabilize the system and drive the system back into R. This technique can significantly improve the sampling efficiency in the training process.

Another technique used in safe RL offline training is to reset the state to its previous value when the state is outside of R. Assume that at k, $\hat{x}_k \in R$. If u_k drives the system state to be outside of R, $\hat{x}_{k+1} \notin R$, the state is reset to its previous state, $\hat{x}_{k+1} = \hat{x}_k$. In such a case, the RL will get a lower reward r_b according to (3.19). Since once the system state is outside of R (if R is the maximum RCIS within $\mathbb{X}$), there is no control action that can drive the system back to R and the system becomes unstable; the interaction between the RL and the system will not further bring much useful information towards learning the optimal control law. By implementing state reset technique, the RL learns from this failure experience and will get second or more chances to learn at the same state $\hat{x}_k$ towards a stable and optimal policy.

Online implementation

After offline training, the safe RL agent learns a control law. However, due to the sampling nature of RL training, the learned control law cannot guarantee the closed-loop stability. To address this issue, in safe RL, each time when the RL agent prescribes a control action, it passes through a safety supervisor as shown in figure 3.4. The safety supervisor uses the system model to calculate the next state $\hat{x}_{k+1}$ based on the prescribed input u_k and the current state x_k. If $\hat{x}_{k+1}$ is within R, then the control action u_k is actually applied to the system; if $\hat{x}_{k+1}$ is outside of R, the RL agent is switched to the training mode again to learn a control action that can maintain the predicted state within R for this specific state; the control law is updated accordingly. A pre-determined maximum number of iterations (maxItr) can be placed to restrict the time that can be used for training at one step in online implementation. When maxItr is sufficiently large, the online training is expected to find a safe action for every state encountered since the RCIS provides the guarantee of the existence of a safe action for all the states within it. To address the issue that the RL cannot find a safe action within maxItr, a backup table (controller) can be used. The stability guaranteed online implementation strategy is summarized in algorithm 3.3.

Algorithm 3.3. Stability guarantee online implementation of safe RL.

Input: x_k, k, $maxItr$
Output: Safe u_k

$notSafe \leftarrow True$, $update = 1$;
while *notSafe* **do**
    Calculate u_k at x_k based on the learned RL policy;
    Based on the model and u_k, predict $\hat{x}_{k+1}$ and r_{k+1};
    if $\hat{x}_{k+1} \in R$ **then**
        $notSafe \leftarrow False$;
    else
        if $update \leq maxItr$ **then**
            Update RL policy with $(x_k, u_k, r_{k+1}, \hat{x}_{k+1})$;
            $update \leftarrow update + 1$;
        else
            Get safe action u_k from the backup table;
            $notSafe \leftarrow False$;
Apply u_k to system (3.1) and obtain x_{k+1};
Reinitialize the algorithm with $k \leftarrow k + 1$

Next, we discuss how to check whether $\hat{x}_{k+1} \in R$ in the presence of process uncertainty ($w \neq 0$). In the presence of w, even if the predicted state based on the nominal model indicates $\hat{x}_{k+1} \in R$, it does not guarantee that the actual state $x_{k+1} \in R$. Therefore, a rigorous approach is required to ensure the safety of the action and cannot solely rely on the predicted state $\hat{x}_{k+1}$ using the nominal model. One approach to address this issue is to consider the worst-case scenario. Specifically, given u_k, we aim to find the w_k value that drives x_k furthest away from R. If in the worst case, $\hat{x}_{k+1}$ is still in R, this implies that the action is safe. Otherwise, u_k is not safe.

To determine whether there is a w_k that can drive the state outside of R, given the form of R in (3.18), the following optimization problem can be solved:

$$J^* = \max_{w_k} \quad \max(A\hat{x}_{k+1} - b) \tag{3.20a}$$

$$\text{s.t.} \quad \hat{x}_{k+1} = f(x_k, u_k, w_k) \tag{3.20b}$$

$$w_k \in \mathbb{W}, \tag{3.20c}$$

where the max(·) operator in the objective function is used to convert the $A\hat{x}_{k+1} - b$ from a vector to a scalar and, more importantly, to select the element with the maximum value, which indicates the boundary with the highest possibility of being violated. The maximum optimization promotes the violation by trying to maximize the largest element in $A\hat{x}_{k+1} - b$ through modifying the decision variable w_k. If the optimal value $J^*>0$, it implies that the prescribed action u_k could result in $\hat{x}_{k+1}$ being outside of R. In this case, u_k is not safe and the RL agent should be penalized and retrained online.

The optimization problem (3.20) is non-smooth due to the presence of the max operator, which makes it unsuitable for gradient-based optimization algorithms that require differentiable objective functions. To overcome this issue, a new variable z is introduced to represent the maximum element in $A\hat{x}_{k+1} - b$, leading to a reformulated optimization problem as follows:

$$J^* = \max_{z, y_i, w_k} z \tag{3.21a}$$

$$\text{s. t.} \ z \leqslant A_i\hat{x}_{k+1} - b_i + My_i, \quad i = 1, \ \dots, \ c \tag{3.21b}$$

$$\sum_{i=1}^{c} y_i = c - 1 \tag{3.21c}$$

$$y_i = \{0, 1\}, \quad i = 1, \dots, c \tag{3.21d}$$

$$\hat{x}_{k+1} = f(x_k, u_k, w_k) \tag{3.21e}$$

$$w_k \in \mathbb{W}, \tag{3.21f}$$

where there are c binary variables y_i corresponding to the c affine constraints in $A\hat{x}_{k+1} - b$, with each y_i indicating whether the ith constraint in (3.21b) is active ($y_i = 0$) or inactive ($y_i = 1$). Constraint (3.21c) forces exactly $c - 1$ constraints to be inactive, with the remaining one being active. Assume that in the optimization, the jth constraint ($A_j\hat{x}_{k+1} - b_j$) has the largest value and its corresponding $y_j = 0$. Then, its corresponding equation in (3.21b) converts to $z \leqslant A_j\hat{x}_{k+1} - b_j$ with

$J^*{=}\max z = A_j\hat{x}_{k+1} - b_j$. Since the optimal value J^* represents the maximum element in $A\hat{x}_{k+1} - b$, the constraint $J^*{\leqslant}A_i\hat{x}_{k+1} - b_i$ is no longer satisfied for the inactive constraints. To ensure that $J^*{\leqslant}A_i\hat{x}_{k+1} - b_i + My_i$ holds for all constraints in (3.21b), an auxiliary positive parameter $M \in \mathbb{R}+$ is introduced and set sufficiently large. The optimization problem (3.21) is a mixed-integer linear problem (MILP). There exist mature and computationally efficient solvers to solve MILPs [63]. After solving the MILP problem, if $J^*{\leqslant}0$, then u_k is safe and can be applied to the system. Otherwise, u_k is not safe and the RL should be retrained online.

3.5.2 Application to the CSTR

The effectiveness of the RCIS-enhanced safe RL is illustrated through controlling the CSTR described in section 3.3.3 as well. The maximum RCIS was computed using the graph-based algorithm discussed earlier and is shown in figure 3.1.

In the offline training of the safe RL, the proximal policy optimization (PPO) [64] was used as the optimization algorithm. In the training, 10,000 episodes and 200 steps per episode were used. The batch size was defined as 10 episodes. The learning rate was 10^{-4} and discounted factor was 0.99. The initial states of all 10,000 episodes were sampled within the RCIS. The reward function was designed as the following:

$$r_{k+1}(x_k, u_k) = \begin{cases} 10,000, & \text{if } \hat{x}_{k+1} \in R \\ -1,000, & \text{otherwise} \end{cases}. \tag{3.22}$$

Based on the aforementioned safe RL training set-up, 20 offline trainings were executed in parallel and the learning performance was calculated as the average of performance over the 20 trainings. The average training reward plot, representing the learning performance, is shown in figure 3.5. In figure 3.5, the horizontal line at

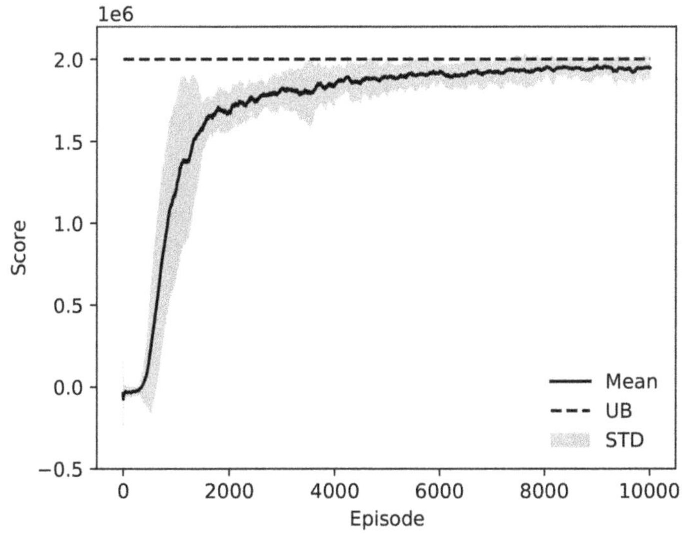

Figure 3.5. Average training score of safe RL.

score 2.0×10^6 represents the maximum score (UB) each episode can achieve if the RL agent maintains the system state within the RCIS all the time. The mean line was calculated based on all 20 trainings and the shaded area shows one standard deviation. From the figure, a clear improvement (higher mean score value, smaller standard deviation) in the training performance can be seen. This signifies the safe RL's ability to effectively learn from the training process.

After assessing the offline training performance, one offline trained safe RL policy was saved and treated as the pre-calculated feedback controller. Then the controller was implemented online following algorithm 3.3. The controller was tested by interacting with the actual CSTR for 10,000 episodes and each episode lasted for 200 steps. During the implementation, uncertainty was introduced. The MILP (3.21) optimization was used to check whether the actions were safe or not.

In all the tests, the online implemented RL successfully kept the system state within the maximum RCIS for all 10,000 episodes. This outcome is expected, as the online implementation ensures a stability guarantee.

This example shows the stability guarantee property of the safe RL, in which the maximum RCIS plays a critical role in providing this guarantee.

3.6 Conclusion

Parts of this section have been reproduced with permission from [62]. Copyright 2023 Elsevier.

This chapter presents graph-based CIS approximation algorithms for general nonlinear systems in the presence of model uncertainty. The applications of CIS in ZEMPC and safe RL are discussed. A CSTR example was used throughout the chapter to illustrate the graph-based CIS approximation algorithms and CIS's critical roles in ZEMPC and safe RL designs.

Bibliography

[1] Nikolopoulou A and Ierapetritou M G 2012 Optimal design of sustainable chemical processes and supply chains: a review *Comput. Chem. Eng.* **44** 94–103

[2] Homer T and Mhaskar P 2017 Constrained control Lyapunov function-based control of nonlinear systems *Syst. Control Lett.* **110** 55–61

[3] Blanchini F 1999 Set invariance in control *Automatica* **35** 1747–67

[4] Mitchell I M, Bayen A M and Tomlin C J 2005 A time-dependent Hamilton–Jacobi formulation of reachable sets for continuous dynamic games *IEEE Trans. Autom. Control* **50** 947–57

[5] Homer T, Mahmood M and Mhaskar P 2020 A trajectory-based method for constructing null controllable regions *Int. J. Robust Nonlinear Control* **30** 776–86

[6] Aubin J P 2009 *Viability Theory* (Boston, MA: Birkhäuser)

[7] Rungger M and Tabuada P 2017 Computing robust controlled invariant sets of linear systems *IEEE Trans. Autom. Control* **62** 3665–70

[8] Rakovic S V, Kerrigan E C, Kouramas K I and Mayne D Q 2005 Invariant approximations of the minimal robust positively invariant set *IEEE Trans. Autom. Control* **50** 406–10

[9] Kerrigan E C 2001 Robust constraint satisfaction: invariant sets and predictive control *PhD Thesis* University of Cambridge, Cambridge

[10] Kolmanovsky I and Gilbert E G 1998 Theory and computation of disturbance invariant sets for discrete-time linear systems *Math. Problems Eng* **4** 317–67

[11] Gilbert E G and Tan K T 1991 Linear systems with state and control constraints: the theory and application of maximal output admissible sets *IEEE Trans. Autom. Control* **36** 1008–20

[12] Bertsekas D 1972 Infinite time reachability of state-space regions by using feedback control *IEEE Trans. Autom. Control* **17** 604–13

[13] Fiacchini M, Alamo T and Camacho E F 2010 On the computation of convex robust control invariant sets for nonlinear systems *Automatica* **46** 1334–8

[14] Alamo T, Cepeda A, Fiacchini M and Camacho E F 2009 Convex invariant sets for discrete-time Lur'e systems *Automatica* **45** 1066–71

[15] Bravo J M, Limón D, Alamo T and Camacho E F 2005 On the computation of invariant sets for constrained nonlinear systems: an interval arithmetic approach *Automatica* **41** 1583–9

[16] Homer T and Mhaskar P 2018 Utilizing null controllable regions to stabilize input-constrained nonlinear systems *Comput. Chem. Eng.* **108** 24–30

[17] Osipenko G S 1983 On a symbolic image of dynamical system *Boundary Value Problems: Interuniv. Collect, Sci. Works, perm* **17** 101–5 (in Russian)

[18] Szolnoki D 2003 Set oriented methods for computing reachable sets and control sets *Discrete Continuous Dyn. Syst.* B*3 361–82*

[19] Eidenschink M 1997 Exploring global dynamics: a numerical algorithm based on the Conley index theory *PhD Thesis* Georgia Institute of Technology

[20] Mischaikow K 2002 Topological techniques for efficient rigorous computation in dynamics *Acta Numer.* **11** 435–77

[21] Rawlings J B 1999 Tutorial: model predictive control technology *Proc. of the 1999 American Control Conf. (Cat. No. 99CH3625)* vol 1 *(San Diego, CA)* (Piscataway, NJ: IEEE) pp 662–76

[22] Rawlings J B, Angeli D and Bates C N 2012 Fundamentals of economic model predictive control *51st IEEE Conf. on Decision and Control (CDC)* (Piscataway, NJ: IEEE) pp 3851–61

[23] Ellis M, Durand H and Christofides P D 2014 A tutorial review of economic model predictive control methods *J. Process Control* **24** 1156–78

[24] Liu S, Mao Y and Liu J 2019 Model predictive control with generalized zone tracking *IEEE Trans. Autom. Control* **64** 4698–704

[25] Liu S and Liu J 2018 Economic model predictive control with zone tracking *Mathematics* **6** 65

[26] Grosman B, Dassau E, Zisser H C, Jovanovič L and Doyle F J 2010 Zone model predictive control: a strategy to minimize hyper- and hypoglycemic events *J. Diabetes Sci. Technol.* **4** 961–75

[27] González A H, Rivadeneira P S, Ferramosca A, Magdelaine N and Moog C H 2020 Stable impulsive zone model predictive control for type 1 diabetic patients based on a long-term model *Optim. Control Appl. Methods* **41** 2115–36

[28] Privara S, Široky J, Ferkl L and Cigler J 2011 Model predictive control of a building heating system: the first experience *Energy Build.* **43** 564–72

[29] Zhang Y, Decardi-Nelson B, Liu J, Shen J and Liu J 2020 Zone economic model predictive control of a coal-fired boiler-turbine generating system *Chem. Eng. Res. Des.* **153** 246–56

[30] Mao Y, Liu S, Nahar J, Liu J and Ding F 2018 Soil moisture regulation of agro-hydrological systems using zone model predictive control *Comput. Electron. Agric.* **154** 239–47

[31] Huang Z, Liu J and Huang B 2023 Model predictive control of agro-hydrological systems based on a two-layer neural network modeling framework *Int. J. Adapt. Control Signal Process.* **37** 1536–58

[32] Scokaert P O M and Rawlings J B 1999 Feasibility issues in linear model predictive control *AIChE J.* **45** 1649–59
[33] Askari M, Moghavvemi M, Almurib H A F and Haidar A M A 2017 Stability of soft-constrained finite horizon model predictive control *IEEE Trans. Ind. Appl.* **53** 5883–92
[34] Heidarinejad M, Liu J and Christofides P D 2012 Economic model predictive control of nonlinear process systems using Lyapunov techniques *AIChE J.* **58** 855–70
[35] Narasingam A and Kwon J S 2017 Development of local dynamic mode decomposition with control: application to model predictive control of hydraulic fracturing *Comput. Chem. Eng.* **106** 501–11
[36] Bangi M S F and Kwon J S 2022 Deep hybrid model-based predictive control with guarantees on domain of applicability *AIChE J.* **69** e18012
[37] Narasingam A, Son S H and Kwon J S 2023 Data-driven feedback stabilisation of nonlinear systems: Koopman-based model predictive control *Int. J. Control* **96** 770–81
[38] Bhadriraju B, Kwon J S and Khan F 2024 A data-driven framework integrating Lyapunov-based MPC and oasis-based observer for control beyond training domains *J. Process Control* **138** 103224
[39] Sitapure N and Kwon J S 2024 Machine learning meets process control: unveiling the potential of LSTMc *AIChE J.* **70** e18356
[40] Sutton R S and Barto A G 2018 *Reinforcement Learning: An Introduction* (Cambridge, MA: MIT Press)
[41] Garcıa J and Fernández F 2015 A comprehensive survey on safe reinforcement learning *J. Mach. Learn. Res.* **16** 1437–80
[42] Osinenko P, Dobriborsci D and Aumer W 2022 Reinforcement learning with guarantees: a review *IFAC-PapersOnLine* **55** 123–8
[43] Gu S, Yang L, Du Y, Chen G, Walter F, Wang J, Yang Y and Knoll A 2022 A review of safe reinforcement learning: methods, theory and applications arXiv: 2205.10330
[44] Kadota Y, Kurano M and Yasuda M 2006 Discounted Markov decision processes with utility constraints *Comput. Math. Appl.* **51** 279–84
[45] Chow Y, Ghavamzadeh M, Janson L and Pavone M 2017 Risk-constrained reinforcement learning with percentile risk criteria *J. Mach. Learn. Res.* **18** 6070–120
[46] Law E L M 2005 Risk-directed exploration in reinforcement learning *MsC Thesis* McGill University
[47] Gehring C and Precup D 2013 Smart exploration in reinforcement learning using absolute temporal difference errors *Proc. of the 2013 Int. Conf. on Autonomous Agents and Multi-agent Systems* pp 1037–44
[48] Zanon M, Gros S and Bemporad A 2019 Practical reinforcement learning of stabilizing economic MPC *18th European Control Conf. (ECC)* (Piscataway, NJ: IEEE) pp 2258–63
[49] Gros S and Zanon M 2021 Reinforcement learning based on MPC and the stochastic policy gradient method *American Control Conf. (ACC)* (Piscataway, NJ: IEEE) pp 1947–52
[50] Decardi-Nelson B and Liu J 2022 Robust economic model predictive control with zone tracking *Chem. Eng. Res. Des.* **177** 502–12
[51] Ma H, Chen J, Eben S, Lin Z, Guan Y, Ren Y and Zheng S 2021 Model-based constrained reinforcement learning using generalized control barrier function *IEEE/RSJ Int. Conf. on Intelligent Robots and Systems (IROS)* (Piscataway, NJ: IEEE) pp 4552–9
[52] Alshiekh M, Bloem R, Ehlers R, Könighofer B, Niekum S and Topcu U 2018 Safe reinforcement learning via shielding *Proc. of the AAAI Conf. on Artificial Intelligence* **vol 32**

[53] Gros S, Zanon M and Bemporad A 2020 Safe reinforcement learning via projection on a safe set: how to achieve optimality? *IFAC-PapersOnLine* **53** 8076–81

[54] Li S and Bastani O 2020 Robust model predictive shielding for safe reinforcement learning with stochastic dynamics *IEEE Int. Conf. on Robotics and Automation (ICRA)* (Piscataway, NJ: IEEE) pp 7166–72

[55] Tabas D and Zhang B 2022 Computationally efficient safe reinforcement learning for power systems *American Control Conf. (ACC)* (Piscataway, NJ: IEEE) pp 3303–10

[56] Osipenko G 2007 *Dynamical Systems, Graphs, and Algorithms Lecture Notes in Mathematics* vol 1889 (Berlin: Springer)

[57] Decardi-Nelson B and Liu J 2021 Computing robust control invariant sets of constrained nonlinear systems: a graph algorithm approach *Comput. Chem. Eng.* **145** 107177

[58] Dellnitz M and Junge O 2002 Set oriented numerical methods for dynamical systems *Handbook of Dynamical Systems* **vol 2** (Amsterdam: Elsevier) pp 221–64

[59] Decardi-Nelson B and Liu J 2022 An efficient implementation of graph-based invariant set algorithm for constrained nonlinear dynamical systems *Comput. Chem. Eng.* **164** 107906

[60] Huang Z, Liu J and Huang B 2023 Generalized robust MPC with zone-tracking *Chem. Eng. Res. Des.* **195** 537–50

[61] Liu S, Mao Y and Liu J 2019 Model-predictive control with generalized zone tracking *IEEE Trans. Autom. Control* **64** 4698–704

[62] Bo S, Yin X and Liu J 2023 Control invariant set enhanced safe reinforcement learning: improved sampling efficiency, guaranteed stability and robustness *Comput. Chem. Engin.* **179** 108413

[63] Sioshansi R *et al* 2017 *Optimization in Engineering* (Cham: Springer International) p 120

[64] Schulman J, Wolski F, Dhariwal P, Radford A and Klimov O 2017 Proximal policy optimization algorithms arXiv:1707.06347

Chapter 4

Machine learning-based multiscale modeling and control of quantum dot manufacturing and their applications

Niranjan Sitapure, Parth Shah and Joseph Sang-Il Kwon

In recent years, the quest for quantum dots (QDs)—semiconducting nanocrystals with customizable optical and optoelectronic properties—has gained significant momentum for their application in next-generation photonic devices. This surge in interest is largely due to their high photoluminescence quantum yield (PLQY), broad color spectrum, adjustable optoelectronic characteristics, and cost-efficient solution-based processing. Additionally, the expanding market for these applications has spurred a demand for rapid and scalable QD production methods, along with the manufacture of related optoelectronic devices. However, the commercialization of QDs faces several challenges: (a) a limited mechanistic understanding of the crystallization kinetics across different QD systems, impeding predictive control over QD size distribution; (b) the lack of a robust paradigm for accurate modeling and scaling up various QD manufacturing processes, such as crystallization and thin-film deposition; and (c) the absence of computationally efficient solutions for the control and optimization of QD processes.

To bridge these knowledge gaps, this work introduces various models to elucidate the mechanism of QD crystal growth, facilitate rapid and scalable manufacturing of QDs and related optoelectronic devices, and establish an effective control framework for different QD processes. Initially, a first-principles kinetic Monte Carlo (kMC) model was developed and experimentally validated to capture the crystallization kinetics of QDs. Subsequently, to address challenges in batch synthesis, continuous QD manufacturing using a plug flow crystallizer (PFC) was demonstrated through a multiscale modeling approach. This methodology was further extended to two-phase slug flow crystallizers (SFCs) via the development of a

doi:10.1088/978-0-7503-6174-3ch4

computational fluid dynamics (CFD)-based multiscale model. Additionally, a highly efficient, data-driven optimal control framework employing a deep neural network (DNN) was formulated to regulate QD crystal size and distribution. It is crucial to note that all the aforementioned models were corroborated by experimental validation.

Despite these advancements, the developed models are highly specific to their intended systems and cannot be readily generalized to other QD systems. To address this limitation, we propose the development of a transformer-based hybrid model. This model aims to harness the exceptional transfer learning capabilities of transformers to achieve better generalization across different QD systems. In summary, the proposed work tackles three major challenges in the QD field—control of QD kinetics, continuous production of QDs, and the design of manufacturing processes for rapid scaling of QD-based devices—by developing a suite of experimentally validated multiscale models and integrating them within an effective control framework.

4.1 Introduction, motivation, and literature review

4.1.1 Motivation

The US Department of Energy has established ambitious goals to lower the cost of photovoltaic energy to approximately 5 cents/kWh by 2030, a significant reduction from the current average of around 15 cents/kWh. This initiative has sparked extensive research focused on the discovery, optimization, and commercialization of high-efficiency solar cells, as evidenced by key studies in the field [1, 2]. In particular, solar cells based on QDs have garnered special interest due to their exceptional quantum efficiency and the ability to fine-tune their absorption/emission spectra, as highlighted by recent advances [3–5]. Additionally, the growing demand for energy-efficient high-resolution displays (HRDs) has driven the exploration of new semiconducting materials capable of nanopatterning to meet high-resolution specifications [6–9]. QDs, with their broad color gamut and high PLQY, have emerged as promising candidates in this space as well [10]. Consequently, the broad spectrum of applications, from solar cells and HRDs to other optoelectronic devices, has stimulated research into various QDs, exploring their properties, manufacturing techniques, and practical applications [11–13].

Research efforts have extensively examined the synthesis of QDs, the impact of processing parameters on their tunable characteristics, and comparisons of synthesis methodologies for improved control over QD size—critical for their photoelectronic properties [14–17]. Despite this, most prior studies have focused on small-scale, lab-based batch synthesis, using a myriad of protocols without providing comprehensive insights into the QD crystallization kinetics crucial for devising more efficient manufacturing techniques [18, 19]. There is a clear demand within the semiconductor industry for a continuous synthesis platform that allows large-scale QD production with precise size control and predictive capabilities. Moreover, the fabrication of QD-based devices, such as solar cells and HRDs, requires sophisticated processing techniques such as lithography and thin-film deposition, areas that remain underexplored from a computational standpoint [20–22]. This highlights a notable gap in the modeling and optimization research necessary for scaling up, improving yields, and achieving better process control in the manufacturing of QD-based optoelectronic thin films. Thus, the sector faces three primary challenges: (a) the need for a deeper mechanistic understanding of QD processes to facilitate the development of enhanced QD models; (b) the absence of established high-fidelity modeling and scaling paradigms for various QD manufacturing processes, including crystallization and thin-film deposition; and (c) the lack of computationally efficient solutions for the control and optimization of these QD processes.

4.1.2 Literature review

Given the complexity of the challenges outlined, it is clear that no single modeling technique or isolated process can offer a comprehensive solution. Instead, a multifaceted approach that combines various modeling, control, and validation techniques is required. This section aims to highlight a selection of works from the existing literature that could be instrumental in addressing these challenges effectively.

Crystallization and size control of QDs

The literature reveals diverse strategies for tuning the size or bandgap of QDs, broadly categorized into three groups. The first category includes studies that have demonstrated the facile tuning of the QD bandgap through halide/anion exchange, offering a straightforward approach to modulate optical properties [23–25]. The second category focuses on the use of various surface ligand species with differing chain lengths and concentrations to influence QD size [26–29]. Here, the ligands act as a shell around the QDs, modulating mass transfer and thus controlling growth kinetics and final QD size. The third group employs hot-injection (HI) techniques, where the reaction temperature is key to controlling QD size [30–33]. Despite the variety of synthesis techniques available, there remains a gap in our understanding of QD crystallization kinetics. For example, the rapid progression of crystallization reactions in HI synthesis complicates size control, leading to significant batch-to-batch size variability. Moreover, while ligand-based size control techniques explore various ligand species and chain lengths, they offer limited insight into the mechanisms through which these ligands influence QD size. Additionally, existing crystallization models, mainly developed for organics and protein crystallization, fail to accurately represent QD growth kinetics [34–37], indicating a need for alternative models that can accurately predict and manage QD size.

Continuous QD manufacturing

Scaling conventional lab-scale QD synthesis to meet market demands is challenging, often resulting in batch-to-batch variations and a broad QD size distribution that can impact device performance. Some efforts have been made towards developing continuous manufacturing techniques for QDs. For instance, a millifluidic platform has been demonstrated for the fast-scalable production of $CsPbBr_3$ QDs within a size range of 5 to 10 nm [38, 39]. Yet, there is a lack of established models to describe continuous QD crystallization and control QD size effectively. Another study introduced an SFC for continuous QD production, utilizing millifluidic SFC to adjust the flow rates of the inert gas and precursor solution for size-tuning $CsPbBr_3$ QDs [40]. This approach suggests that operating an SFC in a stable regime, where gas–liquid slugs alternate, allows each liquid slug to act as an individual batch crystallizer, while the entire system functions as a continuous reactor. However, despite these advancements, the literature is still missing high-fidelity models capable of scaling up and controlling these processes. Existing SFC models do not consider slug-to-slug (S2S) variation or provide appropriate control mechanisms for adjusting air/liquid flow rates to consistently achieve the desired QD size. Therefore, there is an urgent need for the development of a QD crystallization platform that integrates a high-fidelity crystallizer model with an efficient controller, enabling the continuous manufacturing of QDs in both PFCs and SFCs.

Thin-film deposition of QDs

A crucial requirement for many QD-based applications, including solar cells and HRDs, is the formation of QD thin films. Traditionally, spin coating has been the method of choice for fabricating these films, serving as the functional or active layer in devices [41–43]. However, this technique faces several drawbacks such as batch-to-batch variations, low material utilization, and the necessity for a subsequent curing step, all of which hinder the rapid, scalable production of QD-based devices. As a solution, spray coating has been identified as a viable alternative, potentially overcoming these challenges [44–46]. Nonetheless, spray coating introduces its own set of issues, such as the coupling of macroscopic phenomena (e.g. transport and thermal effects) with surface-level interactions (notably, the coffee ring effect), which can lead to uneven thin-film deposition adversely affecting the optoelectronic performance of the QDs [47, 48]. Therefore, the development of an accurate spray coating model for QDs is essential, one that comprehensively accounts for both surface-level and continuum phenomena to facilitate optimal control over thin-film characteristics.

In the realm of nanopatterning, QD thin films have garnered considerable interest for direct integration into HRDs through photolithography [49, 50]. However, the integration process often subjects QD films to damage from the various chemical solvents used during post-processing steps [51, 52]. To circumvent these issues, ligand crosslinking has been explored as a scalable technique for creating stable QD films [53]. Despite the technique's potential, the current literature largely overlooks the detailed kinetics and mechanisms behind the crosslinking reaction. This gap underscores the necessity for a sophisticated model that not only elucidates these aspects but also paves the way for improved nanopatterning methods, thereby enhancing the manufacturing process of QD thin films for future applications.

Multiscale modeling techniques

Addressing the complexities of continuous QD manufacturing and spray coating for thin-film deposition demands a sophisticated multiscale modeling approach.

This approach is essential for integrating crystal growth kinetics, which occurs at the microscopic level, with the macroscopic phenomena of mass and energy balance. In particular, for processes such as PFC and SFC, it is crucial to merge detailed microscopic interactions with overarching system dynamics. Similarly, spray coating involves intricate QD–QD interactions at the microscopic scale that influence thin-film deposition, while also engaging with broader phenomena such as droplet atomization, heat transfer, and evaporation at the macroscopic level.

Given these requirements, a vast majority of QD manufacturing techniques benefit from a fusion of macroscopic models that are universal across systems with specific microscopic interactions unique to QDs. This integration essentially calls for the development of a comprehensive multiscale modeling framework. The literature offers numerous instances where such multiscale approaches have been applied across diverse chemical processes [54, 55], demonstrating their potential to tackle the dual challenges of continuous QD manufacturing and efficient thin-film deposition through spray coating.

Multiscale modeling of crystallization

Christofides and colleagues have made significant contributions to multiscale modeling and control of protein crystallization. Their work successfully integrates kMC simulations at the microscopic level, capturing adsorption, migration, and desorption steps, with macroscopic mass and energy balance equations. Initial efforts led to the development and experimental validation of a high-fidelity kMC model for lysozyme crystallization, demonstrating the model's capability to accurately simulate the crystallization process [34, 56]. Subsequent studies leveraged this kMC model alongside macroscopic equations in a batch crystallizer setting, employing a model predictive controller (MPC) to optimize crystal size distribution (CSD) and even the shape of lysozyme crystals [34, 57, 58]. This approach was later expanded to continuous crystallization processes, integrated with a run-to-run control scheme to enhance control over the CSD [59, 60]. Additionally, these models were enriched by incorporating particle aggregation dynamics, strategies for fine particle removal, and product classification protocols to maximize the productivity of continuous protein crystallizers while minimizing polydispersity [61, 62]. To address the computational demands of these models, another study introduced a multidomain message passing interface (MPI)-powered parallel computation framework. This framework combines problem decomposition, task assignment, and information flow orchestration to achieve a significant speed increase of approximately 40 times, making these complex simulations more feasible [54]. These works by Christofides and his team lay a solid foundation for the development of analogous multiscale models tailored to the unique requirements of QD crystallization processes.

Multiscale modeling of thin-film deposition

In addition to the challenges of thin-film deposition, significant contributions by Christofides and colleagues have demonstrated the integration of surface-level kMC models with gas-phase microscopic models for plasma-enhanced chemical vapor deposition (PECVD) processes, particularly in the manufacturing of silicon solar cells. Their kMC models encompass critical processes such as physisorption, surface migration, hydrogen abstraction, and chemisorption. These models are adeptly combined with two-dimensional gas-phase partial differential equations that simulate chemical reactions, diffusion, and convection [63]. Through this modeling approach, it is possible to achieve thin-film deposition with uniform thickness and an optimized light-trapping surface microstructure by selecting specific wafer surface gratings. Additionally, the application of a run-to-run control strategy, utilizing an exponentially weighted moving average (EWMA) algorithm, has shown promise in minimizing batch-to-batch variability across deposition cycles [64]. The integration of these models with a CFD framework further enhances the accuracy in capturing plasma chemistry and transport phenomena within a 2D/3D axisymmetric reactor geometry, effectively bridging the surface-level wafer domain with the gas-phase [65, 66].

Parallel to these efforts, Sandoval and colleagues have advanced multiscale modeling for thin-film deposition in generalized chemical vapor deposition (CVD) systems. Their approach melds continuum-level transport models with surface-level kMC simulations, leveraging artificial neural networks (ANNs) as surrogates for the multiscale thin-film model. This surrogate model is then integrated into an MPC framework to finely tune thin-film characteristics [67, 68]. The model further evolves into a stochastic MPC framework that incorporates inherent process uncertainties through polynomial series expansion (PSE) and a DNN-based methodology [69, 70]. These groundbreaking studies have laid a solid foundation for the development of a comprehensive multiscale modeling framework tailored for thin-film deposition, offering a blueprint for future advancements in this field.

Kinetic Monte Carlo simulation

The highlighted examples underscore the effectiveness of kMC simulations in capturing the nuances of system-specific microscopic interactions. These simulations serve as a cost-effective and powerful tool for gathering detailed information on the spatiotemporal evolution of various processes, offering a more accessible alternative to molecular dynamics (MD) and other molecular simulation methodologies [71–73]. Significantly, kMC models seamlessly integrate with continuum-level equations, facilitating realistic simulations of chemical reactors, fermenters, pulp digesters, and processes involving crystallization [74]. The versatility of kMC simulations extends beyond simple adsorption, desorption, and migration, accommodating complex phenomena with ease.

For instance, Kwon and colleagues have applied kMC models to intricate biological systems, including the interactions between cholera toxin subunit B (CTB) and various glycan-lectins and their receptors, such as GM1, GM2, and GD1b. These models precisely delineate the binding mechanisms and competitive interactions involved, laying the groundwork for further models that address cellular heterogeneity and complex signaling pathways [75–81]. Such advances inspire applications in QD research, particularly in modeling QD crosslinking interactions.

In the field of pulping and lignin extraction, kMC simulations have been extensively employed for the modeling and control of microscopic properties in both batch and continuous pulp digesters. These simulations offer a powerful framework for capturing the stochastic nature of reaction dynamics at a molecular level, enabling detailed insights into the underlying chemical and physical processes occurring during pulping and delignification. By simulating individual molecular interactions over time, kMC provides a way to predict macroscopic outcomes such as fiber morphology and lignin content, which are essential for optimizing the efficiency and quality of pulp production [82–90].

In the realm of heterogeneous catalysis, kMC simulations provide a comprehensive framework for modeling the microkinetic reaction steps, the interactions between reactants, and the modification of lattice sites to account for degradation reactions. Noteworthy contributions include Kwon and colleagues' development of a kMC model for the nitrogen reduction reaction (NRR), aiming to identify high-performance bimetallic transition metal catalysts, alongside other seminal works addressing catalysis [91–96]. Similarly, Sandoval and colleagues have utilized kMC for modeling surface chemistry in tubular reactors, leading to the development of hybrid kMC-data-driven models for enhanced process simulation [97–101].

Expanding the scope of kMC applications, there has been progress in modeling dendrite-based capacity degradation in lithium-ion batteries, incorporating various electrochemical reactions and the mechanical properties of materials to predict and control dendrite formation [102–104]. These studies collectively highlight the robustness of kMC simulations in modeling a wide spectrum of microscopic phenomena, employing tailored equations to address specific challenges in QD interactions and beyond.

Existing data-driven modeling techniques

As previously noted, while a range of high-fidelity modeling techniques, either standalone or in combination, can address the challenges mentioned, the need for online process monitoring and control underscores the importance of computationally efficient data-driven surrogate models. A survey of the literature reveals that these surrogate models fall into three distinct categories: (a) subspace identification models, which leverage mathematical techniques to identify system dynamics from data; (b) machine learning (ML)-based models, utilizing algorithms that learn from data to predict or classify outcomes; and (c) next-generation hybrid models, which combine elements from traditional and ML-based approaches to enhance prediction accuracy and computational efficiency.

Subspace identification models

Data-driven model order reduction techniques are increasingly becoming a popular strategy for developing surrogate models, offering a streamlined approach to tackling complex system behaviors with high computational efficiency. Among these techniques, proper orthogonal decomposition (POD) stands out for its effectiveness in reducing the dimensionality of datasets, such as spatially varying Young's modulus profiles, to predict fracture propagation dynamics accurately in hydraulic fracturing scenarios, as demonstrated by Kwon and colleagues. The essence of POD lies in isolating the most significant modes within a dataset, effectively capturing the maximum variance with linear combinations of the original state variables [105–107].

Sparse identification of system dynamics (SINDy) represents another significant leap forward. This technique employs a library of basis functions, chosen based on prior knowledge of the process, to distill a compact and interpretable set of equations that characterize dynamic chemical processes. Its emphasis on sparsity ensures that only the most impactful basis functions are utilized, thus maintaining model simplicity and focus [108, 109]. However, the specificity of SINDy models to particular systems can lead to discrepancies when process parameters change. Addressing this, the Operable Adaptive Sparse Identification of Systems (OASIS) model, developed by Kwon and team, integrates multiple SINDy models through a DNN, allowing for the dynamic selection of the most appropriate model under varying process conditions, thus minimizing discrepancies [110–114].

The Koopman operator framework introduces an alternative approach by leveraging basis functions to transform nonlinear systems into a higher-dimensional linear format, simplifying computations and facilitating integration with the MPC. Kwon and colleagues have effectively applied this framework to diverse systems, including reactor operations and hydraulic fracturing processes, demonstrating its broad applicability [115, 116]. A significant advancement within this domain is the Koopman Lyapunov-based MPC (KLMPC), which not only has been shown to ensure closed-loop stability for certain chemical systems but also has been extended to reinforce this stability in nonlinear systems through a bilinear Koopman MPC system [117, 118]. Further developments include an offset-free KLMPC framework, which has shown promising results in regulating batch pulp digesters with high accuracy, addressing the challenge of plant–model mismatch [119, 120]. These advanced data-driven reduced-order modeling techniques present viable solutions to the computational challenges often encountered in QD systems, showcasing the potential for significant efficiency gains in process modeling and control.

Hybrid modeling techniques

While ML techniques offer high accuracy and straightforward model training, they often result in black-box models that lack interpretability and present challenges for numerical integration within online MPC or optimal operation scenarios. This has catalyzed the development of hybrid models, which meld system-agnostic first principles (such as mass and energy balance, transport phenomena) with system-specific data-driven parameters (such as process parameters and kinetic constants). This 'best-of-both-worlds' approach yields models that combine the predictive power of data-driven methods with the interpretability and foundational rigor of first principles [70, 121–124].

Kwon and colleagues have been at the forefront of developing such hybrid models for a variety of complex chemical systems [125, 126]. For instance, one study presents a DNN-based hybrid model that integrates the PKN model of fracture propagation in hydraulic fracturing with dynamically modeled leak-off rates to accurately describe fracture dynamics [127]. Another investigation adapts this hybrid approach to estimate dynamic process parameters (such as rate constants and diffusion coefficients) and incorporate them with macroscopic mass and energy balances for an industrial-scale fermentation process [128, 129]. This model was further evolved to employ a physics-informed neural network (PINN) for elucidating batch kinetics in various lab-scale chemical processes, marrying physical laws with data-driven insights [81, 130]. Ultimately, these hybrid models have been successfully integrated into MPC frameworks, ensuring stability and robust control within their domain of application [131]. These examples underscore the potential of hybrid modeling in enriching system-agnostic equations with specific insights derived from ML models, particularly in the context of QD systems.

4.1.3 Objectives and organization of this chapter

The manufacturing of QDs and their integration into high-performance devices have become increasingly critical in recent years, driven by the push for advanced QD-based technologies. This field faces three primary challenges: (a) devising an accurate model for the crystallization process in the continuous production of QDs; (b) developing high-fidelity models for the various processes involved in creating QD-based devices; and (c) establishing a control framework that leverages computationally efficient surrogate models for both processes. The aim of this doctoral study is to tackle these issues through a synergistic application of microscopic and macroscopic modeling, ML techniques, and advanced control strategies. The organizational structure of this chapter, illustrated in figure 4.1, outlines our methodological approach to addressing these challenges.

Section 4.2 delves into the development and validation of a continuous tubular flow crystallizer designed for the scalable production of QDs. This section highlights the creation and experimental validation of a kMC-based microscopic model for

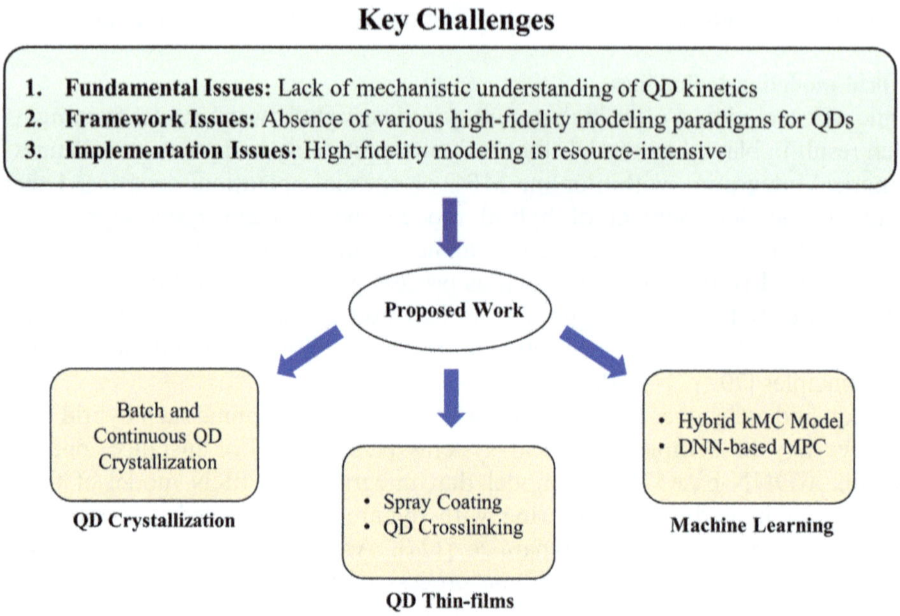

Figure 4.1. Schematic illustration of this chapter.

crystallization, which is further optimized using a nonlinear optimizer for precise control over QD size. Following this, section 4.3 explores the construction of a multiscale model aimed at controlling QD size and CSD within SFCs. This model integrates CFD, crystallization modeling, and DNN to achieve its goals. While these sections focus on addressing the challenge of continuous QD manufacturing, the scope of this chapter does not extend to the detailed exploration of QD thin-film production. For comprehensive insights into this topic, readers are encouraged to consult [132–134].

Moreover, section 4.4 briefly outlines two promising directions for future research, setting the stage for continued exploration in this field. Finally, section 4.5 concludes the chapter, summarizing the key findings and contributions of this work to the domain of QD manufacturing and application development.

4.2 Multiscale modeling and control of tubular crystallizer for continuous QD manufacturing

Parts of this section have been reproduced with permission from [141]. Copyright 2020 Elsevier.

As mentioned earlier, (a) understanding the crystallization kinetics for QDs (i.e. in this case, $CsPbBr_3$ QDs), and (b) development of a tubular crystallizer model and an associated control framework will be a crucial factor for large-scale continuous manufacturing of $CsPbBr_3$ QDs with size-tuning capabilities. These challenges are addressed by proposing a multiscale model for the PFC. Specifically, the macroscopic phase of the PFC is modeled to describe the dynamic spatiotemporal evolution of precursor concentration and QD size using a continuum model for mass balance. The discretized continuum model is integrated with a kMC model to formulate a high-fidelity multiscale model; the kMC simulation describes the physics behind QD synthesis (i.e. the $CsPbBr_3$ unit cell attachment and mass-transfer rate of ligand adsorption and desorption of the ligand), and its dependence on the local precursor concentration and superficial flow velocity. Then, the developed crystallizer model is experimentally validated. Then, an optimizer, which dictates the optimal input profile (i.e. inlet precursor concentration and flow rate) for set-point tracking is designed. More details about this chapter are presented in our previous work [139, 141].

4.2.1 Mathematical modeling of PFC

Process description

We consider a PFC as shown in figure 4.2 that has a length of 10 cm and an internal diameter of 0.1 cm. The PFC has two manipulated inputs, namely, precursor concentration, C_o, and superficial flow velocity, v_z, which can be used to fine-tune the size of $CsPbBr_3$ QDs at the end of the PFC, r_{final}. For an initial precursor concentration of C_o, and superficial flow velocity, v_o, the QD crystals start growing at z_o (the entry location of the PFC), and reach the desired set-point value, $r_{\text{final,sp}}$, at the end of the PFC, z_n. Thus, for an accurate description of QD crystal growth in a PFC, it is important to investigate the spatiotemporal evolution of QD crystal size, $r(t, z)$, and precursor concentration, $C(t, z)$, in the PFC.

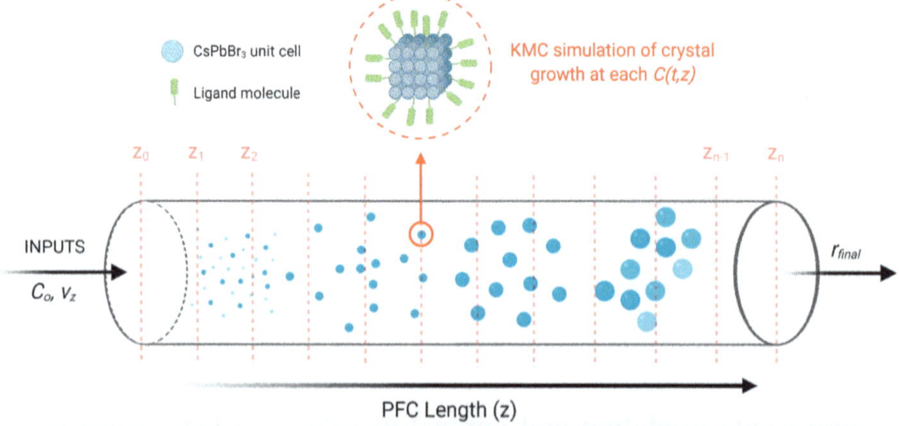

Figure 4.2. Schematic illustration of the proposed multiscale model of a PFC. The inputs to the PFC are the precursor concentration, C_o, and superficial flow velocity, v_z. The QD crystal size at the end of the PFC is denoted by r_{final}. The blue spheres represent the QD crystals growing along the flow direction, and the green extensions around the crystal represent the ligand molecules. (Reproduced with permission from [139]. Copyright 2021 Elsevier.)

Macroscopic model of the PFC

By employing a standard continuum model, and considering spherical nanocrystals at room temperature, we can describe the mass balance within the PFC as follows:

$$\frac{\partial C}{\partial t} = -v_z \frac{\partial C}{\partial z} - \rho_s n_s(t, z) 4\pi r^2(t, z) G_r(C, v_z, r), \tag{4.1}$$

where $r(t, z)$ is the radius of nanocrystals, ρ_s is the density of $CsPbBr_3$ QD crystal, ρ_f is the density of the solution phase, $n_s(t, z)$ is the number of crystals in the PFC, v_z is the superficial flow velocity in the PFC, and G_r is the one-dimensional (1D) crystal growth rate.

Generally, G_r is estimated experimentally and is a nonlinear function of precursor concentration and temperature of the solution phase [37, 135]. However, room-temperature experimental studies show that crystal growth in $CsPbBr_3$ QD synthesis depends on the superficial flow velocity [40, 136]. Further, during $CsPbBr_3$ crystal growth, there are also other phenomena (i.e. ligand-shell formation [137] and a Br rich crystal surface [33]) leading to a size-dependent crystal growth behavior. To this end, an alternative approach to computing $G_r(C, v_z, r)$ is required.

Microscopic description of QD crystal growth

To accurately model the microscopic crystallization kinetics, the following experimental observation needs to be accounted for in the proposed kMC model. First, the rigorous mixing of the precursors results in multiple $PbBr_6$ octahedra and subsequent formation of $CsPbBr_3$ unit cells, which then aggregate to form QD crystals. Second, this is accompanied by the attachment of ligand molecules on the $CsPbBr_3$ crystal surface. Third, recent studies have demonstrated the pivotal role of the superficial flow velocity in influencing the mixing characteristics in a PFC [136, 138].

kMC model development
A kMC model based on a solid-on-solid (SOS) approach, which takes into account the microscopic events (i.e. adsorption, migration, and desorption) occurring in the crystallization process is developed [139, 140].

Surface kinetics
First, the rate of crystal growth, which dictates the attachment of $CsPbBr_3$ unit cell to the crystal nuclei, is given as follows:

$$r_a = NK_o \exp\left(\frac{\Delta\mu_{\text{cry}}}{kT}\right);\ \Delta\mu_{\text{cry}} = k_B T \ln\left(\frac{C_{\text{cry}}}{S}\right), \tag{4.2}$$

where K_0 is the attachment rate at equilibrium ($\Delta\mu_{\text{cry}} = 0$), T is the temperature in kelvins, N is equal to the kMC lattice size ($N = 200$), C_{cry} is the precursor concentration [$CsPbBr_3 0$], S is the solubility of [$CsPbBr_3$], which is a constant, k_B is the Boltzmann constant and $\Delta\mu_{\text{cry}}$ refers to the difference in the chemical potentials of [$CsPbBr_3$] in the solution phase and the crystal.

Second, the rate of ligand attachment is dictated by the molar flux of the ligand molecules from the solution phase to the crystal surface, which is defined as follows:

$$\begin{aligned} J_{\text{lig}} &= h_m\left(C_{\text{lig}} - C_{\text{lig}}^*\right) \\ r_{\text{lig}} &= NJ_{\text{lig}}N_A d^2, \end{aligned} \tag{4.3}$$

where h_m is the convective mass-transfer coefficient, C_{lig} is the concentration of free ligand molecules in the solution phase, C_{lig}^* is the equilibrium concentration of free ligand molecules on the crystal surface, N_A is Avagadro's constant, and d is the edge length of one unit cell of crystalline $CsPbBr_3$.

Third, the loosely bound ligands can detach and dissolve back in the solution, and this is described as follows:

$$r_d(i) = K_o \exp\left(-i\frac{E_b}{k_B T}\right)N_{\text{lig}}(i), \tag{4.4}$$

where E_b is the average binding energy per bond between two ligands, $i \in [0, 2]$ is the number of nearest ligand neighbors, and $N_{\text{lig}}(i)$ is the number of lattice sites occupied by ligand molecules having i nearest neighbors.

Lastly, the total rate, W_{total}, is the sum of all the three individual rates and is defined as $W_{\text{total}} = r_a + r_{\text{lig}} + \sum_{i=0}^{2} r_d(i)$.

kMC event execution
The probability of $CsPbBr_3$ unit cell attachment, ligand attachment, and ligand detachment is represented as P_a, P_{lig}, and P_d, respectively. These probabilities are calculated as shown below:

Table 4.1. Probability conditions for selecting a kMC event. (Reproduced with permission from [139]. Copyright 2021 from Elsevier.)

Probability conditions	Event executed
$0 < \xi_1 \leqslant P_a$	$CsPbBr_3$ unit cell attachment
$P_a < \xi_1 \leqslant P_{\text{lig}} + P_a$	Ligand attachment
$P_a + P_{\text{lig}} < \xi_1 \leqslant P_a + P_{\text{lig}} + P_d$	Ligand detachment

$$P_a = \frac{r_a}{W_{\text{total}}}$$
$$P_{\text{lig}} = \frac{r_{\text{lig}}}{W_{\text{total}}} \tag{4.5}$$
$$P_d = \frac{\sum_{i=0}^{2} r_d(i)}{W_{\text{total}}}.$$

Only one kMC event can be executed at each event execution step. Thus, to determine which event is to be executed, a random number, $\xi_1 \in (0, 1]$, is generated and used in conjunction with table 4.1.

First, if a $CsPbBr_3$ unit cell attachment event is selected, with lattice site $i \in [0, N]$ is chosen at random, then a $CsPbBr_3$ unit cell is deposited on lattice site i and will result in crystal growth. Second, if a ligand attachment event is selected, then lattice site $j \in [0, N]$ is selected at random, and a ligand molecule can successfully attach onto the top layer of lattice site j, as shown in figure 4.3. Lastly, it is important to correlate the virtually executed kMC events to the time elapsed in the real world. Hence, after every kMC event, either successful or unsuccessful, there is a corresponding time increment, Δt, which is given by $\Delta t = \frac{-\log \xi_3}{W_{\text{total}}}$, where $\xi_3 \in (0, 1]$ is a random number. The detailed methodology can be found in our previous work [141].

4.2.2 Experimental validation results

For experimental validation, time-series data of $CsPbBr_3$ QD crystal size were collected using a millifluidic platform [40, 136].

Figure 4.4(a) shows the kMC simulation and the experimental observations at a precursor concentration of 10 mM and varying superficial flow velocities. It is evident from the validation results that the proposed kMC simulation captures the underlying crystallization phenomenon and predicts the experimental observations well (i.e. error is in the range of 2%–15%). More importantly, figure 4.4 suggests that QD crystal size can be effectively controlled by manipulating the precursor concentration and superficial flow velocity.

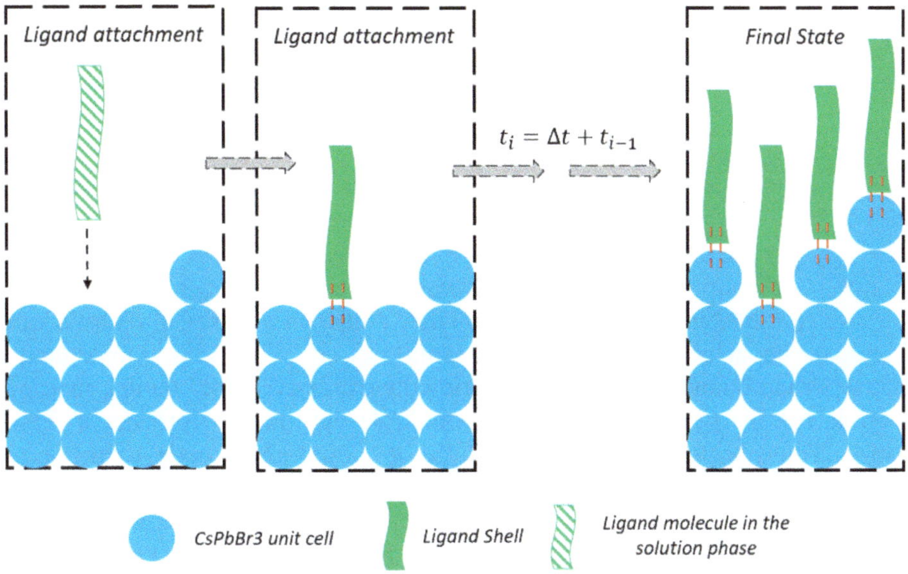

Figure 4.3. Schematic illustration of the ligand attachment event. The green waves schematically show the formation of the ligand-shell on the crystal surface, and the blue spheres represent the individual $CsPbBr_3$ unit cells which build up to form one $CsPbBr_3$ QD crystal. (Reproduced with permission from [139]. Copyright 2021 Elsevier.)

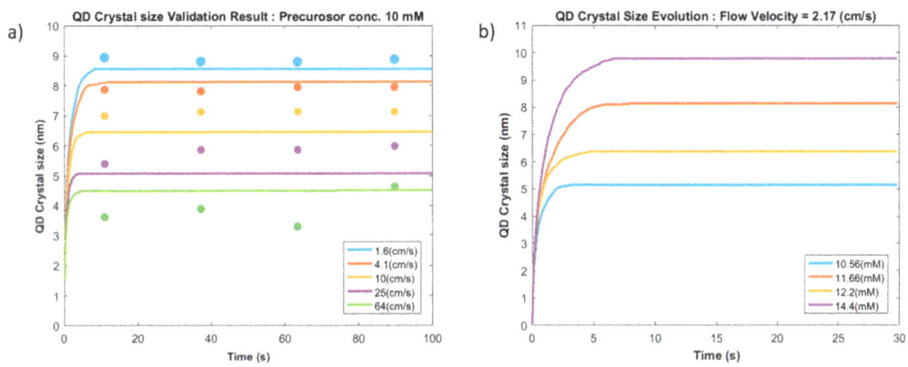

Figure 4.4. (a) Validation results for the proposed kMC simulation, where the solid line represents the KMC simulation and the scatter plot represents the experimental results [138] and (b) KMC simulation results for varying precursor concentrations at a constant superficial flow velocity of 2.17 cm s^{-1}. (Reproduced with permission from [139]. Copyright 2021 Elsevier.)

4.2.3 Optimal operation of PFC

Thus, for establishing a PFC for continuous QD manufacturing, it is paramount to determine the values of C_o and v_o to achieve a minimal offset from the set-point value for r_{final} (i.e. the desired QD size). The formulation shown below presents an optimization problem to compute the optimal inputs, C_o and v_o, for a desired r_{final} value:

$$\begin{aligned}
&\underset{C_o, v_o}{\text{Minimize}} && (r_{\text{final}} - r_{\text{final, sp}})^2 + \lambda u^2 \\
&\text{s.t.} && 5 < C_o(t) \leqslant 15 \ (\text{mol m}^{-3}) \\
& && 0.05 < v_o(t) \leqslant 10 \ (\text{cm s}^{-1}) \\
& \frac{\partial C(t, z)}{\partial t} = -v_z \frac{\partial C(t, z)}{\partial z} && - \rho_s n_s 4\pi r^2(t, z) G_r(C(t, z), v_z, r(t, z)),
\end{aligned} \tag{4.6}$$

where $r_{\text{final,sp}}$ is the desired QD crystal size, u is a vector of input variables (C_0 and v_o), and λ is the input regularization parameter. The optimizer in equation (4.6) is subject to certain constraints, which limits the values of C_o and v_o to realistic values which can be achieved in an experimental set-up [136, 138]. Unfortunately, $G_r(C, v_z, r)$ is obtained by the computationally expensive kMC model, making it infeasible for the optimizer to perform multiple iterations to procure a feasible solution. Thus, we use an ANN to mimic the developed kMC model to increase the efficiency of the optimizer. Next, we can combine the kMC-mimicking ANN with the partial differential equation (PDE) for concentration balance to develop an ANN-based hybrid model that can efficiently describe the dynamics in a PFC.

ANN architecture and training

An ANN with three inputs (i.e. $C(t, z)$, v_z, and $r(t, z)$), a single output, $G_r(C, v_z, r)$, and two hidden layers has been developed. The input data comprise the kMC simulation results of QD crystal growth for varying precursor concentrations and superficial flow velocities, which are averaged over 10 kMC trials. Various ANN architectures with two hidden layers and different numbers of neurons were tested, and it was observed that ANN with 20 neurons gave the least validation error of 0.0011 along with a testing error of 0.0095. Accordingly, figure 4.5 shows the comparison between the proposed ANN, with two hidden layers and ten neurons in each hidden layer, and the kMC simulation. It is evident that the ANN mimics the kMC simulation very well, and is 24 000 times faster than the kMC.

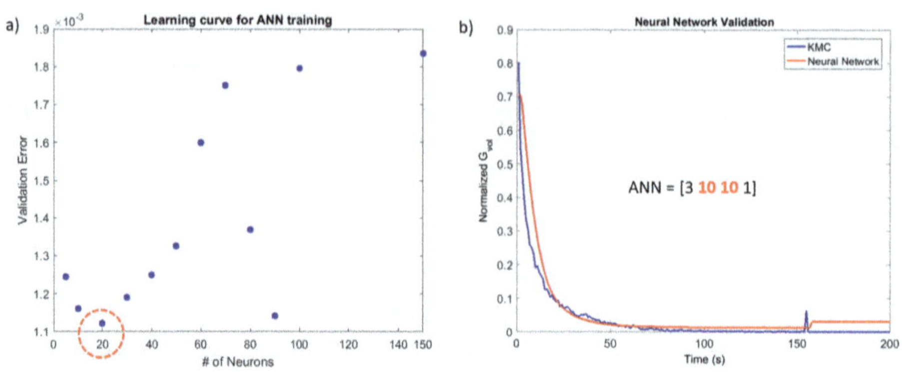

Figure 4.5. (a) Learning curve for ANN training, which shows that the ANN with two hidden layers, each with ten neurons, has the least validation error and (b) comparison of the trained ANN and kMC simulation. (Reproduced with permission from [139]. Copyright 2021 Elsevier.)

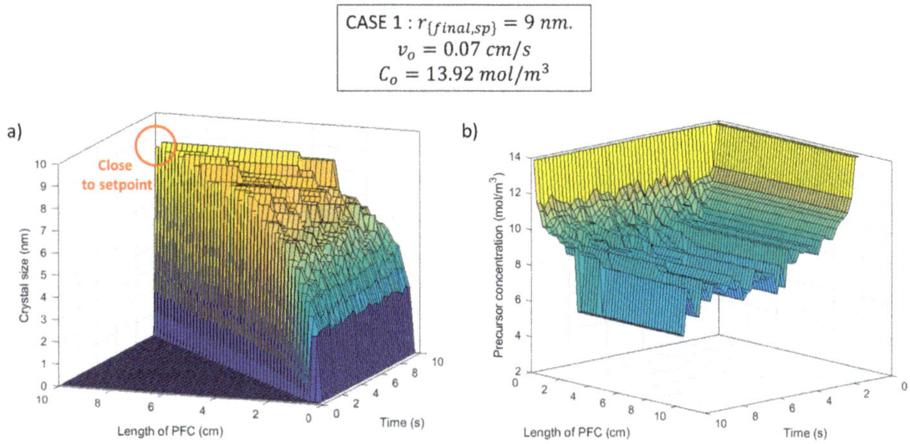

Figure 4.6. (a) Spatiotemporal evolution of QD crystal size for the optimal operation of the PFC for a set-point value of 9 nm,and (b) precursor concentration profile for the optimal operation of the PFC. The relative deviation from the set-point value is ~6.1%. (Reproduced with permission from [139]. Copyright 2021 Elsevier.)

As mentioned earlier, the resulting ANN model simulates the computationally heavy kMC process in an expedited manner and thus it can be integrated with the discretized version of the concentration balance to develop a simple ANN-based hybrid model for the PFC.

Results for QD size control

Figure 4.6(a) shows the QD crystal size evolution in the PFC under optimal operation mode with a set-point of 9 nm. Near the entry of the PFC, and for the first 2–3 s, the QD crystals grow very rapidly before they gradually saturate to terminal size, r_{final}, towards the end of the PFC. Initially, the large precursor concentration and absence of a ligand-shell give a boost to fast crystal growth and then, as precursor concentration drops and ligand-shell formation takes place, the QD crystal growth ceases and reaches a terminal value. Specifically, the relative deviation between r_{final} and the set-point value is ~6.1%. Figure 4.6(b) shows the precursor concentration profile in the PFC. Initially, the precursor concentration drops significantly due to the rapid crystal growth, and then it saturates to a stable value. This happens simultaneously with the QD size saturation (figure 4.6(b)), because if the QD crystals cease to grow any further, additional precursors are not consumed, thereby stabilizing the precursor concentration towards the end of the PFC.

4.3 Multiscale modeling of slug flow crystallizers for QD production

Parts of this section have been reproduced with permission from [146]. Copyright 2021 Elsevier.

The previous chapter focused on the development of a high-fidelity multiscale model for continuous crystallization of QDs using a PFC, and subsequent control of QD size using an ANN-based optimal operation framework. This concept of tubular crystallization can be extended to two-phase SFCs, which can provide access to larger control space by manipulation of three process inputs (i.e. gas and liquid velocity, and solute concentration). However, there is a scarcity of models that accurately describe SFCs by considering (a) the existence of a stable slug regime and (b) considering the effect of S2S variation on CSD. To address the aforementioned two challenges, we developed an ANSYS Fluent-based CFD model to accurately investigate the complexity associated with an SFC. The CFD model was capable of finding a stable slug flow regime in an SFC and analysing the S2S variation. Further, the slug variation results from the CFD simulation were fed to a slug crystallizer model which combines a continuum and a kMC model. The proposed CFD-based multiscale model (i.e. CFD + continuum + kMC) was used to investigate the effect of slug variation on CSD in an SFC and underlined the need to minimize this effect. Thus, an optimization problem was formulated to find the optimal operating condition for the SFC, which will provide good set-point tracking performance while maintaining a narrow CSD. Overall, (a) the CFD model was able to capture the phenomenon of S2S variation and the CFD results were fed to the slug crystallizer model that successfully described the crystallization of QDs in an SFC, and (b) the optimal operation framework exhibited good set-point tracking performance. More details about this chapter are presented in our previous work [146].

4.3.1 CFD-based multiscale modeling of SFC

Process description

Recently, a millifluidic SFC platform was used to fine-tune the size of QDs by manipulating the slug flow velocity and the precursor concentration [24, 142]. Here, the alternating air–liquid slugs are achieved when SFC is operated in a stable slug regime. In this scenario, it was observed that each liquid slug acts as an independent batch crystallizer while the overall SFC can be continuously operated. Based on these recent developments, in this work, we consider an SFC as shown in figure 4.7. The pipe diameter and length are set to be 1.5 mm and 100 cm, respectively. Also, it has an L-shaped junction which has two inlets, one for nitrogen, and the other for the precursor solution. Furthermore, the velocity of gas (nitrogen), v_{gas}, and velocity of liquid (toluene), v_{liq}, are adjusted to investigate the existence of a stable slug flow regime.

CFD modeling of two-phase slug flow

While there exists a plethora of methods for modeling two-phase immiscible fluids, the volume of fluid (VOF) is the most common method [143]. Due to its superior ability to track topological changes in the interface and being naturally conservative, it has been widely applied to a number of different applications [144, 145].

Governing transport equations

Incompressible Navier–Stokes equations, which can be used to model the velocity (u), and pressure (p) for two interacting fluids with constant density, are given as follows:

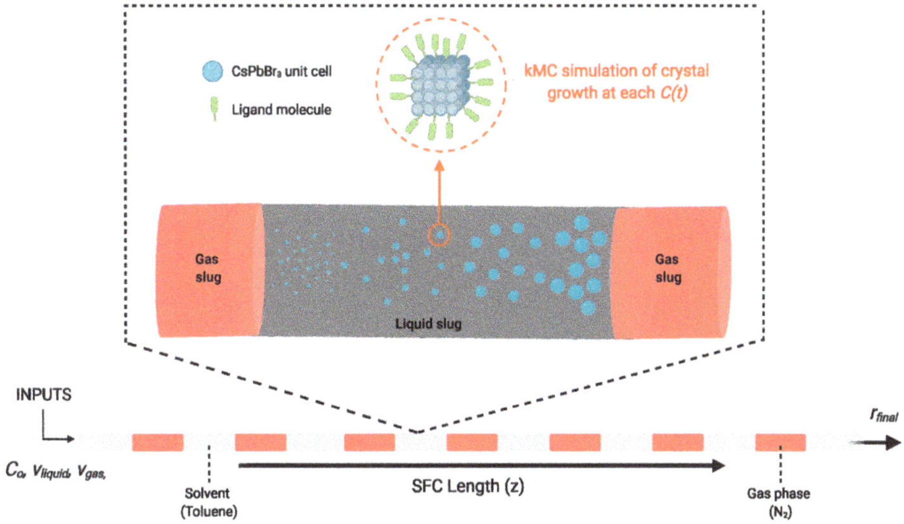

Figure 4.7. Schematic illustration of the multiscale phenomena taking place in an SFC. The blue spheres represent the QD crystals growing along the flow direction, and the green extensions around the crystal represent the ligand molecules. (Reproduced with permission from [146]. Copyright 2021 American Chemical Society.)

$$\frac{\partial \mathbf{u}}{\partial t} + \nabla \cdot (\mathbf{uu}) = -\frac{(\nabla p + \mu \nabla^2 \mathbf{u})}{\rho} + \frac{F_{\mathrm{SF}}}{\rho} + \mathbf{g}$$
$$\nabla \cdot \mathbf{u} = 0, \tag{4.7}$$

where **u** is the fluid velocity, p is the pressure, **g** is the gravitational acceleration, F_{SF} is the continuum surface force (CSF) vector, ρ is the bulk phase density, and μ is the bulk phase viscosity. This formulation assumes an isothermal flow of two immiscible Newtonian fluids. The VOF method tracks the topological changes at the interface by creating an isocontour of a globally defined function (α), which is defined as follows [146, 147]:

$$\frac{\partial \alpha}{\partial t} + \nabla \cdot (\mathbf{u}\alpha) = 0. \tag{4.8}$$

A detailed description of the modeling can be found in [148].

Meshing and numerical schemes

The SFC was fine-meshed with triangular structured cells to ensure numerical convergence. The meshing quality, which can be assessed using various indicators such as skewness and aspect ratio, is shown in figure 4.8. Generally, an aspect ratio close to unity suggests a good quality mesh as shown in figure 4.8(c). Also, atmospheric pressure was maintained at the outlet, and a no-slip boundary condition was used at the walls. The pressure-implicit with the splitting of operators (PISO) was used for pressure velocity coupling, and the numerical convergence

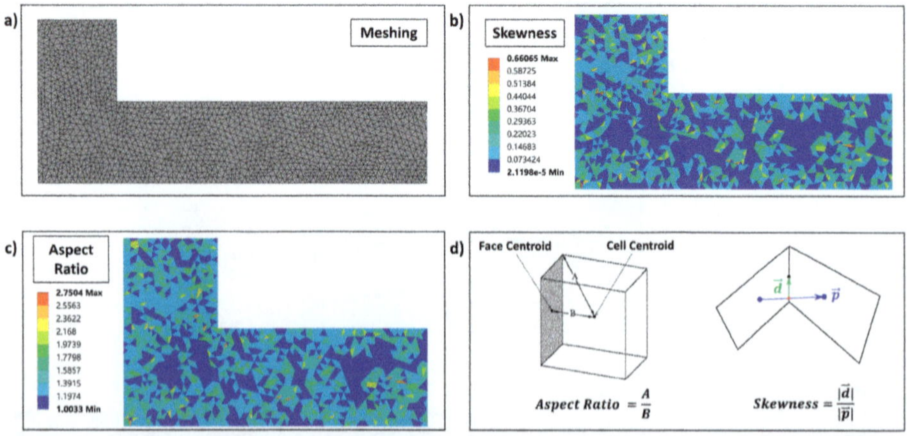

Figure 4.8. Snapshot of the ANSYS Fluent Mesh tool showing (a) the triangular mesh of element size 0.1 mm, (b) the distribution of skewness throughout the mesh, (c) the distribution of aspect ratio throughout the mesh, and (d) an illustration of skewness and aspect ratio. (Reproduced with permission from [146]. Copyright 2021 American Chemical Society.)

Table 4.2. Parameters for the ANSYS Fluent simulation. (Reproduced with permission from [146]. Copyright 2021 American Chemical Society.)

Parameters	Values
Density (toluene)	866 (kg m^{-3})
Density (nitrogen)	1.225 (kg m^{-3})
Interfacial tension (toluene–nitrogen)	0.022 (N m^{-1})
Mesh size	0.1 mm
Number of nodes in mesh	649 605
Number of elements in mesh	314 772

criteria for the continuity equation was set to 0.0001. Lastly, table 4.2 presents the other details of the ANSYS Fluent simulation.

Slug crystallizer model

Macroscopic modeling: mass and energy balance

In the stable slug regime, each liquid slug can be treated as a well-mixed mini-batch crystallizer [149, 150]. Subsequently, the mass and energy balance for slug $i \in [1, N]$ can be described as follows:

$$\begin{aligned} \frac{\mathrm{d}C(i)}{\mathrm{d}t} &= \rho_s n_s(i, t) 4\pi r^2(i, t) G_r(i) \\ \frac{\mathrm{d}T(i)}{\mathrm{d}t} &= -\frac{\rho_s \Delta H_c}{\rho_f C_p} n_s(i, t) 4\pi r^2(i, t) G_r(i) - \frac{U_c a_c}{\rho_f C_p}(T(i) - T_w), \end{aligned} \tag{4.9}$$

where $r(i, t)$ is the average crystal radius in slug i, $C(i)$ is the precursor concentration in slug i, $T(i)$ is the temperature in slug i, ρ_s is the density of $CsPbBr_3$ crystals, ρ_f is the density of the solution phase (toluene), C_p is the specific heat of the solution phase, U_c is the heat transfer coefficient for the heating/cooling jacket, a_c is the surface area of the heating/cooling jacket, T_w is the temperature of the jacket wall, $n_s(i, t)$ is the number of $CsPbBr_3$ crystals in slug i, and $G_r(i)$ is the 1D crystal growth rate in slug i.

Microscopic modeling: kMC simulation

Furthermore, since $CsPbBr_3$ QDs are a relatively new system, there is a scarcity of reliable correlations and empirical rate laws for $G_r(i)$. Therefore, it is of significant interest to develop a microscopic crystal growth model which considers the effect of ligand-shell formation and slug velocity on $G_r(i)$. Our previous work describes the microscopic kMC crystallization model in detail, and is used to compute $G_r(i)$ [141]. Briefly, there are three major events in the kMC simulations, namely, unit cell attachment, ligand attachment, and ligand detachment. The interplay of the aforementioned kMC events simulates the growth of QD crystals and computes $G_r(i)$.

Integration of macroscopic and microscopic models

The solution to equation (4.9) is found by computing $G_r(i)$ for slug i using the kMC simulation, which describes the microscopic phenomenon underlying $CsPbBr_3$ crystal growth. Specifically, since $C(i, t)$ and $G_r(i)$ affect each other, the kMC simulation is performed for each t and slug i to accurately trace the evolution of $C(i, t)$ and $r(i, t)$ as shown in figure 4.9. It is important to note that each slug i has a slightly different slug velocity, which leads to a variation in the evolution of crystal size and precursor concentration; this variation ultimately results in a wide CSD. Lastly, the overall modeling methodology is schematically described in figure 4.9.

4.3.2 Results and discussion

Stable slug regime and experimental validation

To find the stable slug flow regime, CFD simulations were performed under various initial conditions (i.e. different sets of nitrogen and toluene velocities). Specifically, a grid search was performed, where the nitrogen velocity was varied from 0.01 m s^{-1} to 0.1 m s^{-1}, and the toluene velocity was varied from 0.01 to 0.04 m s^{-1}. Then the CFD results, which include the slug regime data and slug velocity, for each of these grid points were tabulated for further analysis.

A stable slug flow was observed under certain initial conditions. For example, figure 4.10 shows the temporal evolution of the nitrogen volume fraction in the SFC when v_{nitrogen} is 0.03 m s^{-1} and v_{toluene} is 0.04 m s^{-1}. Clearly, the figure shows various steps in the formation of a stable slug flow, which are summarized in detail in our previous work [148]. To further confirm the existence of a stable slug flow, the temporal evolution of gas volume fraction at the tracer plane AB is shown in figure 4.11. The oscillating amplitude (y-axis) in figure 4.11 is characteristic of a

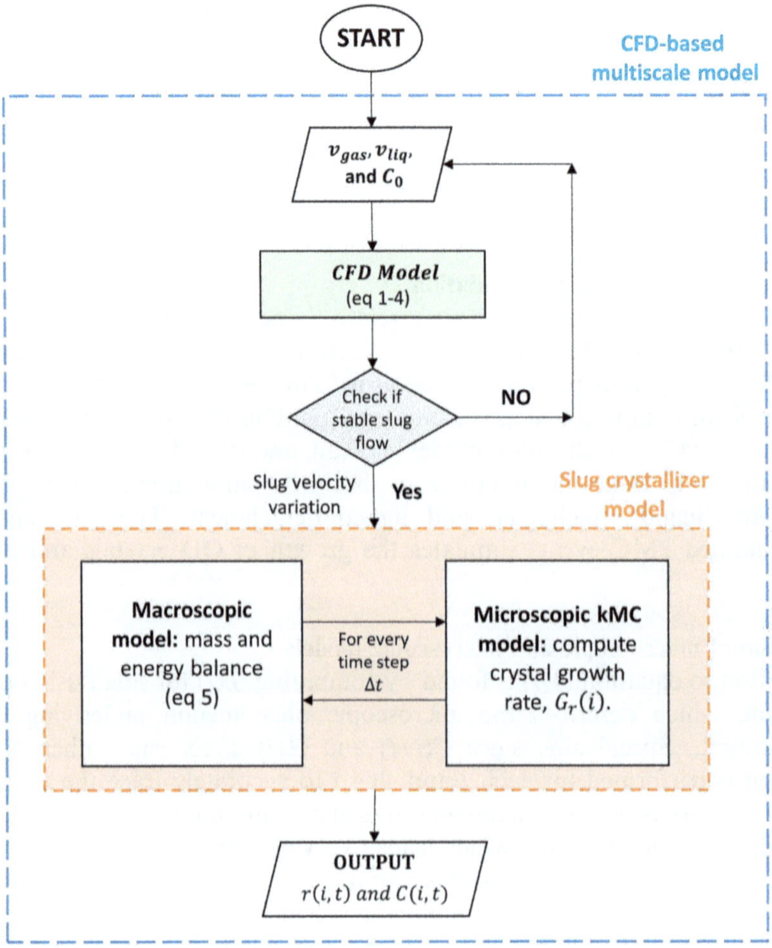

Figure 4.9. Schematic illustration of the workflow employed in the proposed CFD-based multiscale modeling framework. (Reproduced with permission from [146]. Copyright 2021 American Chemical Society.)

stable alternating slug flow condition [146]. Specifically, gas volume fraction oscillated between a magnitude of 1 (gas slug), and 0 (liquid slug), thereby indicating the existence of alternating gas–liquid slugs.

Furthermore, to validate the proposed CFD model, the SFC was operated at a set of fluid velocities, and the corresponding slug velocities were measured experimentally. From figure 4.12 it can be observed that the ANSYS simulation mimics the experimental observations well for a set of fluid velocities (indicated by 'Counts' on the *x*-axis in figure 4.12).

Fitting slug velocity to normal distribution

The phenomenon of S2S can be observed even in the presence of a stable slug flow regime [152]. For example, figure 4.13 shows that the slug velocity peaks (i.e. *y*-axis) are slightly different for each incoming slug. Then, the slug velocity is plotted in

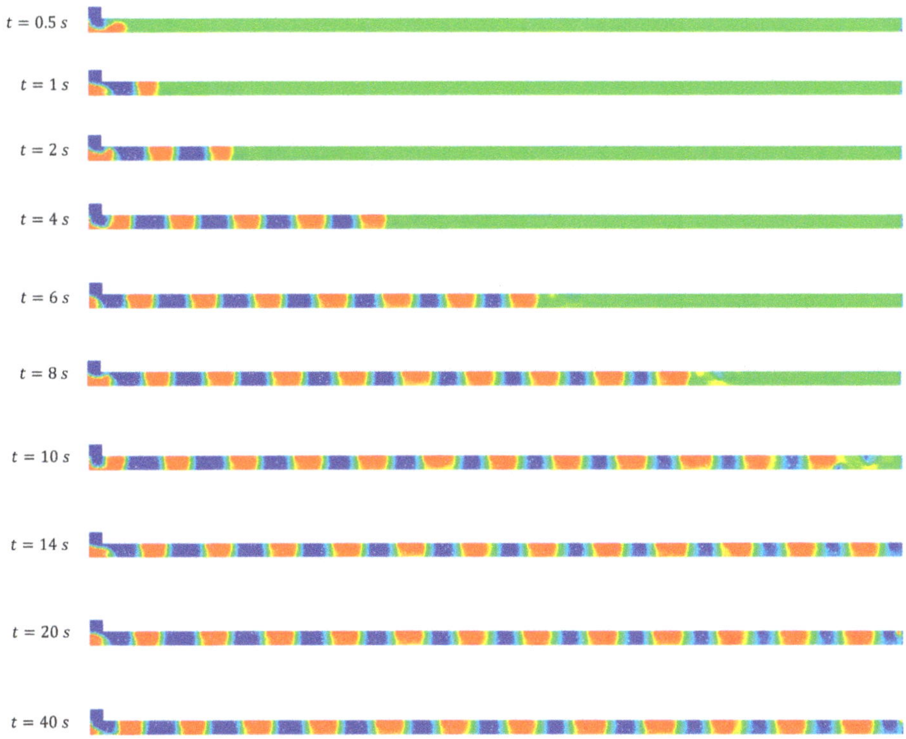

Figure 4.10. Compilation of the CFD simulation snapshots, which shows the contour of nitrogen volume fraction in the SFC. The nitrogen velocity (v_{nitrogen}) was 0.03 m s^{-1} and the toluene velocity (v_{toluene}) was 0.04 m s^{-1}. (Reproduced with permission from [146]. Copyright 2021 American Chemical Society.)

figure 4.14 to see if there is a certain pattern to S2S. It is evident that the slug velocity in an SFC closely follows a normal distribution and thus quantifies this variation.

Lookup table for predicting slug velocity variation

The previous section exhibited the existence of S2S variation and indicated that slug velocity conforms well to a normal distribution. When a similar analysis was performed for all the grid points in the fluid velocity space, it was observed that S2S follows a normal distribution with varying mean (μ) and standard deviation (σ). Furthermore, the surface contour in figure 4.15 is used in the form of a lookup table as it provides the values of average slug velocity (μ), and slug velocity variation (σ) at the required set of fluid velocities. Overall, figure 4.15 serves as a guide to determine the fluid velocities which result in a minimum slug variation, and possibly lead to a homogeneous crystallization behavior across the SFC.

4.3.3 Multivariable optimal operation problem

An optimization problem aimed at finding the optimal input profile (i.e. nitrogen velocity (v_{nitrogen}), toluene velocity (v_{toluene}), and precursor concentration (C_0)), which

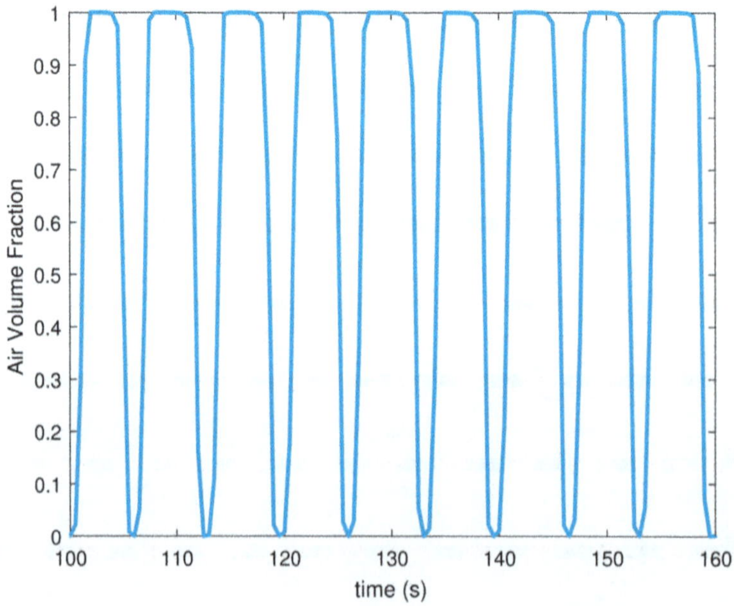

Figure 4.11. Temporal evolution of gas volume fraction in the SFC when operated at a stable slug flow condition. The gas velocity was 0.03 m s^{-1} and the toluene velocity was 0.04 m s^{-1}. (Reproduced with permission from [146]. Copyright 2021 American Chemical Society.)

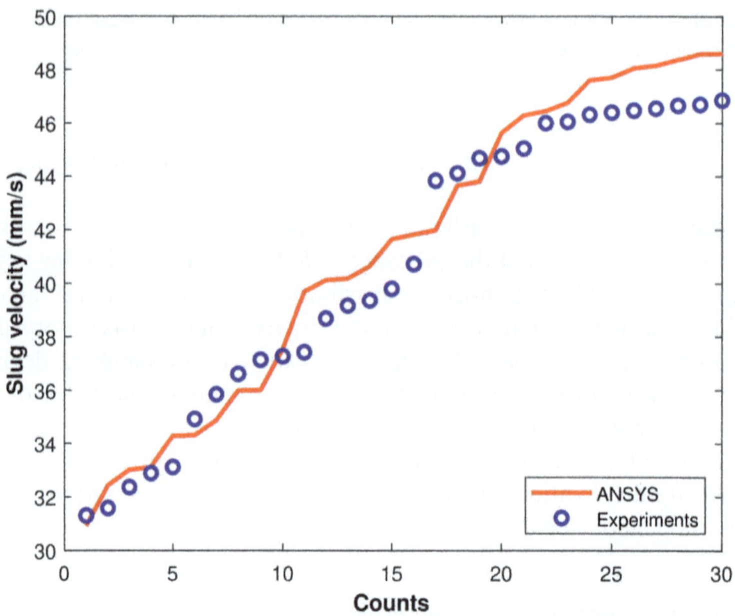

Figure 4.12. Comparison of slug velocity measured from experiments [151] and computed from the CFD simulation for a set of fluid velocities. (Reproduced with permission from [146]. Copyright 2021 American Chemical Society.)

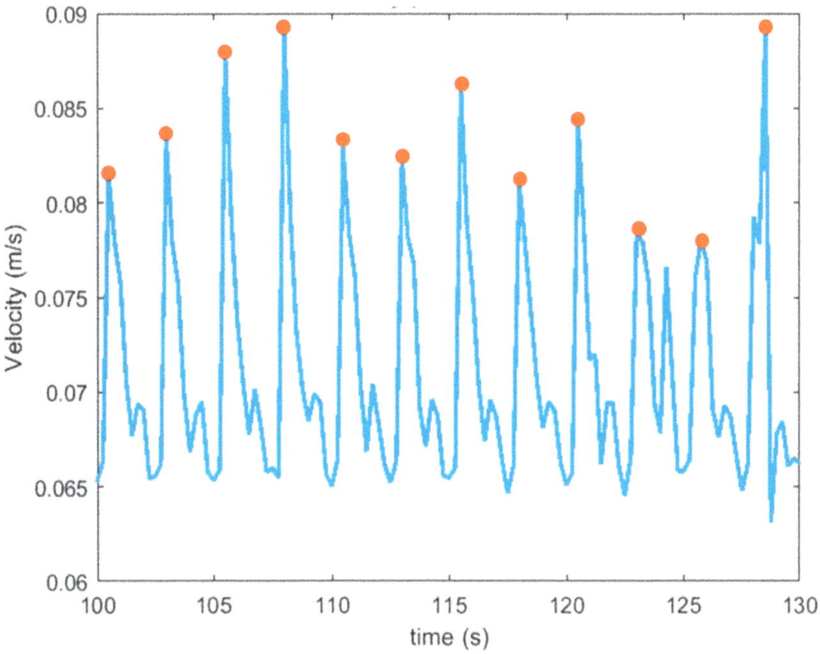

Figure 4.13. Slug velocity at the tracer plane AB showcasing the varying slug velocity for each incoming liquid slug. In this case, v_{nitrogen} was 0.03 m s^{-1} and v_{toluene} was 0.04 m s^{-1}. (Reproduced with permission from [146]. Copyright 2021 American Chemical Society.)

can ensure the good set-point tracking (i.e. QD size) along with a narrow CSD, was constructed as follows:

$$\begin{aligned} \underset{C_o, v_{\text{nitrogen}}, v_{\text{toluene}}}{\text{Minimize}} \quad & (r_{\text{final}} - r_{\text{final, sp}})^2 + \lambda \sigma_{\text{cry}}^2 \\ \text{s.t} \quad & 5 < C_o(t) \leqslant 15 \ (\text{mol m}^{-3}) \\ & v_{\text{nitrogen}}(t),\ v_{\text{toluene}}(t) \in \text{Stable slug regime} \\ & v_{\text{slug}}(i) \in \phi\big(v_{\text{nitrogen}},\ v_{\text{toluene}}\big) \\ & \frac{\partial C(i)}{\partial t} = \rho_s n_{\text{seed}}(i) 4\pi r^2(i,\ t) G_r(i), \end{aligned} \tag{4.10}$$

where $r_{\text{final,sp}}$ is the desired QD crystal size (i.e. set-point), r_{final} is the predicted QD size at the SFC exit, σ_{cry} is the standard deviation of CSD, λ is the regularization parameter, which penalizes a broad CSD, and $\phi(v_{\text{nitrogen}},\ v_{\text{toluene}})$ is the normal distribution of slug velocities as described in the lookup table. Specifically, $\phi(v_{\text{nitrogen}},\ v_{\text{toluene}})$ is used to generate an array of slug velocities for N liquid slugs. This information is fed to the slug crystallizer model, and the crystal size evolution and precursor concentration profile are generated (figure 4.9). Furthermore, since the solution to equation (4.10) would require computation of crystal growth rate, $G_r(i)$, over multiple optimization iterations, the previously developed ANN [141] was utilized to increase the efficiency of the optimizer.

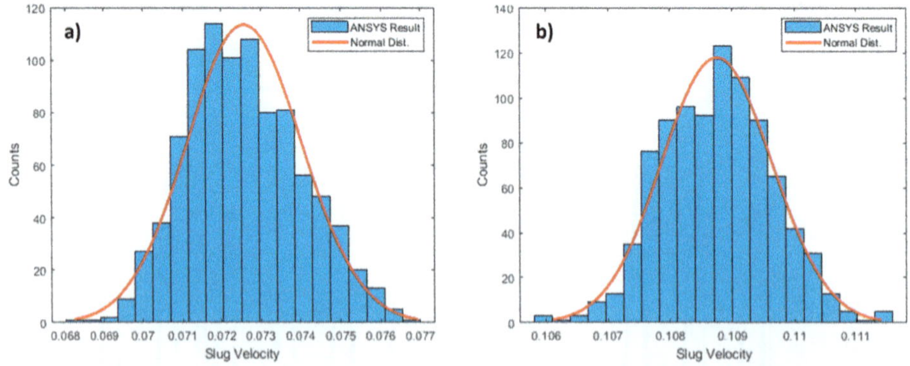

Figure 4.14. Slug velocity variation in the SFC showcasing a normal distribution for different sets of fluid velocities: (a) $v_{\text{nitrogen}} = 0.03\,\text{m s}^{-1}$, $v_{\text{toluene}} = 0.04\,\text{m s}^{-1}$, $\mu = 0.072\,\text{m s}^{-1}$, and $\sigma = 2.1\%$; and (b) $v_{\text{toluene}} = 0.03\,\text{m s}^{-1}$, $v_{\text{nitrogen}} = 0.06\,\text{m s}^{-1}$, $\mu = 0.091\,\text{m s}^{-1}$, and $\sigma = 1.4\%$. Here, μ represents the mean slug flow velocity, and σ represents the standard deviation of the slug velocity. (Reproduced with permission from [146]. Copyright 2021 American Chemical Society.)

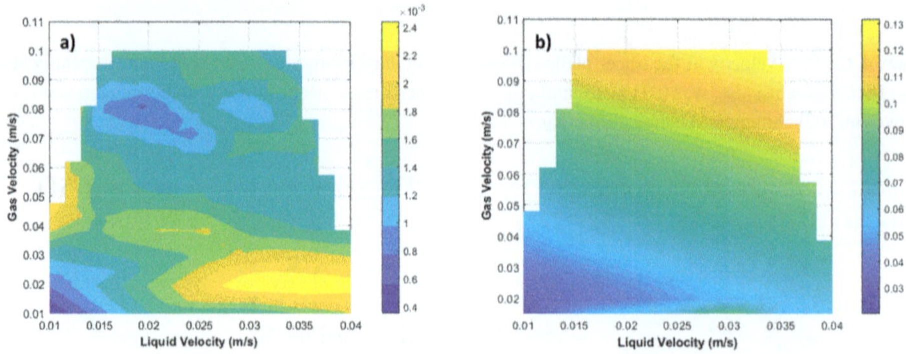

Figure 4.15. Surface contour plot showcasing the spread of (a) variation in slug velocity (σ), and (b) mean slug velocity (μ) across the fluid velocity space. The region with no color corresponds to the fluid velocities at which a stable slug flow condition was not achieved. (Reproduced with permission from [146]. Copyright 2021 American Chemical Society.)

Optimal operation results

The optimal operation problem was solved for a set-point (QD size) value of 9 nm. The solution to the optimal operation problem yields a C_0 value of 13 mM, v_{nitrogen} of $0.082\,\text{m s}^{-1}$, and v_{toluene} of $0.022\,\text{m s}^{-1}$ as the optimal input values, which are indeed in the stable slug regime. Feeding this input to the slug crystallizer model provides the evolution of QD crystal size with time, and the CSD at the exit of the SFC (figure 4.16). Specifically, the average crystal size evolution in the SFC is shown in figure 4.16(a). The light blue region illustrates the variation of crystal size in N different slugs and represents the evolution of QD crystal size. Figure 4.16(b) shows

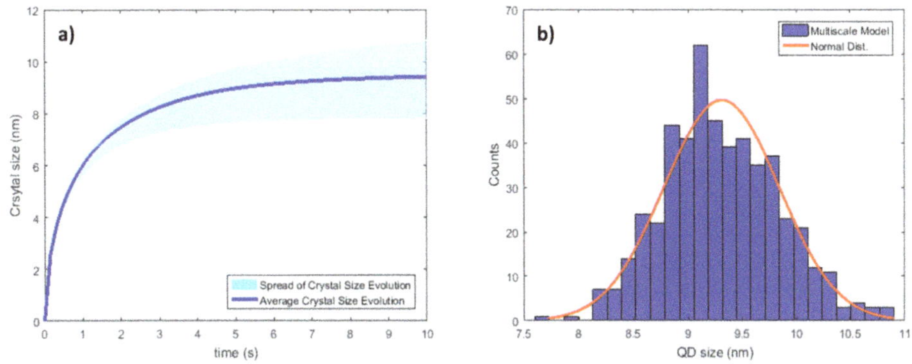

Figure 4.16. (a) Crystal size evolution of $CsPbBr_3$ QDs in SFC and (b) CSD of $CsPbBr_3$ QDs at the SFC exit. The CSD at the SFC exit follows a normal distribution with a mean (μ_{cry}) of 9.38 nm, and a standard deviation (σ_{cry}) of 5.71%. (Reproduced with permission from [146]. Copyright 2021 American Chemical Society.)

the CSD computed at the SFC exit follows a normal distribution with a mean (μ_{cry}) of 9.38 nm and a standard deviation (σ_{cry}) of 5.71%. Furthermore, the relative deviation between μ_{cry} and the set-point value is ~3.5%. Also, a σ_{cry} of 5.71% suggests that the crystal size in N different slugs is not significantly different.

4.4 Future directions

This work is focused on developing a multiscale modeling and control framework for continuous QD production and QD thin-film deposition for various optoelectronic applications. Using the above-developed models as a launchpad, high-impact future works can be formulated to create more generalizable and easily scalable QD models. Specifically, the modeling philosophy followed in the current work has two parts: (a) developing first-principles high-fidelity QD models and then (b) constructing ML-based black-box models that are computationally efficient. Although this strategy can be implemented for accurate modeling and control of QD manufacturing, it can be improved by following the paradigm of hybrid modeling. Specifically, a more intimate coupling between system-agnostic physics (e.g. mass and energy balance equations, population balance, and transport phenomena, etc) with system-specific information (i.e. kinetic parameters, growth and nucleation rate, etc) will result in a hybrid model that is both accurate and computationally efficient. To this end, the following section provides brief details on developing a powerful hybrid model for QD processes and considers a specific example of crystallization (figure 4.17).

4.4.1 Transformer-enhanced hybrid modeling of QD systems

Recent years have seen a tremendous push toward the development of digital twins for different chemical systems that facilitate better online process monitoring, control, and on-the-go process optimization. The majority of the digital twins are

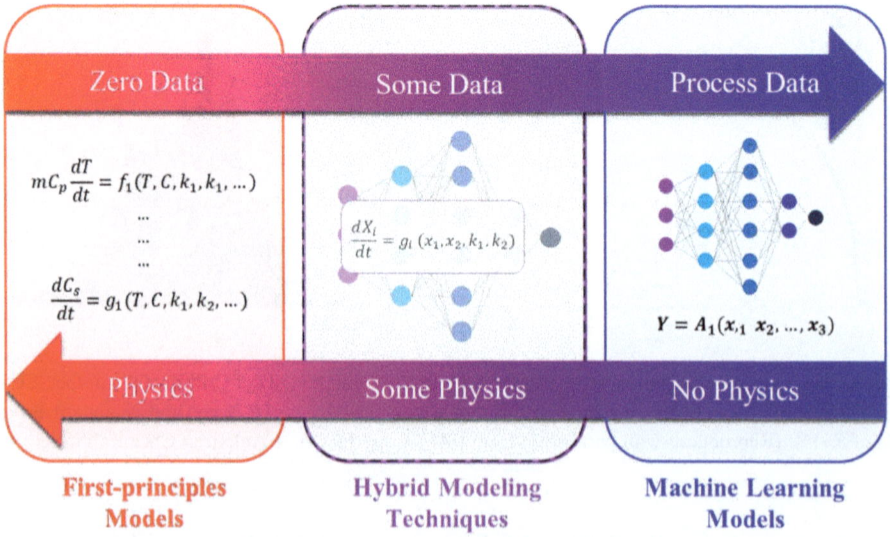

Figure 4.17. Schematic illustration of different modeling paradigms in PSE.

based on a fully data-driven approach such as DNNs along with a few examples of RNNs to account for the time-series dynamics of chemical systems. However, given the hesitance of directly deploying a black-box tool in practice due to safety and operational issues, these models have not seen the light of day. To address this challenge, hybrid models that combine first-principles physics-based dynamics with some data-driven components have increased in popularity as they are considered to be a 'best-of-both-worlds' approach. The first-principles module comprises typical mass and energy balance equations, rate kinetics, and a population balance model (PBM), while the data-driven part estimates the system-specific parameters (i.e. kinetic rate constant, growth and nucleation rate, and other parameters), thereby acting as an input to the first-principles module.

Unfortunately, existing hybrid models (that are largely based on DNNs) require *a priori* information of the kinetic equations which then can be fed with the estimated parameters. This provides a major implementation hurdle as knowing the exact kinetic information of complex systems (i.e. crystallization, fermentation, heterogeneous catalysis) is difficult. For example, in crystallization, knowledge of growth rate (G) and nucleation rate (B) and their dependence on supersaturation, temperature, crystal moments, etc, is difficult to procure. Further, DNN-based hybrid models find it difficult to make accurate time-series predictions in the presence of process uncertainties and noise. Thus, to resolve these two challenges, an alternative paradigm of hybrid modeling that can (a) utilize process data to approximate the underlying kinetic function and (b) accurately capture the short and long-term evolution of system states is required.

Emergence of transformer models

On the lines of transfer learning, very recently, transformer-based large language models (LLMs) and vision transformers have pioneered the advent of disruptive applications including ChatGPT, CodeGPT, Dall-E, and others. These applications act as unified models within the natural language processing (NLP) and computer vision space [153–160]. The key driving force for this revolution has been the advent of transformer networks that utilize (a) a multiheaded self-attention (MAH) mechanism to gain human-like contextual understanding between the words in a given input sentence, (b) positional encoding to embed sequential information efficiently, (c) impressive transfer learning capabilities for rapid fine-tuning for niche tasks, and (d) highly parallelization-friendly architecture (unlike sequential models such as RNN) that empowers training of extraordinarily large models (i.e. 100M+ to 10B+ parameters) with of 10TB+ of textual/image data by enabling maximum use of graphics processing units (GPUs). Transformers utilize attention modules that are trained on large amounts of textual data from multiple sources to learn the underlying grammar, syntax, vocabulary, and nuances of a language. This allows the trained transformer model to be directly used for tasks such as sentiment analysis and fraudulent email detection or fine-tuned for specific tasks such as filtering movie reviews. Unfortunately, utilizing pretrained language models (LLMs) for chemical systems, which involve multivariate time-series data with *floats* and *integers*, is a challenging task with only a few studies in the field. Schweidtmann and colleagues recently converted process flow sheets into input tokens compatible with NLP models, allowing auto-completion of chemical process flow sheets [161]. Meanwhile, AlphaFold is a transformer model that predicts protein structure based solely on its amino acid sequence, providing a groundbreaking solution to the longstanding problem of protein folding [162–164]. Similarly, Venkatasubramanian and Mann demonstrate excellent improvement on existing protein modeling transformers [165]. Lastly, a recent study showcased first-of-a-kind transformer-based MPC implementation for regulation crystal size in an industrial-scale batch crystallizer [166, 167].

Moreover, transformers show a remarkable ability to establish strong correlations between input and outputs, even in the presence of system noise or uncertainties. These models adeptly focus on short-term and long-term dependencies in the evolution of system states [168, 169]. In essence, the attention mechanism performs a scaled-dot product calculation between various input vectors, enabling it to selectively pay attention to significant long-term (e.g. concentration evolution) and short-term (e.g. sudden change in temperature due to control actions) process alterations by assigning higher attention scores to such instances. As a result, the attention mechanism serves as a filtering mechanism to dynamically handle process uncertainties and data noise by effectively dampening weak correlations and amplifying strong interactions between the system states. Given these unique features, transformers can be highly suitable for time-series predictions and control tasks in crystallization. This is especially true given the complex dependencies that exist between system states, which are highly sensitive to short-term control actions. Despite these exciting possibilities, the practical implementation of various transformer architectures for the modeling and control of various chemical processes, especially for hybrid modeling, has yet to be conclusively demonstrated.

Developing TST-based hybrid models

Motivated by these exciting opportunities, a hybrid time-series transformer (TST) model for a chemical system with a specific focus on crystallization can be developed [166, 170]. Specifically, the first-principles module will include system-agnostic dynamics (i.e. mass-energy balance and population balance equations) while the transformer model approximates a system-specific functional form of growth and nucleation dynamics to allow for a dynamic coupling with the first-principles module. Further, since the existing *vanilla*-transformer architecture utilizes a DNN for approximating the nonlinearities between inputs and outputs, it is not well suited for providing multivariate time-series predictions. Thus, long-short-term-memory networks (LSTM) are integrated with the TST architecture to create a novel TST-LSTM framework that combines the remarkable ability of transformers to learn underlying system dynamics and functionality with the superior performance of LSTM on time-series predictions. As a result, the hybrid-TST-LSTM model is comprised of a first-principles module comprised of generalized mass, energy, and PBM, and a TST-LSTM model that utilizes state information (i.e. concentration, temperature, crystal moments, etc) to estimate the functional form of G and B. The developed hybrid model can be trained and tested for simulating various QD crystallization systems developed in the previous sections and can be even integrated with an MPC. Overall, the unique ability of transformers to learn complex system dynamics and functions, coupled with the superior temporal predictions offered by LSTM and first-principles modeling, offers exciting opportunities for the development of accurate hybrid model-based digital twins for various chemical systems. Moreover, the developed framework can be extended to other chemical systems, such as fermentation, reaction engineering, battery systems, and catalysis [128, 171–173]. In light of these promising avenues, the future is indeed bright, and we look forward to seeing transformer-based hybrid models make a profound impact in the chemical industry (figure 4.18).

4.5 Conclusions

Although there has been a huge impetus for research and commercialization of high-performance QDs for various optoelectronic applications (i.e. solar cells and display technology), there are certain challenges that have not been adequately addressed in the literature. For example, (a) there is a lack of understanding of the crystallization kinetics of various QD systems, which hinders the predictive control of the QD size; (b) there are very few demonstrations of fast-scalable manufacturing of QDs as well as QD-based optoelectronics; and (c) there is an absence of computationally efficient solutions for online control and optimization of QD processes (i.e. QD crystallization and thin-film deposition). To address these challenges (figure 4.19), we have developed a continuous QD manufacturing framework using multiscale modeling and advanced process control techniques. First, a multiscale PFC model and CFD-

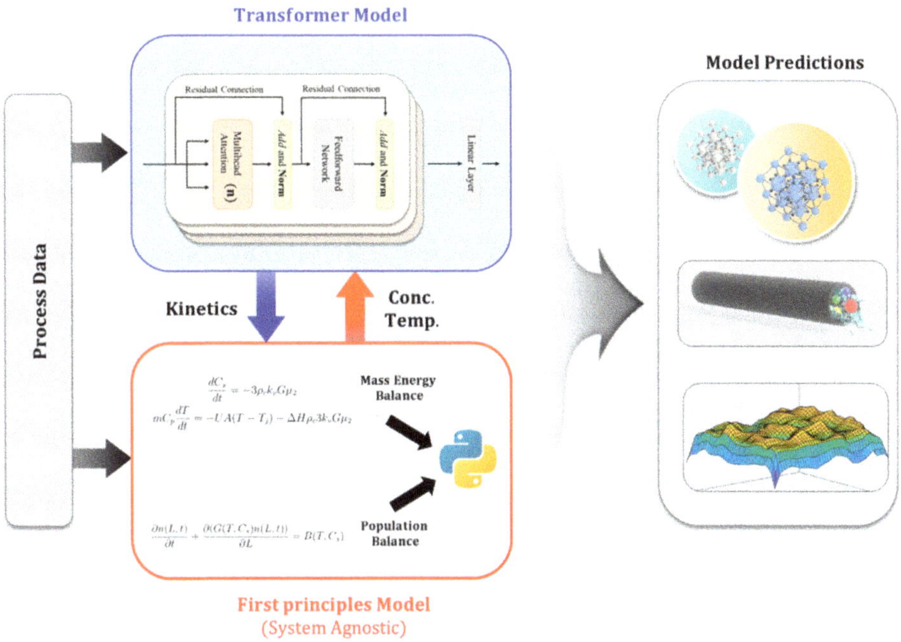

Figure 4.18. Schematic illustration of a transformer-based hybrid model for QD systems.

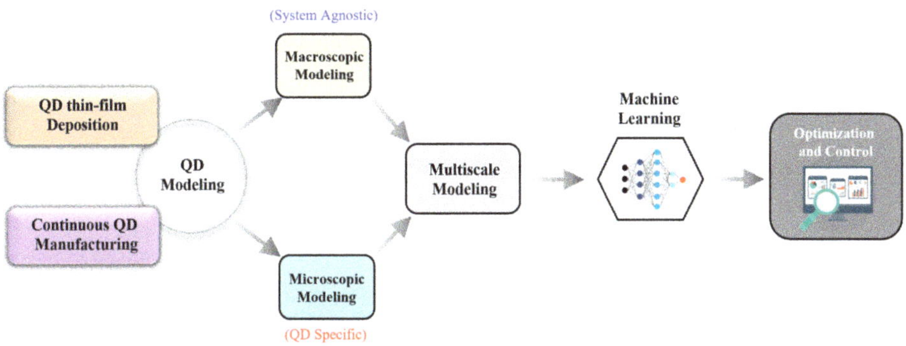

Figure 4.19. Schematic illustration of the QD modeling framework utilized in the current work.

based multiscale SFC were developed to demonstrate set-point tracking of the desired QD size and a narrow CSD. This was done by combining a microscopic kMC model, which incorporates the surface-level phenomenon of unit cell adsorption, ligand attachment, and detachment to describe QD crystallization, with macroscopic mass and energy balance equations. The resulting high-fidelity multiscale model was then surrogated using a DNN to obtain computationally efficient optimal operation of PFC and SFC for regulation QD size (i.e. a key product specification for LEDs and solar cells). Overall, this chapter presented various modeling challenges in QD manufacturing using a combination of multiscale models (i.e. coupled microscopic and macroscopic models), ML, and process control techniques. Moreover, as a natural extension of the developed framework (figure 4.19), a time-series transformer-based hybrid modeling paradigm is suggested as a promising future direction.

Bibliography

[1] Petrus M L, Schlipf J, Li C, Gujar T P, Giesbrecht N, Müller-Buschbaum P, Thelakkat M, Bein T, Hüttner S and Docampo P 2017 Capturing the Sun: a review of the challenges and perspectives of perovskite solar cells *Adv. Energy Mater.* **7** 1700264

[2] Li D, Zhang D, Lim K-S, Hu Y, Rong Y, Mei A, Park N-G and Han H 2021 A review on scaling up perovskite solar cells *Adv. Funct. Mater.* **31** 2008621

[3] Carey G H, Abdelhady A L, Ning Z, Thon S M, Bakr O M and Sargent E H 2015 Colloidal quantum dot solar cells *Chem. Rev.* **115** 12732–63

[4] Ansari M I H, Qurashi A and Nazeeruddin M K 2018 Frontiers, opportunities, and challenges in perovskite solar cells: a critical review *J. Photochem. Photobiol.* C **35** 1–24

[5] Park N-G and Zhu K 2020 Scalable fabrication and coating methods for perovskite solar cells and solar modules *Nat. Rev. Mater.* **5** 333–50

[6] Zhu R, Luo Z, Chen H, Dong Y and Wu S-T 2015 Realizing rec. 2020 color gamut with quantum dot displays *Opt. Express* **23** 23680–93

[7] Zhang X, Wang H-C, Tang A-C, Lin S-Y, Tong H-C, Chen C-Y, Lee Y-C, Tsai T-L and Liu R-S 2016 Robust and stable narrow-band green emitter: an option for advanced wide-color-gamut backlight display *Chem. Mater.* **28** 8493–7

[8] Choi M K, Yang J, Hyeon T and Kim D-H 2018 Flexible quantum dot light-emitting diodes for next-generation displays *NPJ Flex. Electron.* **2** 1–14

[9] Liu Z, Lin C-H, Hyun B-R, Sher C-W, Lv Z, Luo B, Jiang F, Wu T, Ho C-H and Kuo H-C 2020 Micro-light-emitting diodes with quantum dots in display technology *Light: Sci. Appl.* **9** 1–23

[10] Protesescu L, Yakunin S, Bodnarchuk M I, Krieg F, Caputo R, Hendon C H, Yang R X, Walsh A and Kovalenko M V 2015 Nanocrystals of cesium lead halide perovskites ($CsPbX_3$, X= Cl, Br, and I): novel optoelectronic materials showing bright emission with wide color gamut *Nano Lett.* **15** 3692–6

[11] Kramer I J, Minor J C, Moreno-Bautista G, Rollny L, Kanjanaboos P, Kopilovic D, Thon S M, Carey G H, Chou K W and Zhitomirsky D 2015 Efficient spray-coated colloidal quantum dot solar cells *Adv. Mater.* **27** 116–21

[12] Jiang C, Zhong Z, Liu B, He Z, Zou J, Wang L, Wang J, Peng J and Cao Y 2016 Coffee-ring-free quantum dot thin film using inkjet printing from a mixed-solvent system on modified ZnO transport layer for light-emitting devices *ACS Appl. Mater. Interfaces* **8** 26162–8

[13] Yang Z, Wang M, Li J, Dou J, Qiu H and Shao J 2018 Spray-coated $CsPbBr_3$ quantum dot films for perovskite photodiodes *ACS Appl. Mater. Interfaces* **10** 26387–95

[14] Kagan C R, Lifshitz E, Sargent E H and Talapin D V 2016 Building devices from colloidal quantum dots *Science* **33** aac5523

[15] Ramasamy P, Lim D-H, Kim B, Lee S-H, Lee M-S and Lee J-S 2016 All-inorganic cesium lead halide perovskite nanocrystals for photodetector applications *Chem. Commun.* **52** 2067–70

[16] Wei Y, Cheng Z and Lin J 2019 An overview on enhancing the stability of lead halide perovskite quantum dots and their applications in phosphor-converted leds *Chem. Soc. Rev.* **48** 310–50

[17] Liu M, Yazdani N, Yarema M, Jansen M, Wood V and Sargent E H 2021 Colloidal quantum dot electronics *Nat. Electron.* **4** 548–58

[18] Faheem M B, Khan B, Feng C, Farooq M U, Raziq F, Xiao Y and Li Y 2019 All-inorganic perovskite solar cells: energetics, key challenges, and strategies toward commercialization *ACS Energy Lett.* **5** 290–320

[19] Pelayo F, de Arquer G, Talapin D V, Klimov V I, Arakawa Y, Bayer M and Sargent E H 2021 Semiconductor quantum dots: technological progress and future challenges *Science* **373** eaaz8541

[20] Danek M, Jensen K F, Murray C B and Bawendi M G 1996 Synthesis of luminescent thin-film CdSe/ZnSe quantum dot composites using CdSe quantum dots passivated with an overlayer of ZnSe *Chem. Mater.* **8** 173–80

[21] Kovalenko M V 2015 Opportunities and challenges for quantum dot photovoltaics *Nat. Nanotechnol.* **10** 994–7

[22] Kim M R and Ma D 2015 Quantum-dot-based solar cells: recent advances, strategies, and challenges *J. Phys. Chem. Lett.* **6** 85–99

[23] Lou Y, Fang M, Chen J and Zhao Y 2018 Formation of highly luminescent cesium bismuth halide perovskite quantum dots tuned by anion exchange *Chem. Commun.* **54** 3779–82

[24] Abdel-Latif K, Epps R W, Kerr C B, Papa C M, Castellano F N and Abolhasani M 2019 Facile room-temperature anion exchange reactions of inorganic perovskite quantum dots enabled by a modular microfluidic platform *Adv. Funct. Mater.* **29** 1900712

[25] Chiba T, Hayashi Y, Ebe H, Hoshi K, Sato J, Sato S, Pu Y-J, Ohisa S and Kido J 2018 Anion-exchange red perovskite quantum dots with ammonium iodine salts for highly efficient light-emitting devices *Nat. Photon.* **12** 681–7

[26] Pan A, He B, Fan X, Liu Z, Urban J J, Alivisatos A P, He L and Liu Y 2016 Insight into the ligand-mediated synthesis of colloidal $CsPbBr_3$ perovskite nanocrystals: the role of organic acid, base, and cesium precursors *ACS Nano* **10** 7943–54

[27] Sun S, Yuan D, Xu Y, Wang A and Deng Z 2016 Ligand-mediated synthesis of shape-controlled cesium lead halide perovskite nanocrystals via reprecipitation process at room temperature *ACS Nano* **10** 3648–57

[28] Luo B, Pu Y-C, Lindley S A, Yang Y, Lu L, Li Y, Li X and Zhang J Z 2016 Organolead halide perovskite nanocrystals: branched capping ligands control crystal size and stability *Ang. Chem. Int. Ed.* **55** 8864–8

[29] Roo J D, Ibáñez M, Geiregat P, Nedelcu G, Walravens W, Maes J, Martins J C, Driessche I V, Kovalenko M V and Hens Z 2016 Highly dynamic ligand binding and light absorption coefficient of cesium lead bromide perovskite nanocrystals *ACS Nano* **10** 2071–81

[30] Parobek D, Yitong , Qiao T and Son D H 2018 Direct hot-injection synthesis of Mn-doped $CsPbBr_3$ nanocrystals *Chem. Mater.* **30** 2939–44

[31] Yassitepe E *et al* 2016 Amine-free synthesis of cesium lead halide perovskite quantum dots for efficient light-emitting diodes *Adv. Funct. Mater.* **26** 8757–63

[32] Woo J Y, Kim Y, Bae J, Kim T G, Kim J W, Lee D C and Jeong S 2017 Highly stable cesium lead halide perovskite nanocrystals through *in situ* lead halide inorganic passivation *Chem. Mater.* **29** 7088–92

[33] Dong Y, Qiao T, Kim D, Parobek D, Rossi D and Hee Son D 2018 Precise control of quantum confinement in cesium lead halide perovskite quantum dots via thermodynamic equilibrium *Nano Lett.* **18** 3716–22

[34] Kwon J S, Nayhouse M, Christofides P D and Orkoulas G 2013 Modeling and control of protein crystal shape and size in batch crystallization *AIChE J.* **59** 2317–27

[35] Durbin S D and Feher G 1986 Crystal growth studies of lysozyme as a model for protein crystallization *J. Cryst. Growth* **76** 583–92

[36] Durbin S D and Feher G 1991 Simulation of lysozyme crystal growth by the Monte Carlo method *Cryst. Growth* **110** 41–51

[37] Markande A, Nezzal A, Fitzpatrick J, Aerts L and Redl A 2012 Influence of impurities on the crystallization of dextrose monohydrate *J. Cryst. Growth* **353** 145–51

[38] Epps R W, Bowen M S, Volk A A, Abdel-Latif K, Han S, Reyes K G, Amassian A and Abolhasani M 2020 Artificial chemist: an autonomous quantum dot synthesis bot *Adv. Mater.* **32** 2001626

[39] Epps R W and Abolhasani M 2021 Modern nanoscience: Convergence of AI, robotics, and colloidal synthesis *Appl. Phys. Rev.* **8** 041316

[40] Epps R W, Felton K C, Coley C W and Abolhasani M 2018 A modular microfluidic technology for systematic studies of colloidal semiconductor nanocrystals *J. Vis. Exp.* **135** e57666

[41] Zheng J, Zhang M, Lau C F J, Deng X, Kim J, Ma Q, Chen C, Green M A, Huang S and Ho-Baillie A W Y 2017 Spin-coating free fabrication for highly efficient perovskite solar cells *Solar Energy Mater. Solar Cells* **168** 165–71

[42] Pi X, Li Q, Li D and Yang D 2011 Spin-coating silicon-quantum-dot ink to improve solar cell efficiency *Solar Energy Mater. Solar Cells* **95** 2941–5

[43] Lee Y-J, Lee C-J and Cheng C-M 2010 Enhancing the conversion efficiency of red emission by spin-coating CdSe quantum dots on the green nanorod light-emitting diode *Opt. Express* **18** A554–61

[44] Chen M, Yu H, Kershaw S V, Xu H, Gupta S, Hetsch F, Rogach A L and Zhao N 2014 Fast, air-stable infrared photodetectors based on spray-deposited aqueous hgte quantum dots *Adv. Funct. Mater.* **24** 53–9

[45] Bishop J E, Smith J A and Lidzey D G 2020 Development of spray-coated perovskite solar cells *ACS Appl. Mater. Interfaces* **12** 48237–45

[46] Li M, Zhang W, Shao G, Kan H, Song Z, Xu S, Yu H, Jiang S, Luo J and Liu H 2016 Sensitive NO_2 gas sensors employing spray-coated colloidal quantum dots *Thin Solid Films* **618** 271–6

[47] Chen H, Ding X, Pan X, Hayat T, Alsaedi A, Ding Y and Dai S 2018 Reducing the universal ‘coffee-ring effect’ by a vapor-assisted spraying method for high-efficiency $CH_3NH_3PbI_3$ perovskite solar cells *ACS Appl. Mater. Interfaces* **10** 23466–75

[48] Zhang Z, Zhang X, Xin Z, Deng M, Wen Y and Song Y 2013 Controlled inkjetting of a conductive pattern of silver nanoparticles based on the coffee-ring effect *Adv. Mater.* **25** 6714–8

[49] Schramböck M, Andrews A M, Roch T, Schrenk W, Lugstein A and Strasser G 2006 Nano-patterning and growth of self-assembled quantum dots *Microelectron. J.* **37** 1532–4

[50] Xie W, Gomes R, Aubert T, Bisschop S, Zhu Y, Hens Z, Brainis E and Thourhout D V 2015 Nanoscale and single-dot patterning of colloidal quantum dots *Nano Lett.* **15** 7481–7

[51] Liu G, Zhao H, Diao F, Ling Z and Wang Y 2018 Stable tandem luminescent solar concentrators based on CdSe/CdS quantum dots and carbon dots *J. Mater. Chem.* C **6** 10059–66
[52] Lim J, Bae W K, Kwak J, Lee S, Lee C and Char K 2012 Perspective on synthesis, device structures, and printing processes for quantum dot displays *Opt. Mater. Express* **2** 594–628
[53] Oh B M, Jeong Y, Zheng Y J, Young Cho N, Song M, Woo Choi J and Kim J H 2021 Simple one-pot synthesis and high-resolution patterning of perovskite quantum dots using a photocurable ligand *Chem. Commun.* **57** 12824–7
[54] Kwon J S, Nayhouse M and Christofides P D 2015 Multiscale, multidomain modeling and parallel computation: application to crystal shape evolution in crystallization *Ind. Eng. Chem. Res.* **54** 11903–14
[55] Li M and Christofides P D 2005 Multi-scale modeling and analysis of an industrial HVOF thermal spray process *Chem. Eng. Sci.* **60** 3649–69
[56] Nayhouse M, Kwon J S, Christofides P D and Orkoulas G 2013 Crystal shape modeling and control in protein crystal growth *Chem. Eng. Sci.* **87** 216–23
[57] Kwon J S, Nayhouse M, Christofides P D and Orkoulas G 2014 Protein crystal shape and size control in batch crystallization: comparing model predictive control with conventional operating policies *Ind. Eng. Chem. Res.* **53** 5002–14
[58] Nagpal S, Sitapure N, Gagnon Z and Kwon J S 2024 Advancing crystal growth prediction: an adaptive kMC model spanning multiple regimes *Chem. Eng. Sci.* **299** 120472
[59] Kwon J S, Nayhouse M, Orkoulas G and Christofides P D 2014 Crystal shape and size control using a plug flow crystallization configuration *Chem. Eng. Sci.* **119** 30–9
[60] Kwon J S, Nayhouse M, Orkoulas G, Ni D and Christofides P D 2015 Run-to-run-based model predictive control of protein crystal shape in batch crystallization *Ind. Eng. Chem. Res.* **54** 4293–302
[61] Kwon J S, Nayhouse M, Christofides P D and Orkoulas G 2013 Modeling and control of shape distribution of protein crystal aggregates *Chem. Eng. Sci.* **104** 484–97
[62] Kwon J S, Nayhouse M, Orkoulas G and Christofides P D 2014 Enhancing the crystal production rate and reducing polydispersity in continuous protein crystallization *Ind. Eng. Chem. Res.* **53** 15538–48
[63] Crose M, Kwon J S, Nayhouse M, Ni D and Christofides P D 2015 Multiscale modeling and operation of PECVD of thin film solar cells *Chem. Eng. Sci.* **136** 50–61
[64] Crose M, Kwon J S, Tran A and Christofides P D 2017 Multiscale modeling and run-to-run control of PECVD of thin film solar cells *Renew. Energy* **100** 129–40
[65] Crose M, Zhang W, Tran A and Christofides P D 2018 Multiscale three-dimensional CFD modeling for PECVD of amorphous silicon thin films *Comput. Chem. Eng.* **113** 184–95
[66] Crose M, Zhang W, Tran A and Christofides P D 2019 Run-to-run control of PECVD systems: application to a multiscale three-dimensional CFD model of silicon thin film deposition *AIChE J.* **65** e16400
[67] Kimaev G and Ricardez-Sandoval L A 2019 Nonlinear model predictive control of a multiscale thin film deposition process using artificial neural networks *Chem. Eng. Sci.* **207** 1230–45
[68] Kimaev G and Ricardez-Sandoval L A 2020 Artificial neural network discrimination for parameter estimation and optimal product design of thin films manufactured by chemical vapor deposition *J. Phys. Chem.* C **124** 18615–27

[69] Rasoulian S and Ricardez-Sandoval L A 2016 Stochastic nonlinear model predictive control applied to a thin film deposition process under uncertainty *Chem. Eng. Sci.* **140** 90–103

[70] Chaffart D and Ricardez-Sandoval L A 2018 Optimization and control of a thin film growth process: a hybrid first principles/artificial neural network based multiscale modelling approach *Comput. Chem. Eng.* **119** 465–79

[71] Pahari S, Moon J, Akbulut M, Hwang S and Kwon J S 2021 Model predictive control for wormlike micelles (WLMs): application to a system of CTAB and NaCl *Chem. Eng. Res. Des.* **174** 30–41

[72] Pahari S, Liu S, Lee C H, Akbulut M and Kwon J S 2022 SAXS-guided unbiased coarse-grained Monte Carlo simulation for identification of self-assembly nanostructures and dimensions *Soft Matter* **18** 5282–92

[73] Pahari S, Bhadriraju B, Akbulut M and Kwon J S 2021 A slip-spring framework to study relaxation dynamics of entangled wormlike micelles with kinetic Monte Carlo algorithm *J. Colloid Interface Sci.* **600** 550–60

[74] Chaffart D and Ricardez-Sandoval L A 2022 A three dimensional kinetic Monte Carlo defect-free crystal dissolution model for biological systems, with application to uncertainty analysis and robust optimization *Comput. Chem. Eng.* **157** 107586

[75] Lee D, Singla A, Wu H-J and Kwon J S 2018 An integrated numerical and experimental framework for modeling of CTB and GD1B ganglioside binding kinetics *AIChE J.* **64** 3882–93

[76] Choi H-K, Lee D, Singla A, Kwon J S and Wu H-J 2019 The influence of heteromultivalency on lectin–glycan binding behavior *Glycobiology* **29** 397–408

[77] Lee D, Green A, Wu H-J and Kwon J S 2021 Hybrid PDE-kMC modeling approach to simulate multivalent lectin-glycan binding process *AIChE J.* **67** e17453

[78] Worstell N C, Singla A, Saenkham P, Galbadage T, Sule P, Lee D, Mohr A, Kwon J S, Cirillo J D and Wu H-J 2018 Hetero-multivalency of *Pseudomonas aeruginosa* lectin LecA binding to model membranes *Sci. Rep.* **8** 8419

[79] Lee D, Mohr A, Kwon J S and Wu H-J 2018 Kinetic Monte Carlo modeling of multivalent binding of CTB proteins with GM1 receptors *Comput. Chem. Eng.* **118** 283–95

[80] Lee D, Jayaraman A and Kwon J S 2020 Identification of cell-to-cell heterogeneity through systems engineering approaches *AIChE J.* **66** e16925

[81] Lee D, Jayaraman A and Kwon J S 2020 Development of a hybrid model for a partially known intracellular signaling pathway through correction term estimation and neural network modeling *PLoS Comput. Biol.* **16** e1008472

[82] Choi H-K and Kwon J S 2019 Multiscale modeling and control of kappa number and porosity in a batch-type pulp digester *AIChE J.* **65** e16589

[83] Choi H-K and Kwon J S 2020 Multiscale modeling and multiobjective control of wood fiber morphology in batch pulp digester *AIChE J.* e16972

[84] Choi H-K and Kwon J S 2019 Modeling and control of cell wall thickness in batch delignification *Comput. Chem. Eng.* **128** 512–23

[85] Son S H, Choi H-K and Kwon J S 2021 Application of offset-free Koopman-based model predictive control to a batch pulp digester *AIChE J.* **67** e17301

[86] Son S H, Choi H-K and Kwon J S 2020 Multiscale modeling and control of pulp digester under fiber-to-fiber heterogeneity *Comput. Chem. Eng.* **143** 107117

[87] Shah P, Choi H-K and Kwon J S 2023 Achieving optimal paper properties: a layered multiscale KMC and LSTM-ANN-based control approach for kraft pulping *Processes* **11** 809

[88] Kim J, Pahari S, Ryu J, Zhang M, Yang Q, Geun Yoo C and Kwon J S 2024 Advancing biomass fractionation with real-time prediction of lignin content and MWD: a KMC-based multiscale model for optimized lignin extraction *Chem. Eng. J.* **479** 147226

[89] Pahari S, Kim J, Choi H-K, Zhang M, Ji A, Geun Yoo C and Kwon J S 2023 Multiscale kinetic modeling of biomass fractionation in an experiment: understanding individual reaction mechanisms and cellulose degradation *Chem. Eng. J.* **467** 143021

[90] Lee C H, Kim J, Ryu J, Won W, Yoo C G and Kwon J S 2024 Lignin structure dynamics: advanced real-time molecular sensing strategies *Chem. Eng. J.* **487** 150680

[91] Lee C H, Pahari S, Sitapure N, Barteau M A and Kwon J S 2023 Investigating high-performance non-precious transition metal oxide catalysts for nitrogen reduction reaction: a multifaceted DFT-kMC-LSTM approach *ACS Catal.* **13** 8336–46

[92] Lee C H, Pahari S, Sitapure N, Barteau M A and Kwon J S 2022 DFT-kMC analysis for identifying novel bimetallic electrocatalysts for enhanced NRR performance by suppressing HER at ambient conditions via active-site separation *ACS Catal.* **12** 15609–17

[93] Stamatakis M, Chen Y and Vlachos D G 2011 First-principles-based kinetic Monte Carlo simulation of the structure sensitivity of the water-gas shift reaction on platinum surfaces *J. Phys. Chem.* C **115** 24750–62

[94] Kimaev G, Chaffart D and Ricardez-Sandoval L A 2020 Multilevel Monte Carlo applied for uncertainty quantification in stochastic multiscale systems *AIChE J.* **66** e16262

[95] Salciccioli M, Stamatakis M, Caratzoulas S and Vlachos D G 2011 A review of multiscale modeling of metal-catalyzed reactions: mechanism development for complexity and emergent behavior *Chem. Eng. Sci.* **66** 4319–55

[96] Li J, Croiset E and Ricardez-Sandoval L 2015 Carbon nanotube growth: first-principles-based kinetic Monte Carlo model *J. Catal.* **326** 15–25

[97] Chaffart D and Ricardez-Sandoval L A 2018 Robust optimization of a multiscale heterogeneous catalytic reactor system with spatially-varying uncertainty descriptions using polynomial chaos expansions *Can. J. Chem. Eng.* **96** 113–31

[98] Chaffart D and Ricardez-Sandoval L A 2017 Robust dynamic optimization in heterogeneous multiscale catalytic flow reactors using polynomial chaos expansion *J. Process Control* **60** 128–40

[99] Li J, Liu G, Ren B, Croiset E, Zhang Y and Ricardez-Sandoval L 2019 Mechanistic study of site blocking catalytic deactivation through accelerated kinetic Monte Carlo *J. Catal.* **378** 176–83

[100] Chaffart D, Shi S, Ma C, Lv C and Ricardez-Sandoval L A 2022 A moving front kinetic Monte Carlo algorithm for moving interface systems *J. Phys. Chem.* B **126** 2040–59

[101] Chaffart D, Shi S, Ma C, Lv C and Ricardez-Sandoval L A 2023 A semi-empirical force balance-based model to capture sessile droplet spread on smooth surfaces: a moving front kinetic Monte Carlo study *Phys. Fluids* **35** 032109

[102] Sitapure N, Lee H, Ospina-Acevedo F, Balbuena P B, Hwang S and Kwon J S 2021 A computational approach to characterize formation of a passivation layer in lithium metal anodes *AIChE J.* **67** e17073

[103] Lee H, Sitapure N, Hwang S and Kwon J S 2021 Multiscale modeling of dendrite formation in lithium-ion batteries *Comput. Chem. Eng.* **153** 107415

[104] Hwang G, Sitapure N, Moon J, Lee H, Hwang S and Kwon J S 2022 Model predictive control of lithium-ion batteries: development of optimal charging profile for reduced intracycle capacity fade using an enhanced single particle model (SPM) with first-principled chemical/mechanical degradation mechanisms *Chem. Eng. J.* **435** 134768

[105] Narasingam A and Kwon J S 2017 Development of local dynamic mode decomposition with control: application to model predictive control of hydraulic fracturing *Comput. Chem. Eng.* **106** 501–11

[106] Narasingam A, Siddhamshetty P and Kwon J S 2018 Handling spatial heterogeneity in reservoir parameters using proper orthogonal decomposition based ensemble Kalman filter for model-based feedback control of hydraulic fracturing *Ind. Eng. Chem. Res.* **57** 3977–89

[107] Narasingam A, Siddhamshetty P and Kwon J S 2017 Temporal clustering for order reduction of nonlinear parabolic PDE systems with time-dependent spatial domains: application to a hydraulic fracturing process *AIChE J.* **63** 3818–31

[108] Narasingam A and Kwon J S 2018 Data-driven identification of interpretable reduced-order models using sparse regression *Comput. Chem. Eng.* **119** 101–11

[109] Brunton S L, Proctor J L and Kutz J N 2016 Discovering governing equations from data by sparse identification of nonlinear dynamical systems *Proc. Natl Acad. Sci.* **113** 3932–7

[110] Bhadriraju B, Narasingam A and Kwon J S 2019 Machine learning-based adaptive model identification of systems: application to a chemical process *Chem. Eng. Res. Des.* **152** 372–83

[111] Bhadriraju B, Bangi M S F, Narasingam A and Kwon J S 2020 Operable adaptive sparse identification of systems: application to chemical processes *AIChE J.* **66** e16980

[112] Bhadriraju B, Kwon J S and Khan F 2021 Risk-based fault prediction of chemical processes using operable adaptive sparse identification of systems (OASIS) *Comput. Chem. Eng.* **152** 107378

[113] Bhadriraju B, Kwon J S and Khan F 2023 An adaptive data-driven approach for two-timescale dynamics prediction and remaining useful life estimation of Li-ion batteries *Comput. Chem. Eng.* **175** 108275

[114] Bhadriraju B, Kwon J S and Khan F 2024 A data-driven framework integrating Lyapunov-based MPC and OASIS-based observer for control beyond training domains *J. Process Control* **138** 103224

[115] Narasingam A and Kwon J S 2019 Koopman Lyapunov-based model predictive control of nonlinear chemical process systems *AIChE J.* **65** e16743

[116] Narasingam A and Kwon J S 2020 Application of Koopman operator for model-based control of fracture propagation and proppant transport in hydraulic fracturing operation *J. Process Control* **91** 25–36

[117] Narasingam A and Kwon J S 2020 Closed-loop stabilization of nonlinear systems using Koopman Lyapunov-based model predictive control *59th IEEE Conf. on Decision and Control (CDC), Jeju Island, Republic of Korea* (Piscataway, NJ: IEEE) pp 704–9

[118] Narasingam A, Son S H and Kwon J S 2023 Data-driven feedback stabilisation of nonlinear systems: Koopman-based model predictive control *Int. J. Control* **96** 770–81

[119] Son S H, Choi H-K, Moon J and Kwon J S 2022 Hybrid Koopman model predictive control of nonlinear systems using multiple EDMD models: an application to a batch pulp digester with feed fluctuation *Control Eng. Pract.* **118** 104956

[120] Son S H, Narasingam A and Kwon J S 2022 Development of offset-free Koopman Lyapunov-based model predictive control and mathematical analysis for zero steady-state

offset condition considering influence of Lyapunov constraints on equilibrium point *J. Process Control* **118** 26–36
[121] Hassanpour H, Mhaskar P, House J M and Salsbury T I 2020 A hybrid modeling approach integrating first-principles knowledge with statistical methods for fault detection in HVAC systems *Comput. Chem. Eng.* **142** 107022
[122] Ghosh D, Hermonat E, Mhaskar P, Snowling S and Goel R 2019 Hybrid modeling approach integrating first-principles models with subspace identification *Ind. Eng. Chem. Res.* **58** 13533–43
[123] Pahari S, Shah P and Kwon J S 2024 Unveiling latent chemical mechanisms: hybrid modeling for estimating spatiotemporally varying parameters in moving boundary problems *Ind. Eng. Chem. Res.* **63** 1501–14
[124] Pahari S, Shah P and Kwon J S 2024 Achieving robustness in hybrid models: a physics-informed regularization approach for spatiotemporal parameter estimation in PDEs *Chem. Eng. Res. Des.* **204** 292–302
[125] Shah P, Pahari S, Bhavsar R and Kwon J S 2024 Hybrid modeling of first-principles and machine learning: A step-by-step tutorial review for practical implementation *Comput. Chem. Eng.* **194** 108926
[126] Kwon J S-I 2024 Adding big data into the equation *Nat. Chem. Eng.* **1** 724
[127] Bangi M S F and Kwon J S 2020 Deep hybrid modeling of chemical process: application to hydraulic fracturing *Comput. Chem. Eng.* **134** 106696
[128] Shah P, Sheriff M Z, Bangi M S F, Kravaris C, Kwon J S, Botre C and Hirota J 2022 Deep neural network-based hybrid modeling and experimental validation for an industry-scale fermentation process: identification of time-varying dependencies among parameters *Chem. Eng. J.* **441** 135643
[129] Shah P, Sheriff M Z, Bangi M S F, Kravaris C, Kwon J S, Botre C and Hirota J 2023 Multi-rate observer design and optimal control to maximize productivity of an industry-scale fermentation process *AIChE J.* **69** e17946
[130] Bangi M S F, Kao K and Kwon J S 2022 Physics-informed neural networks for hybrid modeling of lab-scale batch fermentation for β-carotene production using saccharomyces cerevisiae *Chem. Eng. Res. Des.* **179** 415–23
[131] Bangi M S F and Kwon J S 2023 Deep hybrid model-based predictive control with guarantees on domain of applicability *AIChE J.* **69** e18012
[132] Yang J *et al* 2022 Nondestructive photopatterning of heavy-metal-free quantum dots *Adv. Mater.* **34** 2205504
[133] Sitapure N, Kwon T H, Lee M, Kim B S, Kang M S and Kwon J S 2022 Modeling ligand crosslinking for interlocking quantum dots in thin-films *J. Mater. Chem.* C **10** 7132–40
[134] Sitapure N and Kwon J S 2022 Neural network-based model predictive control for thin-film chemical deposition of quantum dots using data from a multiscale simulation *Chem. Eng. Res. Des.* **183** 595
[135] Ridder B J, Majumder A and Nagy Z K 2014 Population balance model-based multi-objective optimization of a multisegment multiaddition (MSMA) continuous plug-flow antisolvent crystallizer *Ind. Eng. Chem. Res.* **53** 4387–97
[136] Epps R W, Volk A A, Abdel-Latif K and Abolhasani M 2020 An automated flow chemistry platform to decouple mixing and reaction times *React. Chem. Eng.* **5** 1212–7

[137] Cho J, Jin H, Sellers D G, Watson D F, Son D H and Banerjee S 2017 Influence of ligand shell ordering on dimensional confinement of cesium lead bromide ($CsPbBr_3$) perovskite nanoplatelets *J. Mater. Chem.* C **5** 8810–8

[138] Epps R W, Felton K C, Coley C W and Abolhasani M 2017 Automated microfluidic platform for systematic studies of colloidal perovskite nanocrystals: towards continuous nano-manufacturing *Lab Chip* **17** 4040–7

[139] Sitapure N, Qiao T, Son D and Kwon J S 2020 Kinetic Monte Carlo modeling of the equilibrium-based size control of $CsPbBr_3$ perovskite quantum dots in strongly confined regime *Comput. Chem. Eng.* **139** 106872

[140] Sitapure N, Qiao T, Son D H and Kwon J S 2020 Modeling and size control of $CsPbBr_3$ perovskite quantum dots *IEEE American Control Conf. (ACC) (Denver, CO)* (Piscataway, NJ: IEEE) pp 4331–6

[141] Sitapure N, Epps R, Abolhasani M and Kwon J S 2020 Multiscale modeling and optimal operation of millifluidic synthesis of perovskite quantum dots: towards size-controlled continuous manufacturing *Chem. Eng. J.* **413** 127905

[142] Abdel-Latif K, Bateni F, Crouse S and Abolhasani M 2020 Flow synthesis of metal halide perovskite quantum dots: from rapid parameter space mapping to AI-guided modular manufacturing *Matter* **3** 1053–86

[143] Annaland M V, Dijkhuizen W, Deen N G and Kuipers J A M 2006 Numerical simulation of behavior of gas bubbles using a 3D front-tracking method *AIChE J.* **52** 99–110

[144] Bothe D, Koebe M, Wielage K and Warnecke H-J 2003 VOF-simulations of mass transfer from single bubbles and bubble chains rising in aqueous solutions *Fluids Engineering Division Summer Meeting* vol 36 975 pp 423–9

[145] Benson D J 2002 Volume of fluid interface reconstruction methods for multi-material problems *Appl. Mech. Rev.* **55** 151–65

[146] Grzybowski H and Mosdorf R 2014 Modelling of two-phase flow in a minichannel using level-set method *J. Phys.* **530** 012049

[147] Soulaimani A and Saad Y 1998 An arbitrary Lagrangian–Eulerian finite element method for solving three-dimensional free surface flows *Comput. Methods Appl. Mech. Eng.* **162** 79–106

[148] Sitapure N, Epps R W, Abolhasani M and Kwon J S 2021 CFD-based computational studies of quantum dot size control in slug flow crystallizers: handling slug-to-slug variation *Ind. Eng. Chem. Res.* **60** 4930–41

[149] Rasche M L, Jiang M and Braatz R D 2016 Mathematical modeling and optimal design of multi-stage slug-flow crystallization *Comput. Chem. Eng.* **95** 240–8

[150] Su M and Gao Y 2018 Air-liquid segmented continuous crystallization process optimization of the flow field, growth rate, and size distribution of crystals *Ind. Eng. Chem. Res.* **57** 3781–91

[151] Kerr C B, Epps R W and Abolhasani M 2019 A low-cost, non-invasive phase velocity and length meter and controller for multiphase lab-in-a-tube devices *Lab Chip* **19** 2107–13

[152] Wang X, Guo L and Zhang X 2006 Development of liquid slug length in gas–liquid slug flow along horizontal pipeline: experiment and simulation *J. Chem. Eng.* **14** 626–33

[153] OpenAI 2023 *GPT-4 technical report* 2303.08774 https://arxiv.org/abs/2303.08774

[154] Brown T, Mann B, Ryder N, Subbiah M, Kaplan J D, Dhariwal P, Neelakantan A, Shyam P, Sastry G and Askell A 2020 Language models are few-shot learners *Proc. Int. Conf.*

Neural Inform. Process. Syst. 33 1877–901 https://papers.nips.cc/paper/2020/hash/1457c0d6bfcb4967418bfb8ac142f64a-Abstract.html

[155] Dosovitskiy A, Beyer L, Kolesnikov A, Weissenborn D, Zhai X, Unterthiner T, Dehghani M, Minderer M, Heigold G and Gelly S 2020 An image is worth 16x16 words: transformers for image recognition at scale arXiv: 2010.11929

[156] Shoeybi M, Patwary M, Puri R, LeGresley P, Casper J and Catanzaro B 2019 Megatron-LM: training multi-billion parameter language models using model parallelism arXiv: 1909.08053

[157] Narayanan D *et al* 2021 Efficient large-scale language model training on GPU clusters using megatron-LM *Proc. of the Int. Conf. for High Performance Computing, Networking, Storage and Analysis* pp 1–15

[158] Radford A, Wu J, Child R, Luan D, Amodei D and Sutskever I 2019 Language models are unsupervised multitask learners *OpenAI blog* **1** 9

[159] Liu Z, Lin Y, Cao Y, Hu H, Wei Y, Zhang Z, Lin S and Guo B 2021 Swin transformer: hierarchical vision transformer using shifted windows *Proc. of the IEEE/CVF Int. Conf. on Computer Vision* pp 10012–22

[160] Rombach R, Blattmann A, Lorenz D, Esser P and Ommer B 2022 High-resolution image synthesis with latent diffusion models *Proc. of the IEEE/CVF Conf. on Computer Vision and Pattern Recognition* pp 10684–95

[161] Vogel G, Balhorn L S and Schweidtmann A M 2023 Learning from flowsheets: a generative transformer model for autocompletion of flowsheets *Comput. Chem. Eng.* **171** 108162

[162] Jumper J, Evans R, Pritzel A, Green T, Figurnov M, Ronneberger O, Tunyasuvunakool K, Bates R, Žídek A and Potapenko A 2021 Highly accurate protein structure prediction with AlphaFold *Nature* **596** 583–9

[163] Evans R, O'Neill M, Pritzel A, Antropova N, Senior A, Green T, Žídek A, Bates R, Blackwell S and Yim J 2021 Protein complex prediction with AlphaFold-Multimer *BioRxiv* https://www.biorxiv.org/content/10.1101/2021.10.04.463034v2

[164] Senior A W, Evans R, Jumper J, Kirkpatrick J, Sifre L, Green T, Qin C, Žídek A, Nelson A W R and Bridgland A 2020 Improved protein structure prediction using potentials from deep learning *Nature* **577** 706–10

[165] Mann V and Venkatasubramanian V 2021 Predicting chemical reaction outcomes: a grammar ontology-based transformer framework *AIChE J.* **67** e17190

[166] Sitapure N and Kwon J S 2023 Exploring the potential of time-series transformers for process modeling and control in chemical systems: an inevitable paradigm shift? *Chem. Eng. Res. Des.* **194** 461–77

[167] Sitapure N and Kwon J S 2023 A unified approach for modeling and control of crystallization of quantum dots (QDs) *Digit. Chem. Eng.* **6** 100077

[168] Vaswani A, Shazeer N, Parmar N, Uszkoreit J, Jones L, Gomez A N, Kaiser L and Polosukhin I 2017 Attention is all you need *Proc. Int. Conf. Neural Inform. Process. Syst.* **30** 6000–10 https://papers.nips.cc/paper_files/paper/2017/hash/3f5ee243547dee91fbd053c1c4a845aa-Abstract.html

[169] Devlin J, Chang M-W, Lee K and Toutanova K 2018 BERT: pre-training of deep bidirectional transformers for language understanding arXiv: 1810.04805

[170] Sitapure N and Kwon J S 2023 CrystalGPT: enhancing system-to-system transferability in crystallization prediction and control using time-series-transformers *Comput. Chem. Eng.* **177** 108339

[171] Li M 2017 Li-ion dynamics and state of charge estimation *Renew. Energy* **100** 44–52

[172] Nascimento R G, Corbetta M, Kulkarni C S and Viana F A C 2021 Hybrid physics-informed neural networks for lithium-ion battery modeling and prognosis *J. Power Sources* **513** 230526
[173] Zheng Z, Stephens R M, Braatz R D, Alkire R C and Petzold L R 2008 A hybrid multiscale kinetic Monte Carlo method for simulation of copper electrodeposition *J. Comput. Phys.* **227** 5184–99

IOP Publishing

High-Performance Computing and Artificial Intelligence in Process Engineering

Mingheng Li and Yi Heng

Chapter 5

The rise of time-travelers: are transformer-based models the key to unlocking a new paradigm in surrogate modeling for dynamic systems?

Joseph Sang-Il Kwon

5.1 Introduction

Over the past couple of years, significant disruptions have unfolded across natural language processing (NLP), computer vision, and adaptive AI models. These advancements, exemplified by applications such as ChatGPT, Codex, DALL-E, and ChatSonic, showcase human-like contextual understanding. Facilitating these developments are sophisticated large language models (LLMs) such as BERT, Megatron-LM, and GPT series models [1–7]. The driving force behind this revolution is the emergence of transformer networks. These networks leverage multiheaded self-attention mechanisms, positional encoding, and impressive transfer learning capabilities, enabling rapid fine-tuning for niche tasks. Moreover, their highly parallelizable architecture empowers the training of exceptionally large models with tens of billions of parameters, utilizing vast amounts of data efficiently.

These advancements have enabled the creation of adaptive chatbots, automated labeling, image generation, and code development tasks. This progress is primarily attributed to the efficient transfer learning capabilities exhibited by transformers, allowing models to grasp underlying language nuances and syntax from diverse textual data sources. Prior state-of-the-art approaches such as recurrent neural networks (RNNs) and long short-term memory (LSTM) models struggled to keep pace due to inefficiencies in handling large datasets and parallelization. The swift progress in NLP and computer vision is thus owed to the superior transfer learning capabilities and parallelization-friendly architecture of transformers.

Likewise, transformers are poised to revolutionize chemical systems, offering high-value applications such as accurate process monitoring, advanced process

doi:10.1088/978-0-7503-6174-3ch5

control, and optimization [8, 9]. For instance, transformer models trained on process data could learn system dynamics and be fine-tuned for new applications, potentially transforming the field. However, integrating these capabilities into chemical systems poses challenges, with only a few studies exploring this domain. Despite these challenges, recent efforts demonstrate the feasibility of leveraging transformer models for chemical systems tasks. For example, Schweidtmann and team utilized SFILES notation to adapt chemical process diagrams for NLP model interpretation, specifically fine-tuning a GPT-2 model to predict connections in these diagrams, aiding in their automatic completion [10]. Additionally, AlphaFold, a leading-edge transformer model, has transformed protein structure prediction with its amino acid sequence-based approach, solving a long-standing challenge in protein folding [10, 11]. Venkatasubramanian and Mann have also made significant contributions to protein structure prediction using transformer models [12]. The application of advanced transformer models for forecasting in chemical systems, such as predicting the end-of-life and state-of-charge (SOC) for various battery systems, remains a relatively unexplored area with few examples [13, 14]. Despite the potential transformative impact of transformer models on chemical system modeling, the concrete implementation of first-generation transformer models in this field is still in its early stages and has not been fully realized.

Addressing this gap, this work aims to develop a novel transformer model tailored for multivariate time-series prediction in chemical systems. The proposed time-series transformer (TST), as shown in figure 5.1, incorporates multiple encoder blocks with attention heads, enabling it to capture long-range dependencies and changes in system states effectively. The study evaluates various aspects of TSTs, including model size and training epochs, to inform future development. Furthermore, while this work represents a foundational step in TST development, it sets the stage for more sophisticated transformer models tailored to specific chemical operations or serving as large-scale cognitive models processing information from multiple sources. Future directions include leveraging transfer learning for modeling new process equipment and developing plant-level transformer models for fault monitoring and prevention.

The subsequent sections of the chapter are structured as follows. Initially, we provide an in-depth analysis of the construction of different TST models and their fundamental computational framework. Following this, we present the outcomes of time-series prediction using TSTs (i.e. CrystalGPT), along with their integration into hybrid models, wherein they are combined with first-principles models. For illustration purposes, we examine their performance in modeling a batch crystallization process, showcasing the capabilities of CrystalGPT and TST-based hybrids. Subsequently, we delve into a discussion regarding the potential extensive impact of TSTs in chemical systems. Finally, we conclude with a summary.

5.2 Time-series transformers

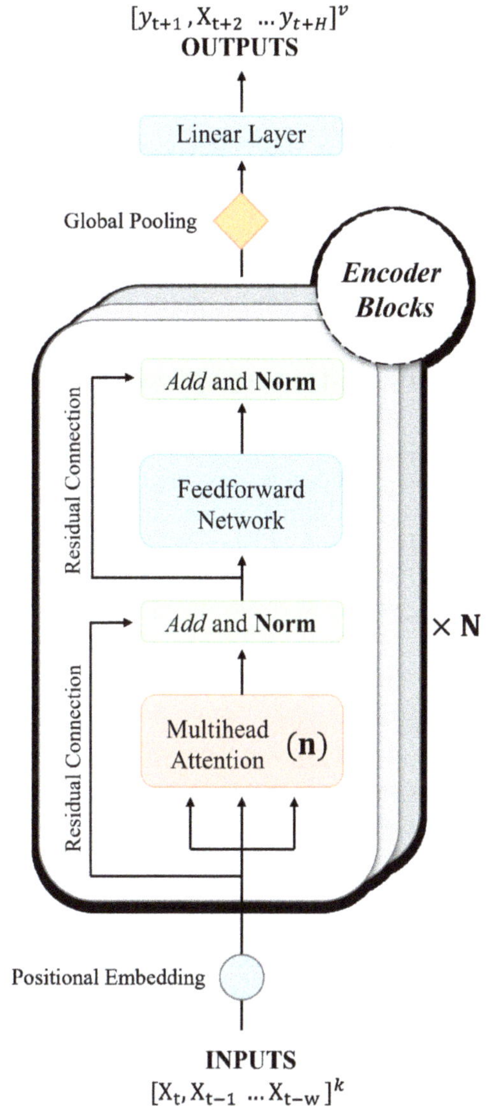

Figure 5.1. Generalized TST framework with N encoder blocks and n attention heads. (Reproduced with permission from [15]. Copyright 2023 Elsevier.)

5.2.1 Operation of encoder–decoder transformers

The functioning of a sophisticated vanilla transformer architecture, commonly employed for tasks in NLP, revolves around encoder–decoder blocks, which consist of identical sub-layers and incorporate a globally pooled output layer. The input sequence undergoes a series of transformations including truncation/padding, tokenization, positional encoding, self-attention, and cross-attention mechanisms. These transformations ultimately result in the contextualized embedding of the source sequence. Notably, the initial input tensor (X_{RAW}) is projected into a higher-dimensional space (d_{model}) and subsequently processed through a sinusoidal positional encoder (PE) to systematically identify each word's position within the sequence, generating the tensor X_{PE}.

Subsequently, X_{PE} is fed into a stack of encoder blocks, each equipped with multiple attention heads (MAHs) to compute 'self-attention' scores, which are then processed by a feed-forward network (FFN). The computation of attention scores by each MAH is crucial for capturing semantic relationships, context, and the relative importance of every word, resulting in contextually enriched embeddings. Utilizing the query–key–value ([**Q**,**K**,**V**]) approach for calculating attention scores, the results from the various MAHs are combined to form the tensor X_{EN}.

Following this, X_{EN} is forwarded to a stack of decoder blocks, each equipped with its own set of MAHs and independent input (X_{DEC}) to facilitate the computation of 'cross-attention'. This comprehensive framework enhances the transformer's ability to focus on high-value cross-attention scores, facilitating a seamless transition and connection between the source and target sequences of words, ultimately resulting in text generation that closely resembles human-like communication. For a more comprehensive understanding, we encourage readers to explore the existing literature on transformer models [1, 2, 5].

5.2.2 TST architecture

Drawing inspiration from the encoder–decoder architecture prevalent in NLP-transformers, we have developed a novel TST framework tailored specifically for time-series prediction tasks, as detailed in our previous works [15, 16]. Here, we provide a concise overview of this architecture. Initially, time-series data from chemical systems, encompassing state information spanning the current and past W time steps (i.e. the source sequence), is encapsulated in a k-dimensional input tensor denoted as $[X_{t-W}, X_{t-W+1}, \dots, X_t]$. Subsequently, it is elevated to a higher-dimensional space (d_{model}), representing the internal hidden dimension of a TST. In the next step, PE terms based on the time variable t_i are incorporated and utilized as follows:

$$\begin{aligned} \text{Even position: } \mathrm{PE}_{(t_i,\, 2j)} &= \sin\left(\frac{t_i}{10\,000^{2j/d}}\right) \\ \text{Odd position: } \mathrm{PE}_{(t_i,\, 2j+1)} &= \cos\left(\frac{t_i}{10\,000^{2j/d}}\right), \end{aligned} \tag{5.1}$$

where t_i is the time at location i in the input sequence, and $j \in \mathbb{R}^{d_{\mathrm{model}}}$ is the feature dimension. By incorporating PE, the transformer generates sine and cosine waves with varying wavelengths, enabling nuanced attention to different positions in the input sequence, thereby enhancing performance [1, 15]. Each unique input value at position i and dimension j is assigned a unique identifier, either $\mathrm{PE}_{(i,\, 2j)}$ or $\mathrm{PE}_{(i,\, 2j+1)}$, depending on whether the time step occupies an even or odd position within the source sequence. Furthermore, the output of the TST is perceived as a sequence with a prediction horizon of H and dimension v, denoted as $[y_{t+1}, y_{t+2}, \dots, y_{t+H}]$.

This sophisticated framework, enriched with various parameters, empowers the transformer to discern and adapt to temporal dependencies, ultimately providing a potent sequence-to-sequence prediction capability. The TST architecture comprises several encoder–decoder blocks, each housing n attention heads (figure 5.2).

Moreover, the attention mechanism in each attention head is based on the [**Q**,**K**,**V**] mechanism, symbolizing the system's state at position i with an observation horizon W (i.e. $X_{t-i} \in [X_{t-W}, X_{t-W+1}, \dots, X_t]$). The internal computations within an attention layer are presented as follows:

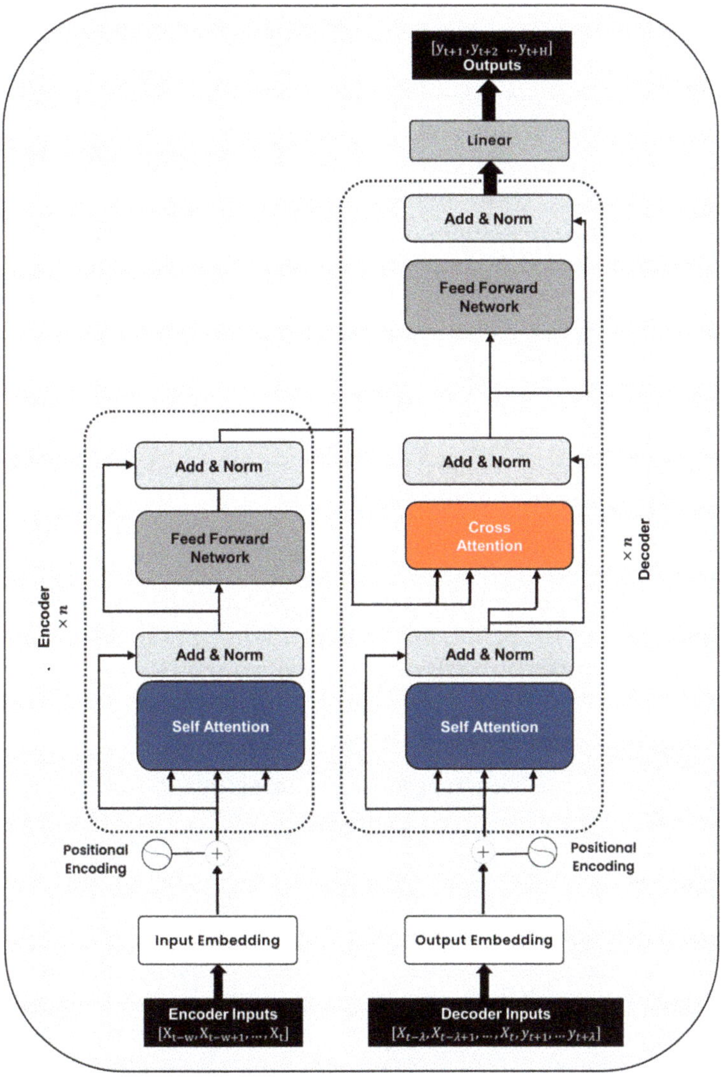

Figure 5.2. Generalized architecture for an encoder–decoder TST. (Reproduced with permission from [88]. Copyright 2023 American Chemical Society.)

$$
\begin{aligned}
A_{P,n} &= \sum_{i}^{k} \lambda_{n,i} \mathbf{V} \\
\lambda_{n,i} &= \frac{\exp\left(\mathbf{Q}^T \mathbf{K}_i / \sqrt{D_k}\right)}{\sum_{j=1}^{k} \exp\left(\mathbf{Q}^T \mathbf{K}_j / \sqrt{D_k}\right)} \\
\sum_{i=1}^{k} \lambda_{n,i} &= 1.
\end{aligned}
\tag{5.2}
$$

Here, $A_{P,n}$ represents the attention value for head n in encoder block P. Additionally, $\mathbf{Q} \in \mathbb{R}^{D_k}$, $\mathbf{K} \in \mathbb{R}^{D_k}$, and $\mathbf{V} \in \mathbb{R}^{D_v}$ denote queries, keys, and values, respectively, where D_k and D_v are the dimensions of the keys and values. Furthermore, the attention score ($\lambda_{n,i}$) indicates the relative importance between different words in the input sequence. The softmax calculation in equation (5.2) ensures that the scaled sum of attention scores equals 1 for each input (i.e. $\sum_{i=1}^{k} \lambda_{n,i} = 1$). During the self-attention process within the encoder block, **Q**, **K**, and **V** all derive from the same input tensor, X_{PE}. Conversely, during the cross-attention calculation in the decoder block, **Q** stems from X_{DEC}, whereas **K** and **V** originate from the processed output of the encoder stack, denoted by X_{EN}. Moreover, MAHs are trained on the same **[Q,K,V]** to facilitate automatic learning of diverse features from the input data. The output from MAHs represents a combination of these different features:

$$
\mathrm{MAH}(Q, K, V) = \mathrm{Concat}[\mathrm{head}_1, \mathrm{head}_2, \ldots, \mathrm{head}_n]. \tag{5.3}
$$

Finally, the attention scores obtained from each encoder or decoder block undergo a positional encoding process, presented as follows:

$$
\mathrm{FFN}(\sigma_{i+1}) = \mathrm{ReLU}(\sigma_i \theta_i + b_i), \tag{5.4}
$$

where σ_i denotes the intermediate state from previous layers, and θ_i and b_i are trainable parameters of the neural network. For a more in-depth understanding of these computations, readers are referred to the literature [15–18].

5.3 Utilizing time-series transformers

Parts of this section have been reproduced with permission from [16]. Copyright 2023 Elsevier.

5.3.1 CrystalGPT

5.3.1.1 Data generation and visualization

The aim of this section is to introduce CrystalGPT, a TST-based model designed to effectively learn from N diverse crystal systems. This model is intended to offer accurate predictions across a wide range of operating conditions and to serve as a baseline predictor for a new (N + 1)th crystal system. To achieve this, we focused on industrially relevant batch sugar crystallization systems as representative case studies for training and implementing CrystalGPT. We compiled crystallization kinetics data for dextrose, sucrose, and lactose crystal systems [19–21] as the foundation for developing 20 unique synthetic crystal systems. These systems encompass a diverse array of process scenarios and represent various polymorphs or derivative crystal systems of the traditional sugar systems. By including more than 20 unique crystal systems, we ensure comprehensive coverage of the parameter space, enabling the TST model to effectively discern, assimilate, and apply structural similarities across these systems.

To this end, we formulated a generalized crystal growth rate (G) as follows:

$$G\ (\mathrm{m\ s^{-1}}) = a_G \exp\left(\frac{-b_G}{RT}\right)(S - 1)^{c_G}, \tag{5.5}$$

where R is the universal gas constant, T is the temperature, and S is the supersaturation. The kinetic parameters are represented by a_G, b_G, and c_G. Similarly, the nucleation rate (B) is expressed as follows:

$$B\ (\#\ (\mathrm{kg \cdot s})^{-1}) = a_B M_T^{b_B}(S - 1)^{c_B}. \tag{5.6}$$

Here, M_T signifies the suspension density, while a_B, b_B, and c_B stand for kinetic parameters. To generate an array of N distinct crystal systems, these kinetic parameters (i.e. $[a_B, b_B, c_B]$ and $[a_G, b_G, c_G]$) were varied using a Gaussian distribution:

$$p_i(x) = \frac{1}{\sigma_i\sqrt{2\pi}} \exp\left(-\frac{1}{2}\left(\frac{x - \overline{p_i}}{\sigma_i}\right)^2\right), \tag{5.7}$$

where $p_i \in [a_B, b_B, c_B, a_G, b_G, c_G]$ represents a kinetic parameter with a mean value of $\overline{p_i}$ and a standard deviation of σ_i. To ensure clarity, the mean value of all process parameters (p_i) was derived from experimentally verified values for different sugar systems. The standard deviation σ_i was assumed to be 20% for each p_i to accommodate significant variations in growth and nucleation rates.

Consequently, the Gaussian distribution of each kinetic parameter, when considered collectively, generates a distribution of growth and nucleation rates, as depicted in figure 5.3. It is essential to note that the spread of this distribution in figure 5.3 influences system-to-system (S2S) transferability, delineating the variable space accessible for the TST model to leverage TL capabilities. Specifically, a narrow distribution encompassing only five crystal systems would impede the TST's capacity to collectively learn transitions between different systems. Conversely, having over 50 systems, each with ample data points, allows the TST to more effectively grasp the underlying relationships between system states, even across a diverse spectrum of systems. Keeping this in consideration, random process parameters (i.e. $[a_B, b_B, c_B, a_G, b_G, c_G]$) for 20 distinct crystal systems, each with a unique pair of (G_i, B_i), were drawn from figure 5.3 to compile the training dataset.

For each unique pair of (G_i, B_i), we consider over 5000 operating conditions, incorporating the following variations for each simulation: (a) an arbitrarily selected jacket cooling curve (i.e. $T_j(t) \in [5, 40]$ °C); (b) solute concentration (i.e. $C_o \in [0.6, 0.9]$ kg/kg); and (c) seed loading (i.e. $M_{T,o} \in [0, 20]$ %), measured in

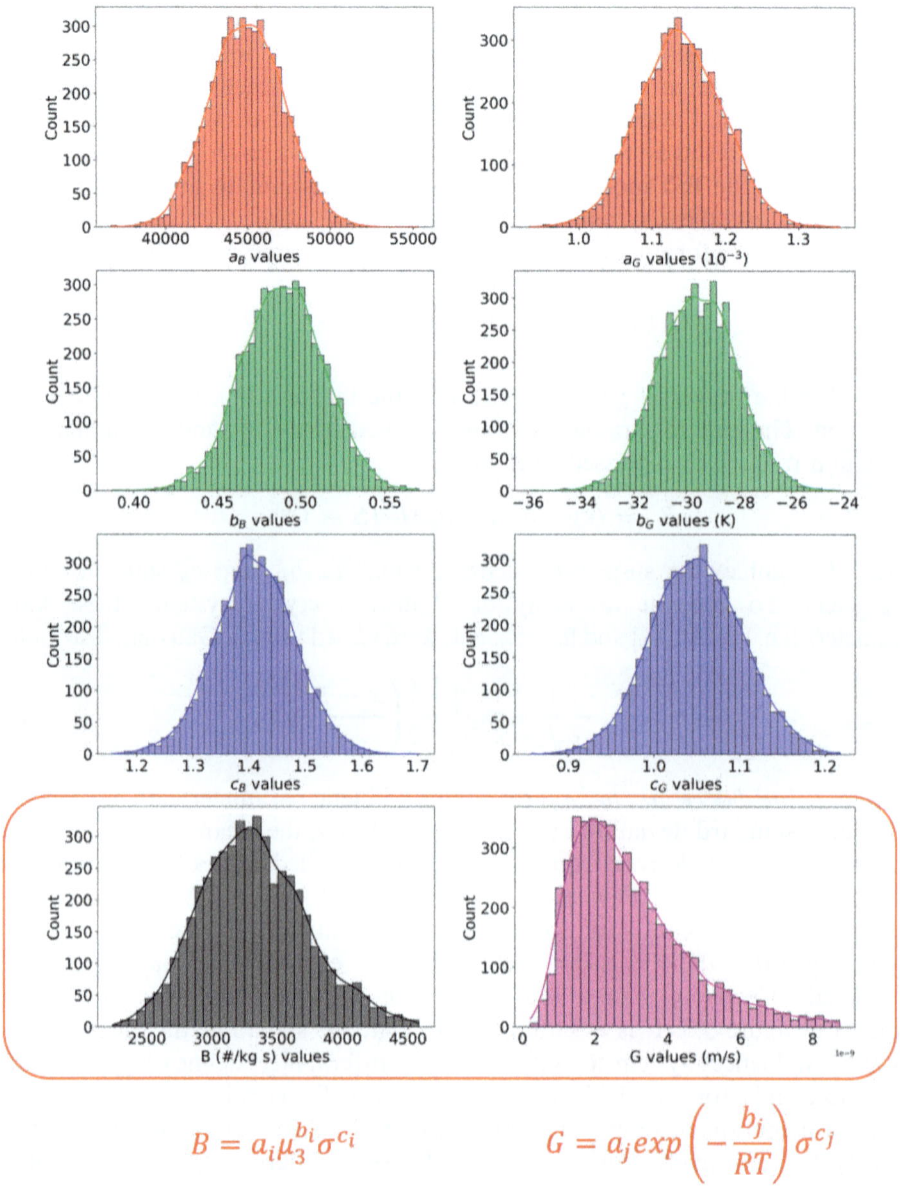

Figure 5.3. The distribution of kinetic parameters (p_i), leading to the corresponding G_i and B_i values. (Reproduced with permission from [16]. Copyright 2023 Elsevier.)

%w/w, with a varying size for these seed crystals (i.e. $\bar{L}_o \in [10, 150]$ μm). This extensive variation in process conditions facilitates data generation across all possible operating regimes, resulting in a comprehensive dataset that aids the TST model in effectively learning the intricate interactions between system states. Subsequently, a population balance model (PBM) model is integrated with mass and energy balance equations (MEBEs) to simulate the aforementioned 20 different crystal systems for each of the 5000+ operating conditions within a jacketed batch crystallizer. Specifically, we utilize a batch crystallization simulator previously developed in-house. While a detailed description of the simulations can be found in our previous works [15, 22], we offer a brief overview here.

After determining the desired crystallization kinetics (i.e. G_i and B_i), the evolution of the crystal size distribution (CSD) can be monitored using a PBM. This model employs a population density function, $n(L, t)$, which is governed by the following equation:

$$\frac{\partial n(L, t)}{\partial t} + \frac{\partial (G(T, C_s)n(L, t))}{\partial L} = B(T, C_s), \tag{5.8}$$

where $n(L, t)$ denotes the number of crystals of size L at time t, $B(T, C_s)$ represents the total nucleation rate, and $G(T, C_s)$ denotes the crystal growth rate. Subsequently, the PBM is integrated with MEBEs, as described below [23]:

$$\begin{aligned} \frac{\mathrm{d}C_s}{\mathrm{d}t} &= -\, 3\rho_c k_v G\mu_2 \\ mC_p\frac{\mathrm{d}T}{\mathrm{d}t} &= -\, UA\left(T - T_j\right) - \Delta H\rho_c 3k_v G\mu_2, \end{aligned} \tag{5.9}$$

where C_s represents the solute concentration, μ_2 is the second moment of crystallization, k_v denotes the shape factor, ρ_c stands for the crystal density, and C_p denotes the heat capacity of the crystallization slurry. Additionally, m denotes the total mass of the slurry, UA represents the area-weighted heat transfer coefficient, T_j stands for the jacket temperature, and ΔH denotes the heat of crystallization.

These simulations were conducted simultaneously using Python's concurrent. futures package, enabling the generation of a dataset comprising over 10 million data points. Subsequently, the data were randomly shuffled and divided into training, validation, and testing sets in a ratio of 70:15:15. Moreover, the input tensor encompasses six process states (i.e. $\left[T_j, C_s, T, \bar{L}, M_T, \text{time}\right]$) for the current and preceding W time steps, while the output tensor includes four states (i.e. $[C_s, T, \bar{L}, M_T]$) for the subsequent H time steps.

Remark 1: Christofides and colleagues have done extensive research on multiscale modeling and control of protein crystallization by coupling microscopic-level kinetic Monte Carlo (kMC) with macroscopic mass and energy balance equations. Specifically, one of the initial studies focuses on the development and experimental validation of a high-fidelity kMC model, which considers several microscopic steps (i.e. adsorption, migration, and desorption), of lysozyme crystallization [24, 25]. The next few studies focused on utilizing the kMC model in conjunction with macroscopic mass and energy balance equations for a batch crystallizer. It was then further utilized to develop a model predictive controller (MPC) [26–30]. The MPC regulates not only the CSD but also the shape of lysozyme crystals [25, 31]. This work further extended to continuous crystallization and was also integrated with a run-to-run control scheme to provide better control over the CSD [32, 33]. Further, the above models were coupled with particle aggregation dynamics, fine removal strategies, and product classification protocols to maximize the productivity of continuous protein crystallizers while reducing the polydispersity [34, 35]. Finally, to handle the issue of computational expense of the above models, one of the studies focuses on the development of a multidomain message passing interface (MPI)-powered parallel computation framework that utilizes a combination of problem decomposition, task assignment, and information flow orchestration techniques to achieve a speed boost of ~40 times [36]. The above works provide a perfect starting point for developing similar multiscale models for quantum dot (QD)-specific crystallization systems.

5.3.1.2 Results and discussion

This section serves two primary objectives: first, it presents the testing results of the CrystalGPT model to underscore its S2S transferability, juxtaposed with the state-of-the-art (SOTA) LSTM model; second, it delves into a case study, exemplifying the practical utility of the CrystalGPT model.

S2S transferability

A dataset comprising approximately 10 million data points, drawn randomly from 20 distinct systems operating under varied conditions, was utilized for training, validation, and testing. Table 5.1 delineates the normalized mean square error (NMSE) values for different CrystalGPT models. Notably, all models exhibit low NMSE values ranging from approximately 5×10^{-4} to 10×10^{-4} for both the $W = 12$ and $W = 50$ cases. Additionally, expanding the TST size marginally impacts the NMSE value up to a certain threshold. However, beyond this point, particularly with large model sizes (e.g. Mega model), there is a decline in predictive performance. This phenomenon can be attributed to the need for an optimal balance between the number of model parameters (N_p) and the dataset size ($\mathbb{D}$) to achieve desirable predictive accuracy [15].

Testing on the $(N + 1)$th crystal system

The preceding results underscore CrystalGPT's efficacy in accurately characterizing a diverse set of 20 crystal systems across various operational scenarios. To further confirm its S2S transferability, we evaluated its performance on an entirely novel,

Table 5.1. Performance comparison of different CrystalGPT models. The top-performing models for each case are highlighted in **bold**. (Reproduced with permission from [16]. Copyright 2023 Elsevier.)

		Base	Big	Large	Mega
# Parameters		1.1M	4.5M	10M	100M
NMSE (10^{-4})	Training	8.93	8.93	**8.27**	9.38
$W = 12$	Validation	9.31	**9.01**	9.14	9.68
	Testing	9.33	**9.29**	9.35	9.69
NMSE (10^{-4})	Training	5.25	6.13	**4.17**	6.38
$W = 50$	Validation	6.21	6.55	**6.10**	6.58
	Testing	6.3	6.61	**6.15**	6.62

unseen twenty-first crystal system. This new system, characterized by unique growth rate (G_i) and nucleation rate (B_i) values, underwent simulation across 2500 operating conditions. Parameters for this twenty-first system were drawn from the distributions illustrated in figure 5.3, aligning structurally with the previous 20 systems in terms of growth and nucleation rates, as well as the PBM and MEBEs, albeit with distinct system parameters.

Two scenarios were considered for this assessment: (a) baseline predictions from CrystalGPT without additional training on data from the twenty-first system, and (b) fine-tuning CrystalGPT using process data from the twenty-first system, followed by performance testing. It is noteworthy that only 2500 operating conditions were considered for the new $(N+1)$th system evaluation. This decision stems from two factors: first, since our aim is to test CrystalGPT's predictive capabilities on a completely new and unseen $(N+1)$th crystal system, we require only a testing dataset rather than a larger dataset for training, validation, and testing. Second, we intend to analyse the impact of model fine-tuning for the new $N+1$ system, necessitating an adequate amount of training data. However, given that CrystalGPT has already been trained on a comprehensive dataset comprising 20 different systems, a smaller dataset for fine-tuning purposes suffices. Hence, considering these factors, simulation data for 2500 (instead of 5000) operating conditions were utilized for the $(N+1)$th system evaluation.

The NMSE values for CrystalGPT, under both the baseline and fine-tuned scenarios, are provided in table 5.2. The results indicate consistent high prediction accuracy across both scenarios. CrystalGPT demonstrates superior performance compared to the LSTM model, achieving NMSE values around 10^{-4}, which is 8 to 10 times lower. Notably, even in the baseline scenarios where it lacks prior exposure to the new crystal system, CrystalGPT exhibits remarkably low NMSE values (1.5×10^{-4}), showcasing its robust transfer learning capabilities. This capability stems from CrystalGPT's ability to leverage shared structural features within the PBM and MEBEs, which describe both the initial 20 systems and the new, previously unseen twenty-first system.

Multiheaded attention empowers S2S transferability

The multiheaded attention mechanism plays a pivotal role in enhancing the transferability of TST-based models across N or $N+1$ different systems. Figure 5.4 illustrates a schematic comparison between a TST model and a

Table 5.2. A comparison between the baseline predictions and the fine-tuned predictions for the twenty-first crystal system. (Reproduced with permission from [16]. Copyright 2023 Elsevier.)

	NMSE (10^{-4})	LSTM	Base	Big	Large
$W = 12$	Baseline	10.72	1.56	1.46	**1.13**
	Fine-tuning	9.05	1.15	**1.05**	1.09
$W = 50$	Baseline	9.74	**1.45**	1.52	1.54
	Fine-tuning	8.85	1.04	1.09	**0.98**

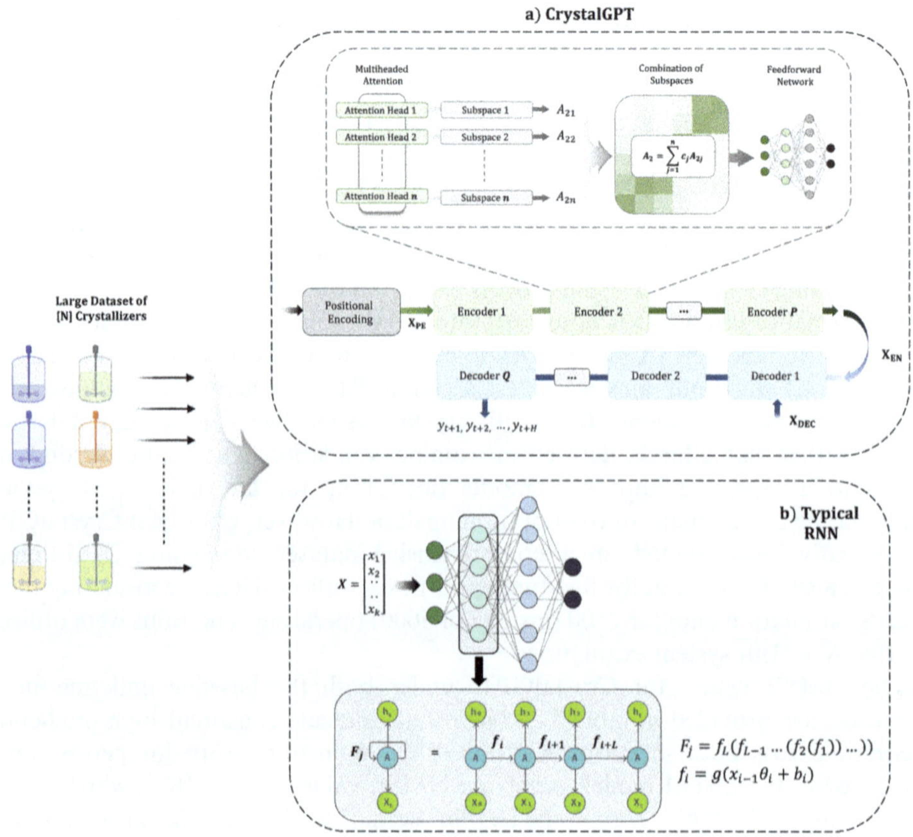

Figure 5.4. Comparison between LSTM, a type of RNN, and CrystalGPT. (Reproduced with permission from [16]. Copyright 2023 Elsevier.)

conventional RNN, highlighting the distinctive features of TSTs that contribute to their superior predictive performance.

In essence, TSTs leverage multiple attention heads within each encoder and decoder block, coupled with an feed-forward network (FFN). These attention heads operate in parallel, processing the same input tensor to compute attention values A_{ij} for each head i in encoder–decoder block j. These attention values are then combined using trainable network parameters c_j and passed through an FFN with ReLU or sigmoid activation functions to capture input–output correlations. This iterative process across encoder–decoder blocks yields intricate representations of the input–output relationship. For any given system $K \in [0, N]$, the parallel nature of the multiple attention heads allows the TST to deconstruct a unified mapping function between inputs and outputs into various constituents or subspaces. These subspaces are collectively weighted to represent the unified mapping function. During training with input data for N systems, each possessing k features, individual attention heads discern different interdependencies among system states. This disassembles the unified mapping function into constituent parts and shared subspace models between different system states, such as the relationships between temperature (T) and solute concentration (C_s), or between C_s and mean crystal size ($\bar{L}$). Moreover, throughout model training, TST parameters learn to adjust the weights of attention values (c_j) across different local regions within the entire state space. For instance, input state information from system A, one of the N systems, is associated with attention score weights within a specific local region, while a different region corresponds to system B, another member of the same set of N systems. Consequently, the TST model becomes adept at mapping between different systems by associating them with corresponding attention score weights.

Second, the distributed internal framework described above plays a crucial role in maintaining high accuracy when extending the TST model to a new, $(N + 1)$th system. Even when the TST encounters state information from an unfamiliar system during input, information it has not seen during training, its various attention heads can identify and focus on the structural similarities between the interdependencies of the system states. These similarities can then be combined using a unique set of weights (c_j) to accurately depict the state dynamics for the $(N + 1)$th system. For instance, consider a hypothetical TST where one attention head captures the relationship between temperature (T) and solute concentration (C_s), represented by a solubility curve, while another attention head might focus on the correlation between the growth rate G and C_s (i.e. the growth rate equation). During training, the TST model learns these distinct representations along with the various possible combinations of attention values from different heads. When confronted with state information from a new system, structurally similar to the previous N systems, the TST model can aggregate attention scores from n different heads for P encoders and Q decoders. This process results in a unique combination of attention values that accurately describes the dynamics of the $(N + 1)$th system.

Furthermore, the capabilities described above can be further enhanced by adjusting the width and depth of the TST [37]. The width of a transformer model is primarily influenced by the number of attention heads and the size of the FFN. A greater number of attention heads enables the capturing of both prominent and subtle system interdependencies, which can then be aggregated to gain a more comprehensive understanding of the input–output relationship. However, the performance gains achieved by increasing the number of attention heads will eventually level off. This phenomenon occurs when the existing n attention heads already adequately represent the dominant modes of system dynamics, resulting in diminishing returns for additional heads. Conversely, the depth of a TST model is determined by the number of encoder–decoder blocks. As each encoder–decoder block comprises n attention heads in parallel, connected to a simple FFN layer, increasing the number of encoder–decoder blocks allows each successive FFN layer to capture more nonlinearity in the input–output relationship [38, 39]. In conclusion, the width and depth of TSTs are critical hyperparameters that significantly influence the S2S transferability of the TST model.

In contrast, an RNN comprises multiple fully connected layers of recurrent neurons. Specifically, the output from each recurrent neuron can be denoted as F_j, representing a convolution of L different neurons. Each neuron is governed by an activation function f_i that encompasses L time steps from the source sequence. During training, when training an RNN for a given Kth system, the model aims to discover a single unified mapping nonlinear function (ϕ) capable of delineating the correlation between inputs and outputs. While this approach proves highly effective for a specific system K, extending the model training to encompass N different systems presents significant challenges in finding a single unified mapping function ϕ. Furthermore, the RNN model processes all time steps in a sequential manner, assigning equal importance to all state inputs. This method inefficiently manages temporal information, often resulting in issues such as vanishing or exploding gradients over large window sizes. Given these characteristics, if a certain RNN model computes an approximate function, extending it to a new $(N + 1)$th system does not yield accurate results. This discrepancy arises because the unified mapping function ϕ treats time steps of different states with equal importance and cannot be deconstructed into various subspaces to collectively provide a new and unique representation of system states. These representations are vital for potentially describing the dynamics of a $(N + 1)$th system (figure 5.5).

Visualization of the attention mechanism

As previously explained, the calculation of attention scores within encoder and decoder blocks is a pivotal aspect of TSTs. Figure 5.6 illustrates the fluctuations in self-attention scores between the input jacket temperature (T_j) and the internal crystallizer temperature (T). In region A, there are moderate attention scores, corresponding to incremental changes in the jacket temperature profile (T_j). Conversely, in region B, there are elevated attention scores observed during sharp declines in T_j, which directly influence T due to the energy balance equation detailed in equation (5.9). Region C, where changes in T_j are relatively subtle, resulting in minor shifts in T, is associated with lower attention scores.

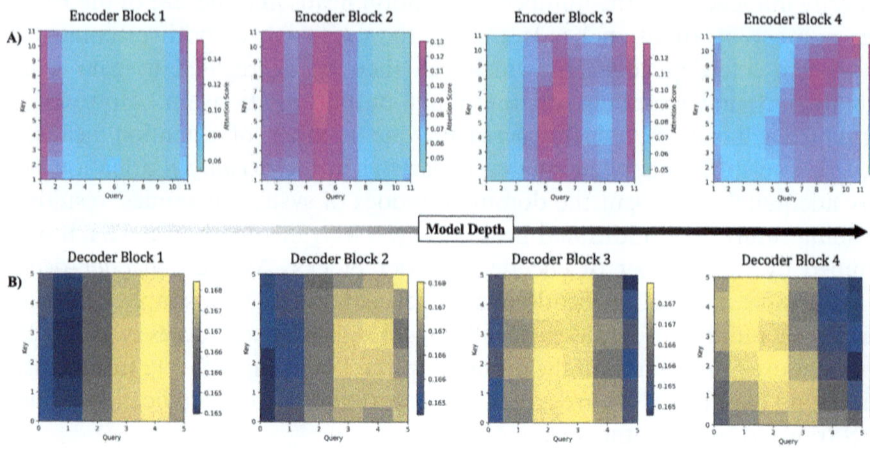

Figure 5.5. Visual representation of self-attention and cross-attention scores within the encoder and decoder blocks, respectively, for the CrystalGPT-Base model with a window size of $W = 12$. (Reproduced with permission from [16]. Copyright 2023 Elsevier.)

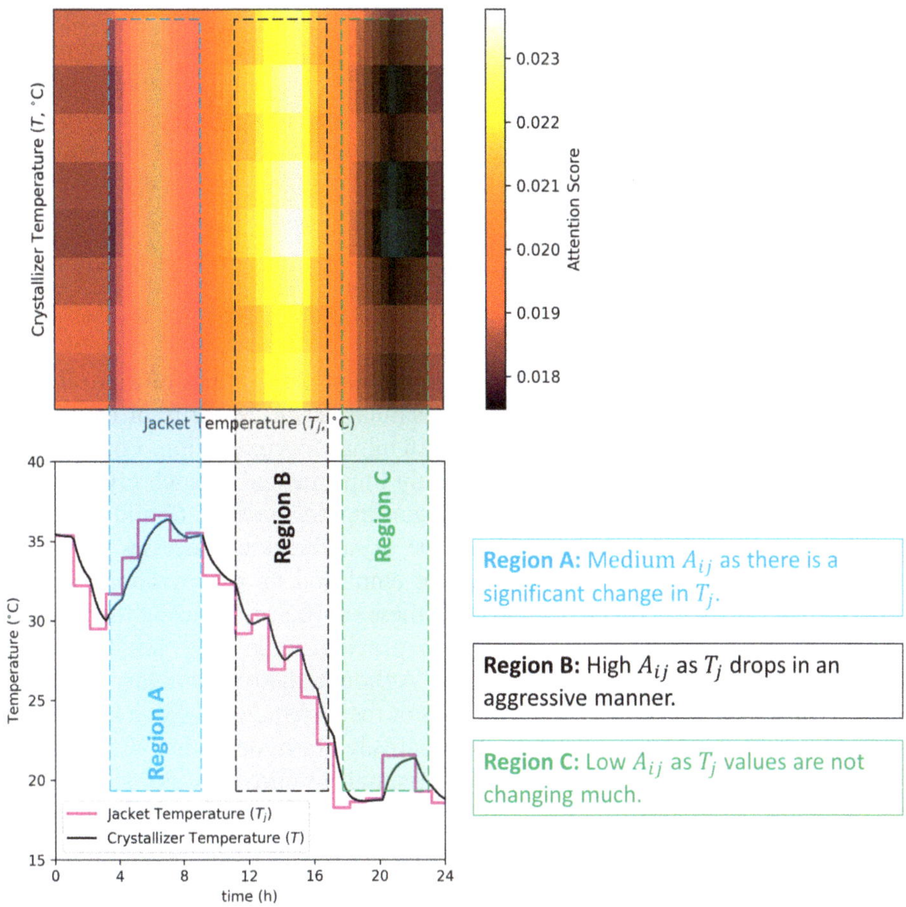

Figure 5.6. Illustration of self-attentions corresponding to the relationship between the jacket temperature input, T_j, and the crystallizer temperature, T, in the CrystalGPT-based model with a window size of $W = 50$. The notation A_{ij} represents the attention score. (Reproduced with permission from [16]. Copyright 2023 Elsevier.)

5.3.2 TST-based hybrid modeling approaches

In this section, we delve into the examination of two distinct methodologies for TST-based hybrid models, outlining their requirements, construction, and training techniques.

5.3.2.1 Problem description

This section delves into the application and effectiveness of TST-based hybrid models, with a particular emphasis on the batch crystallization of dextrose to illustrate their potential. The main goal of these models is to deliver precise predictions of the system's behavior while ensuring the smallest possible deviation between the actual plant operations and the model's predictions. For this purpose, two key sources of data are employed. The first consists of comprehensive process data $(X(t) \in [T_j(t), C_s(t), T(t), \bar{L}(t), \mu_0(t), \mu_1(t), \mu_2(t), \mu_3(t), t])$, covering more than 500 different operational conditions, which are based on the actual kinetics of dextrose crystallization. The second source addresses a common challenge faced by those in the field: the exact kinetics of dextrose crystallization might not be known, yet an estimated version of these kinetics is available and can be used as a point of reference:

$$\begin{aligned} G_{\text{approx}}\left(\text{m s}^{-1}\right) &= 1 \times 10^{-8}(S-1)^2 \\ B_{\text{approx}}\left(\# \,(\text{kg}\cdot\text{s})^{-1}\right) &= 1 \times 10^{5}(S-1)^2. \end{aligned} \tag{5.10}$$

In this context, two methods are available to estimate the kinetics of a novel crystal system: (a) applying the kinetic data from a similar existing system, or (b) initiating fundamental experimental research to derive an initial approximation. For example, a food manufacturing company that has recently implemented a batch crystallizer for dextrose crystallization faces uncertainty regarding the precise G and B kinetics. Nevertheless, available kinetic data from the crystallization processes of fructose, maltose, or other comparable sugars can be employed as a provisional guideline. Given the resemblance between dextrose and these substances, utilizing the kinetics of fructose, maltose, or others as a proxy can provide preliminary insights into the expected behavior and progression over time. Although these preliminary simulations might not perfectly reflect the actual dynamics of the system, they offer a foundational understanding of the process evolution. Alternatively, the undertaking of experimental investigations or trial runs that require minimal resources could yield a basic approximation of the dextrose's G and B kinetics. Subsequently, these preliminary kinetic estimates can be integrated into the first-principles (FP) model (comprising PBM and MEBEs) to simulate batch crystallization. Due to the lack of precise kinetic data, discrepancies between the predicted states and the actual system dynamics, known as plant–model mismatches, are anticipated. To address this discrepancy, this study proposes a series hybrid modeling strategy. This strategy aims to adjust the TST model's kinetic parameters $[k_g, k_b, p_g, p_b]$ to reduce the difference between the plant model and the actual system, with the initial kinetic estimates serving as starting conditions as depicted in the referenced figure configuration schematic.

5.3.2.2 Gathering process data

To validate the predictive efficiency of various hybrid model configurations, assembling a detailed set of process data is crucial. For this purpose, the process of dextrose crystallization, which has been extensively studied, is selected as an exemplary system. This dataset is created by simulating the batch crystallization of dextrose, relying on the specific G and B kinetics detailed in existing research [21]. In essence, due to the lack of directly accessible industrial process data for batch crystallizers, we resort to a detailed simulation, which is comparable to a virtual experiment, of dextrose crystallization. This simulation, which incorporates precise growth and nucleation kinetics, generates synthetic process data for the purpose of training and evaluating the model.

This dataset encompasses simulations under more than 500 different operational scenarios, capturing a wide array of critical variables: (a) the jacket cooling curve, with temperatures varying from 5 °C to 40 °C; (b) solute concentrations ranging from 0.6 to 0.9 kg/kg; and (c) seed loadings from 0% to 20% (in %w/w), along with a variety of seed crystal sizes from 10 μm to 150 μm. Additionally, the specific crystallization kinetics for dextrose, as cited from the literature, are employed as the 'true kinetics' for further discussions and analyses within this study [21]:

$$\begin{aligned} G\ (\mathrm{m\ s^{-1}}) &= 1.14 \times 10^{-3} \exp\left(\frac{-29\ 549}{RT}\right)(S-1)^{1.05} \\ B\ (\#\ (\mathrm{kg} \cdot \mathrm{s})^{-1}) &= 4.50 \times 10^{4}(S-1)^{1.41}. \end{aligned} \tag{5.11}$$

The combined dataset was segmented into three parts: 35 000 data points for training, 15 000 for validation, and 10 000 for testing. Each data point encompasses information across nine process states (specifically, $[T_j, C_s, T, \bar{L}, \mu_0, \mu_1, \mu_2, \mu_3, t]$) over both current and preceding W time steps. This broad spectrum of process conditions enables the generation of data spanning all conceivable operational regimes, thus compiling a dataset that significantly enhances the ability of the series hybrid model to deduce kinetic parameters. Essentially, the hybrid model's goal is to deliver precise predictions of the model's performance, even in the absence of complete knowledge about the true growth and nucleation (G and B) kinetics of dextrose crystallization (figure 5.7).

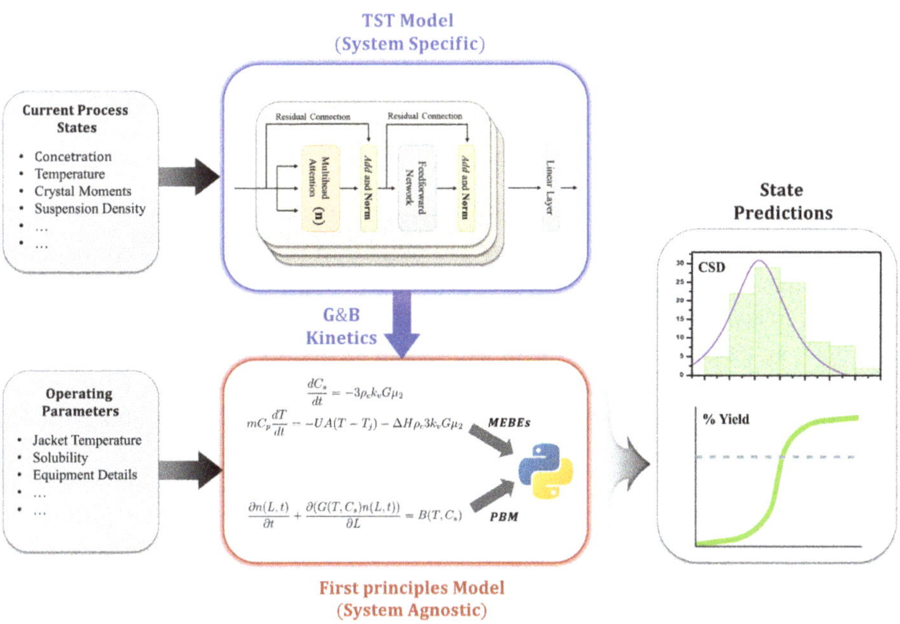

Figure 5.7. A schematic illustration of the TST-based hybrid model in a series configuration. (Reproduced with permission from [88]. Copyright 2023 American Chemical Society.)

Remark 2: In this work, macroscopic models are employed for data generation primarily for demonstration purposes. However, it is typically essential for multi-scale problems, such as crystallization, to incorporate both macroscopic and microscopic models to accurately capture the complex dynamics involved. kMC simulations emerge as a versatile and cost-effective technique for modeling system-specific microscopic interactions, offering a granular view of processes' spatiotemporal evolution at a fraction of the cost of molecular dynamics (MD) or other molecular simulation tools. These simulations adeptly integrate with continuum-level mass-energy balance and population balance equations, enabling realistic modeling of chemical reactors, fermenters, pulp digesters, and crystallization processes [40–43]. The scope of kMC extends well beyond simple adsorption, desorption, and migration steps to encompass complex biological systems and heterogeneous catalysis. Notable examples include the development of kMC models for studying the interaction between cholera toxin subunit B and its receptors, significantly advancing our understanding of cellular heterogeneity and signaling pathways [44–50]. Models based on kMC simulations have found a broad range of applications in the modeling and control of pulp digesters and lignin extraction [51–59]. In the field of catalysis, kMC has been instrumental in creating microkinetic reaction models for the nitrogen reduction reaction, aiding in the identification of high-performance bimetallic transition metal catalysts, among other industrially relevant systems [60–65]. Furthermore, kMC has been applied to surface chemistry modeling within tubular catalytic reactors and to addressing moving boundary problems in microscopic interfacial systems, contributing to the development of hybrid kMC-data-driven models for various processes [66–70]. Additionally, kMC simulations have been extended to the microscopic modeling of battery systems, particularly for studying dendrite-based capacity degradation in lithium-ion batteries, taking into account different electrochemical reactions and mechanical properties of battery materials [71–73].

5.3.2.3 Series hybrid model

Generally, a hybrid model combines a universal FP model (for instance, PBM + MEBEs for crystallization, reactor models + MEBEs for fermentation, among others) with a domain-specific ML model (such as a TST model for predicting crystallization kinetics or Arrhenius-type reaction kinetics) [74, 75]. In a fully integrated series set-up, the current system states along with additional operational conditions or process parameters are inputted into an ML model (here, a TST) to forecast the domain-specific kinetics for crystallization (our case study), which then informs an FP model to simulate the crystallization process, as depicted in a referenced figure. More precisely, the TST model receives state information from the current and preceding W time steps (i.e. $[X_{t-W}, X_{t-W+1}, \dots, X_t]$, where $X_t = [T_j, C_s, T, \bar{L}, \mu_i, M_T, t]$). It outputs the four kinetic parameters for G and B kinetics (i.e. $\hat{z} \in [k_g, k_b, p_b, p_g]$). These parameters are then applied in a specific equation to calculate the instantaneous values of G and B at time t, which are subsequently used in equations to simulate the next step in batch crystallization, thus producing state predictions ($\hat{y}$).

The predicted states ($\hat{y}$) are then evaluated against the actual state evolution (y) to calculate the NMSE, denoted as e. During the training phase of the series hybrid model, it is essential for the prediction errors to be propagated backward through the entirety of the hybrid model. This process updates the parameters (θ) of the TST model, leveraging the automatic differentiation feature of the autograd function found within the PyTorch framework. Essentially, PyTorch creates a computational linkage among all parameters of the TST model and the variables of the FP model. This linkage enables the use of automatic numerical differentiation to efficiently distribute the error (e) throughout the hybrid model. Initially, the partial derivative of e with respect to $\hat{y}$ is calculated and then propagated backward through the FP model to determine the partial derivative of e with respect to $\hat{z}$ (the output of the TST model). Following the calculation of $\frac{\partial e}{\partial \hat{z}}$, the autograd system undertakes backpropagation through the FFNs of both the encoder and decoder segments, akin to standard deep neural network (DNN) operations, adjusting the TST model's parameters (θ) in the process. Further elaboration on training such hybrid models is available in the existing research literature [50, 76–86].

Remark 3: While hybrid models integrate the first-principles model with a data-driven model, there are other different types of such integrations. One of the widely used approaches is physics-informed neural networks (PINNs), which are neural networks that incorporate known physical laws, typically in the form of partial differential equations (PDEs), into their training process. Instead of relying solely on data, PINNs use these physical laws as constraints, ensuring that the network's predictions are consistent with established scientific principles [87–89]. This allows PINNs to solve complex physics problems with fewer data and improve the model's interpretability and accuracy when simulating physical systems.

5.3.2.4 Results and discussion for series hybrid models

Following the outlined training procedure and using the specified dataset, the TST-based hybrid models underwent phases of training, validation, and testing, as detailed in table 5.3 [90]. The NMSE measures the discrepancy between the series hybrid model's predictions and the actual process states, serving as an indicator of the plant–model mismatch (NMSE $= \|\hat{y} - y\|^2$). This evaluation extends to a comparison between two versions of the TST model, namely the Base and the Large, alongside examining the impact of varying the window size W within the range of [1, 12]. The selection of W is pivotal, influencing the model's capacity to interpret context and apply its attention mechanism effectively. A larger W endows the TST model with an expanded perspective, pulling in data from a wider array of preceding time steps. Such an approach enhances the model's ability to discern long-standing trends and dependencies within the dataset, invaluable for recognizing the nuanced effects of procedural adjustments and the lag in temperature response, among other phenomena. For example, alterations to the jacket temperature in batch crystallization processes manifest delayed impacts on the crystallizer's internal temperature, subsequently influencing the supersaturation levels critical to G and B kinetics. Conversely, a scenario showcasing swift crystal growth initially could reduce solute concentration,

Table 5.3. Performance comparison of series hybrid models. The best-performing model is highlighted in **bold**. (Reproduced with permission from [88]. Copyright 2023 American Chemical Society.)

TST model size		Base	Large
# Parameters		2.5M	6.5M
NMSE (10^{-4})	Training	45	56
$W = 1$	Validation	81	73
	Testing	86	93
NMSE (10^{-4})	Training	52	56
$W = 6$	Validation	66	60
	Testing	78	72
NMSE (10^{-4})	Training	**28**	25
$W = 12$	Validation	**55**	50
	Testing	**59**	65

diminishing G and B kinetics at later stages. Conversely, opting for a smaller W concentrates on immediate contexts and prioritizes recent events, which is beneficial for data exhibiting rapid temporal changes or when the latest updates are paramount for precise predictions. For instance, frequent changes in the jacket temperature necessitate a focus on the most immediate data points. Essentially, the TST model's attention mechanism plays a key role by evaluating the significance of each time step within the selected window, allocating greater weight to moments deemed more crucial, thereby enabling the model to navigate through temporal dependencies and spotlight significant trends within the time-series data. Further exploration into the nuances of varying window sizes is documented in the literature [15, 16, 91].

In this context, table 5.3 illustrates that the NMSE values for the training, validation, and testing phases are reduced by 10%–15% when using a window size of $W = 6$, and by 25%–30% with $W = 12$, in comparison to a window size of $W = 1$. The $W = 1$ scenario, commonly employed in simpler DNN-based hybrid models, only leverages current state information without incorporating any contextual insights or recognizing trends, patterns in operating conditions, and the progression of states. Furthermore, the differences in performance between the Base and Large models are minimal, indicating that an oversized TST model with a large number of parameters might not be necessary. The dimension of the Base model aligns well with the sizes of other transformer-based models for chemical processes discussed in the literature [10, 12, 16, 17].

Figure 5.8 provides a visual representation of the series hybrid model's prediction accuracy, displaying a set of state predictions made under a randomly selected operational scenario. This includes the CSD at the conclusion of the crystallization process (i.e. $t = 24$ hr), along with the temporal progression of solute concentration, average crystal size ($\bar{L}$), and the temperature within the crystallizer as forecasted by the series hybrid model, compared against the actual process dynamics. The close correspondence between the model's predictions for both the Base and Large configurations and the real values, as evidenced by an R^2 value exceeding 0.99 for

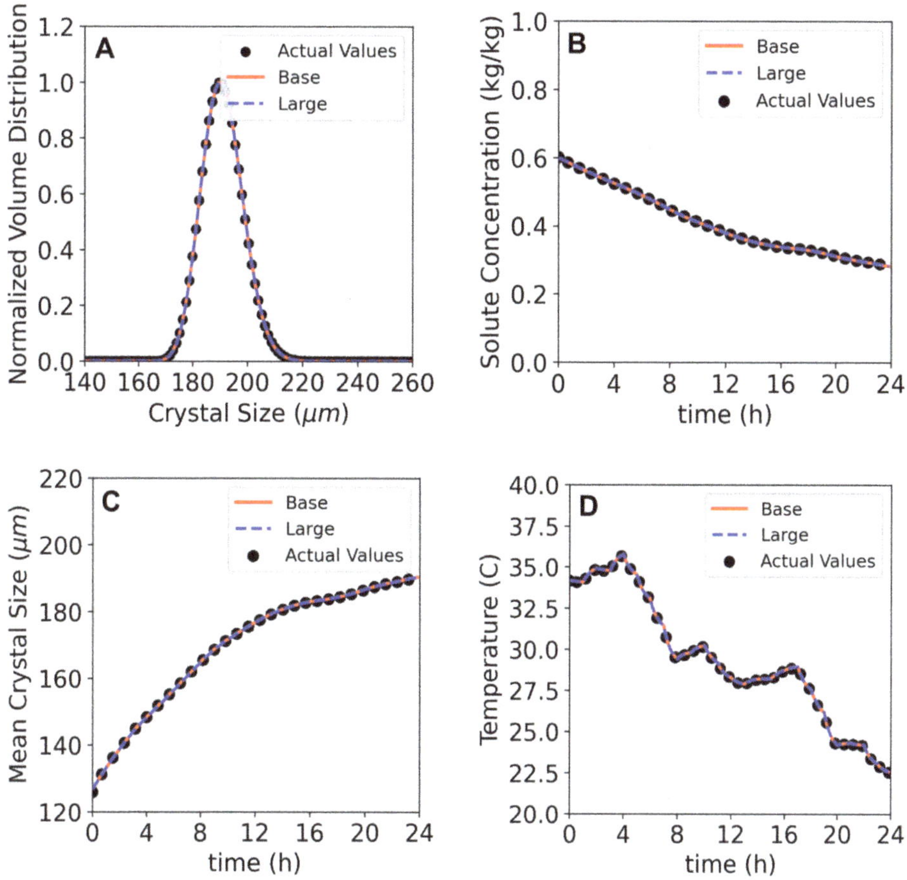

Figure 5.8. The series hybrid model's predictions of (a) the final CSD, (b) solute concentration, (c) mean crystal size ($\bar{L}$), and (d) crystallizer temperature were compared with the actual system dynamics under a randomly selected operating condition. (Reproduced with permission from [88]. Copyright 2023 American Chemical Society.)

all state variables, underscores the model's exceptional predictive accuracy. This observation is further supported by table 5.3, which highlights the negligible difference in predictive capabilities between the Base and Large TST models, a comparison that is vividly confirmed by the data presented in figure 5.8.

In the series hybrid model, the TST model is tasked with identifying patterns within state dynamics and associating them with the kinetic parameters, specifically $\hat{z} \in [k_g, k_b, p_b, p_g]$ from the G and B equations. The state data, encompassing both current and past W time steps, generally exhibits a more consistent trend than the error correction factors, thereby easing the TST model's job in uncovering the relationships between various system states and kinetic parameters. As the TST model becomes increasingly adept at accurately forecasting the G and B kinetics, integrating these predictions into the FP model (PBM + MEBEs) yields state estimations that closely align with the actual observed values (figure 5.9).

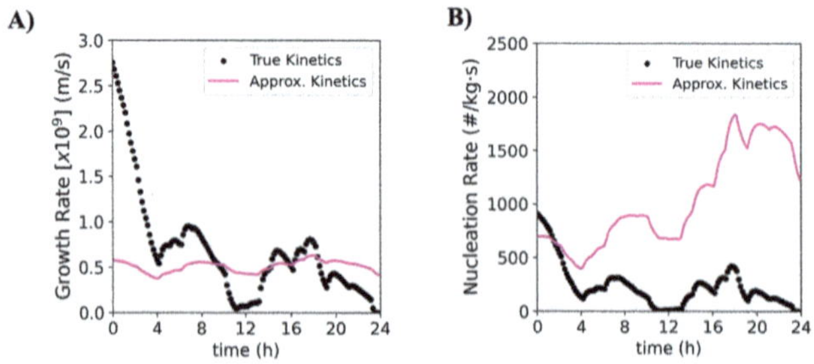

Figure 5.9. Comparison of the true dextrose kinetics (followed by the process data) and approximate kinetics considered for the case study (5.10). (Reproduced with permission from [88]. Copyright 2023 American Chemical Society.)

5.4 Insights and applications of transformer models in chemical systems

Reflecting on the predictive capabilities of TSTs, this section outlines several critical observations and explores potential high-value applications for transformer models across different chemical systems.

5.4.1 Advancements in multiscale modeling through TST models

The advent of attention-based TST models offers promising enhancements to multiscale modeling, which amalgamates physical principles across different scales to elucidate the mechanisms behind observable phenomena. Multiscale modeling has been effectively employed in diverse particulate systems. For instance, Kwon *et al* crafted a multiscale model for the continuous crystallization of QDs, adeptly merging the micro-level growth kinetics of QDs with macro-level dynamics, such as mass and energy balances [92–95]. Similarly, Christofides and his team formulated several multiscale models for protein crystallization by integrating micro-scale crystal growth rates and shapes with macro-scale aggregation dynamics, alongside mass and energy balances. [32, 34–36, 96]. Although general frameworks for multiscale modeling exist, the specificity of interactions across scales—such as surface kinetics, microstructural morphology, and macro-level mass transport—demands a tailored approach for each system to attain precise modeling accuracy. In this scenario, attention-based TST models, with their capacity to focus on significant correlations among system states across scales, can pinpoint vital interaction nodes between these scales. For example, they might link the surface kinetics of a catalyst with the macroscopic concentration gradients in a reactor. The attention scores derived from the model's encoder and decoder highlight the interplay between micro and macro process variables, offering valuable insights for the multiscale modeling of novel systems. This targeted approach allows for the identification and integration of essential variables across scales, ensuring a more coherent model assembly. Moreover, understanding key variable couplings can inform strategies for simplifying complex multiscale models through coarse-graining or reduction techniques. If a system involves n interconnected process variables across scales, identifying the crucial p variables for model simplification can lead to the development of computationally efficient, yet dynamically comprehensive, reduced-order models.

5.4.2 Replacing existing data-driven system identification approaches

Data-driven model order reduction techniques, such as proper orthogonal decomposition (POD) and sparse identification of system dynamics (SINDy), represent pivotal approaches in developing surrogate models for chemical processes. The POD technique effectively captures dominant modes in datasets to predict dynamics in moving boundary problems such as hydraulic fracturing by reducing the dimensionality of variable profiles [97–99]. SINDy, on the other hand, identifies sparse and interpretable equations from a library of basis functions for dynamic chemical processes, addressing plant–model mismatches through the operable adaptive sparse identification of systems (OASIS) models, which conglomerate individual SINDy models using deep neural networks for adaptable process condition modeling [100–106].

Moreover, the Koopman operator framework offers a method to linearize nonlinear systems for easier computation and integration with MPC, extending its applicability to reactor and hydraulic fracking systems. The development of Koopman Lyapunov-based MPC (KLMPC) and bilinear Koopman MPC systems have further enhanced stability and accuracy in chemical system regulation, mitigating plant–model mismatches [107–112]. These advanced data-driven methodologies facilitate overcoming computational hurdles in quantum dots and other chemical process applications, showcasing the breadth and adaptability of reduced-order modeling techniques in addressing complex system dynamics.

Incorporating TSTs into these frameworks adds substantial value by enhancing their predictive accuracy and adaptability. Time-series transformers, with their ability to process sequential data, can capture temporal dependencies and patterns in complex datasets more effectively than traditional methods. This capability makes them particularly suitable for forecasting and anomaly detection in dynamic systems, where understanding the sequence of events is crucial.

5.4.3 Harnessing transfer learning in chemical engineering: a new era

A notable breakthrough with large-scale NLP-transformer models, such as the predecessors of ChatGPT such as GPT-2, is their capacity for transfer learning, which was previously unattainable with RNN and LSTM-based models in a practical sense [113, 114]. The GPT-2 model, with its approximately 125 million parameters trained on extensive textual datasets (including WebCrawl, books, Wikipedia, etc), has mastered the intricacies of English grammar, syntax, and vocabulary. This foundation enables its straightforward adaptation or direct application to tasks such as fraud detection in emails or sentiment analysis, as evidenced by recent research [115, 116].

Looking ahead, the vast repositories of operational data from chemical processing units (such as pressure, temperature, flow rate, and concentration data) could serve to train expansive TST models. Imagine aggregating state data from over 20

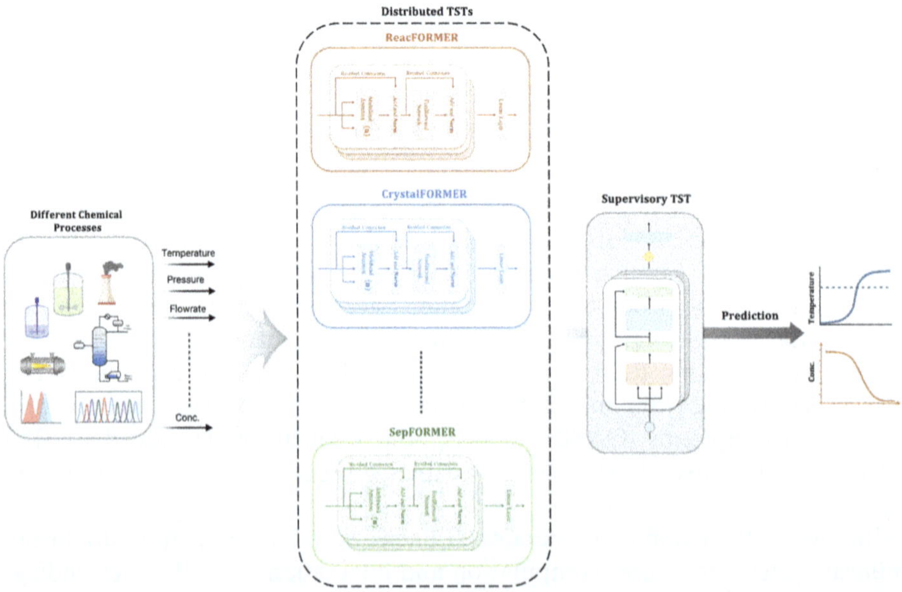

Figure 5.10. Schematic illustration of a distributed TST framework for plant-wide time-series predictions. (Reproduced with permission from [15]. Copyright 2023 Elsevier.)

chemical reactors, each with unique kinetics, into a single TST model, say 'ReacFORMER', which could then be adeptly applied to predict behaviors in a new, twenty-first reactor. This feasibility stems from the model's potential to internalize the fundamental principles of mass and energy balances, along with reaction kinetics, through its elaborate attention mechanisms and encoder layers, facilitating knowledge transfer to unfamiliar reactors.

Moreover, this paradigm can expand across various chemical processes by crafting specialized TST models for distinct process categories (e.g. reactors, crystallization, distillation, separation processes), potentially leading to the creation of models like ReacFORMER, CrystalFORMER, and SepFORMER. Drawing from the principles of distributed MPC systems, these tailor-made TST models could converge into a unified distributed TST framework [117], offering comprehensive predictions across an entire plant, as depicted in figure 5.10. While this vision is yet to be realized, future explorations by academic or industry experts with access to extensive datasets could illuminate the path forward, unlocking new frontiers in chemical process modeling and optimization.

5.4.4 Integrating multiple data sources with transformer models

Recent advancements in modified LLMs have showcased their ability to seamlessly integrate and process data from diverse sources, including text, numbers, and images. A prime example is the GPT-4 model, which has demonstrated this multi-modal data processing capability [118]. Similarly, ACT-1, developed by AdeptLabs, is adept at handling multisource information in real-time and adapting its actions accordingly for task execution. These models emulate human cognitive processes, where various types of information are attentively analysed and synthesized to form a unified understanding of a situation. Traditional deep learning models such as DNNs, CNNs, and RNNs often struggle with this level of integration, as they typically learn from singular data modalities in isolation [119, 120]. The challenge of integrating disparate machine learning models for comprehensive plant-wide applications becomes evident when considering the efficiency, complexity, and potential for misalignment among individually tailored models for different processes.

In this context, multisource aggregative transformer models (MATs) could significantly impact areas such as plant-wide fault diagnosis and preventive maintenance, as depicted in figure 5.11. Specifically, MATs can consolidate a variety of data types, including (a) time-series data from different plant units such as reactors, separators, and condensers; (b) binary alarm system statuses; and (c) textual operation reports and control room inputs. These aggregated data can feed into a sophisticated transformer network capable of diagnosing faults and potentially initiating adaptive control measures to avert failures, showcasing the potential of transformer models to bring about a new level of integration and intelligence in industrial settings.

Most chemical plants possess high-frequency data that are yet to be fully explored and utilized for developing comprehensive models and applications, as depicted in figure 5.11. Inspired by the transformative advancements in LLMs and transformer-based models for computer vision, a range of impactful transformer models can be designed for various chemical processes. These models can harness (a) exceptional transfer learning capabilities, (b) an architecture conducive to parallel processing of

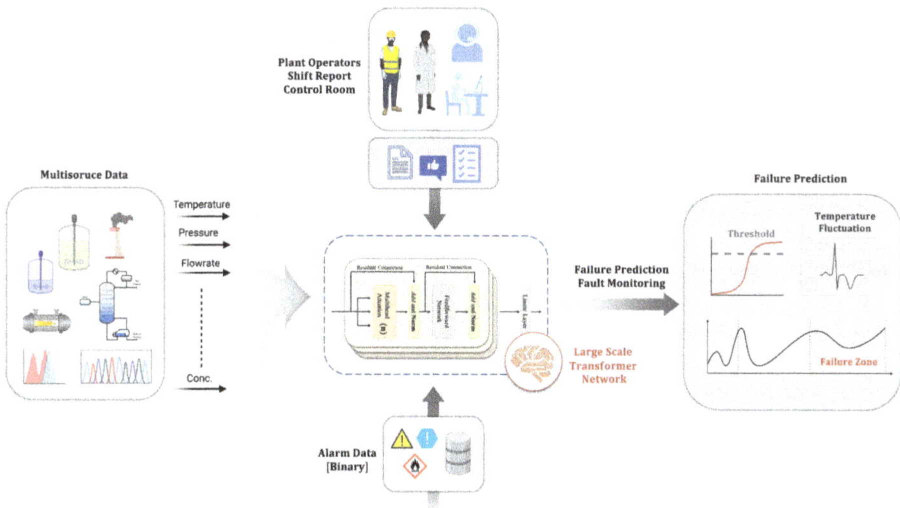

Figure 5.11. Schematic illustration of employing MATs for plant-wide fault prognosis and failure prevention. (Reproduced with permission from [15]. Copyright 2023 Elsevier.)

extensive multisource data, and (c) a selection of architectural designs suited for specific or overarching plant operations. In essence, the advent of diverse transformer models heralds a significant shift towards employing AI-based solutions for both small and large-scale chemical systems, marking the beginning of a long-anticipated revolution in the field.

5.5 Conclusions

The breakthroughs in LLMs and transformer technology for computer vision contrast sharply with the nascent state of transformer models in time-series forecasting, particularly within chemical engineering. This gap prompted the creation and evaluation of several TST models, marking a pioneering effort in complex multivariate time-series analysis for chemical processes. Notably, the newly developed TST models (e.g. CrystalGPT), as well as TST-based hybrid models, demonstrated remarkable predictive accuracy in batch crystallization processes. Comparative analysis of different TST models provided insights into the trade-offs between model complexity and training duration, pinpointing opportunities for enhancement. Despite being at an early stage, these initial TST models show significant promise for further development with additional data. Crucially, this research lays the groundwork for future advancements in (a) custom TST architectures tailored to specific chemical processing challenges, (b) scalable transformers that leverage transfer learning for process innovation, and (c) comprehensive MATs for integrated process monitoring and control across entire plants.

Bibliography

[1] Vaswani A, Shazeer N, Parmar N, Uszkoreit J, Jones L, Gomez A N, Kaiser Ł and Polosukhin I 2017 Attention is all you need *Adv. Neural Inform. Process. Syst.* **30** arXiv:1706.03762

[2] Devlin J, Chang M-W, Lee K and Toutanova K 2018 BERT: Pre-training of deep bidirectional transformers for language understanding arXiv: 1810.04805

[3] Shoeybi M, Patwary M, Puri R, LeGresley P, Casper J and Catanzaro B 2019 Megatron-LM: training multi-billion parameter language models using model parallelism arXiv: 1909.08053

[4] Radford A, Wu J, Child R, Luan D, Amodei D and Sutskever I 2019 Language models are unsupervised multitask learners *OpenAI blog* **1** 9

[5] Brown T, Mann B, Ryder N, Subbiah M, Kaplan J D, Dhariwal P, Neelakantan A, Shyam P, Sastry G and Askell A 2020 Language models are few-shot learners *Adv. Neural Inform. Process. Syst.* **33** 1877–901

[6] Liu Z, Lin Y, Cao Y, Hu H, Wei Y, Zhang Z, Lin S and Guo B 2021 Swin transformer: hierarchical vision transformer using shifted windows *Proc. of the IEEE/CVF Int. Conf. on Computer Vision* 10012–22

[7] Rombach R, Blattmann A, Lorenz D, Esser P and Ommer B 2022 High-resolution image synthesis with latent diffusion models *Proc. of the IEEE/CVF Conf. on Computer Vision and Pattern Recognition* 10674–85

[8] Venkatasubramanian V 2019 The promise of artificial intelligence in chemical engineering: is it here, finally? *AIChE J.* **65** 466–78

[9] Venkatasubramanian V and Mann V 2022 Artificial intelligence in reaction prediction and chemical synthesis *Curr. Opin. Chem. Eng.* **36** 100749

[10] Vogel G, Balhorn L S and Schweidtmann A M 2023 Learning from flowsheets: a generative transformer model for autocompletion of flowsheets *Comput. Chem. Eng.* **171** 108162

[11] Jumper J *et al* 2021 Highly accurate protein structure prediction with AlphaFold *Nature* **596** 583–9

[12] Mann V and Venkatasubramanian V 2021 Predicting chemical reaction outcomes: a grammar ontology-based transformer framework *AIChE J.* **67** e17190

[13] Hannan M A *et al* 2021 Deep learning approach towards accurate state of charge estimation for lithium-ion batteries using self-supervised transformer model *Sci. Rep.* **11** 19541

[14] Chen D, Hong W and Zhou X 2022 Transformer network for remaining useful life prediction of lithium-ion batteries *IEEE Access* **10** 19621–8

[15] Sitapure N and Kwon J S 2023 Exploring the potential of time-series transformers for process modeling and control in chemical systems: an inevitable paradigm shift? *Chem. Eng. Res. Des.* **194** 461–77

[16] Sitapure N and Kwon J S 2023 CrystalGPT: enhancing system-to-system transferability in crystallization prediction and control using time-series-transformers *Comput. Chem. Eng.* **177** 108339

[17] Wen Q, Zhou T, Zhang C, Chen W, Ma Z, Yan J and Sun L 2022 Transformers in time series: a survey arXiv: 2202.07125

[18] Zeng A, Chen M, Zhang L and Xu Q 2022 Are transformers effective for time series forecasting? arXiv: 2205.13504

[19] Shi Y, Liang B and Hartel R W 1990 Crystallization kinetics of alpha-lactose monohydrate in a continuous cooling crystallizer *J. Food Sci.* **55** 817–20

[20] Ouiazzane S, Messnaoui B, Abderafi S, Wouters J and Bounahmidi T 2008 Estimation of sucrose crystallization kinetics from batch crystallizer data *J. Cryst. Growth* **310** 798–803

[21] Markande A, Nezzal A, Fitzpatrick J, Aerts L and Redl A 2012 Influence of impurities on the crystallization of dextrose monohydrate *J. Cryst. Growth* **353** 145–51

[22] Sitapure N and Kwon J S 2023 A unified approach for modeling and control of crystallization of quantum dots (QDs) *Digit. Chem. Eng.* **6** 100077

[23] Worlitschek J and Mazzotti M 2004 Model-based optimization of particle size distribution in batch-cooling crystallization of paracetamol *Cryst. Growth Des.* **4** 891–903

[24] Nayhouse M, Kwon J S, Christofides P D and Orkoulas G 2013 Crystal shape modeling and control in protein crystal growth *Chem. Eng. Sci.* **87** 216–23

[25] Kwon J S, Nayhouse M, Christofides P D and Orkoulas G 2013 Modeling and control of protein crystal shape and size in batch crystallization *AIChE J.* **59** 2317–27

[26] Zhao T, Zheng Y, Gong J and Wu Z 2022 Machine learning-based reduced-order modeling and predictive control of nonlinear processes *Chem. Eng. Res. Des.* **179** 435–51

[27] Agyeman B T, Nouri M, Appels W M, Liu J and Shah S L 2024 Learning-based multi-agent MPC for irrigation scheduling *Control Eng. Pract.* **147** 105908

[28] Xiao M, Hu C and Wu Z 2023 Modeling and predictive control of nonlinear processes using transfer learning method *AIChE J.* **69** e18076

[29] Zhao T, Zheng Y and Wu Z 2023 Feature selection-based machine learning modeling for distributed model predictive control of nonlinear processes *Comput. Chem. Eng.* **169** 108074

[30] Sitapure N and Kwon J S 2024 Machine learning meets process control: Unveiling the potential of LSTMc *AIChE J.* **70** e18356

[31] Kwon J S, Nayhouse M, Christofides P D and Orkoulas G 2014 Protein crystal shape and size control in batch crystallization: comparing model predictive control with conventional operating policies *Ind. Eng. Chem. Res.* **53** 5002–14

[32] Kwon J S, Nayhouse M, Orkoulas G and Christofides P D 2014 Crystal shape and size control using a plug flow crystallization configuration *Chem. Eng. Sci.* **119** 30–9

[33] Kwon J S, Nayhouse M, Orkoulas G, Ni D and Christofides P D 2015 Run-to-run-based model predictive control of protein crystal shape in batch crystallization *Ind. Eng. Chem. Res.* **54** 4293–302

[34] Kwon J S, Nayhouse M, Christofides P D and Orkoulas G 2013 Modeling and control of shape distribution of protein crystal aggregates *Chem. Eng. Sci.* **104** 484–97

[35] Kwon J S, Nayhouse M, Orkoulas G and Christofides P D 2014 Enhancing the crystal production rate and reducing polydispersity in continuous protein crystallization *Ind. Eng. Chem. Res.* **53** 15538–48

[36] Kwon J S, Nayhouse M and Christofides P D 2015 Multiscale, multidomain modeling and parallel computation: application to crystal shape evolution in crystallization *Ind. Eng. Chem. Res.* **54** 11903–14

[37] Levine Y, Wies N, Sharir O, Bata H and Shashua A 2020 Limits to depth efficiencies of self-attention *Adv. Neural Inform. Process. Syst.* **33** 22640–51

[38] Xu H, Liu Q, van Genabith J, Xiong D and Zhang J 2019 Lipschitz constrained parameter initialization for deep transformers arXiv: 1911.03179

[39] Song X, Wu Z, Huang Y, Su D and Meng H 2020 SpecSwap: a simple data augmentation method for end-to-end speech recognition *Interspeech 2020 (Shanghai)* pp 581–5

[40] Pahari S, Moon J, Akbulut M, Hwang S and Kwon J S 2021 Model predictive control for wormlike micelles (WLMs): application to a system of CTAB and NaCl *Chem. Eng. Res. Des.* **174** 30–41

[41] Pahari S, Liu S, Lee C H, Akbulut M and Kwon J S 2022 SAXS-guided unbiased coarse-grained Monte Carlo simulation for identification of self-assembly nanostructures and dimensions *Soft Matter* **18** 5282–92

[42] Pahari S, Bhadriraju B, Akbulut M and Kwon J S 2021 A slip-spring framework to study relaxation dynamics of entangled wormlike micelles with kinetic monte carlo algorithm *J. Colloid Interface Sci.* **600** 550–60

[43] Chaffart D and Ricardez-Sandoval L A 2022 A three dimensional kinetic Monte Carlo defect-free crystal dissolution model for biological systems, with application to uncertainty analysis and robust optimization *Comput. Chem. Eng.* **157** 107586

[44] Lee D, Singla A, Wu H-J and Kwon J S 2018 Dynamic modeling of binding kinetics between GD1b ganglioside and CTB *2018 Annual American Control Conf. (ACC)* (Piscataway, NJ: IEEE) pp 1999–2004

[45] Choi H-K, Lee D, Singla A, Kwon J S and Wu H-J 2019 The influence of heteromultivalency on lectin–glycan binding behavior *Glycobiology* **29** 397–408

[46] Lee D, Green A, Wu H-J and Kwon J S 2021 Hybrid PDE-kMC modeling approach to simulate multivalent lectin-glycan binding process *AIChE J.* **67** e17453

[47] Worstell N C, Singla A, Saenkham P, Galbadage T, Sule P, Lee D, Mohr A, Kwon J S, Cirillo J D and Wu H-J 2018 Hetero-multivalency of *Pseudomonas aeruginosa* lectin LecA binding to model membranes *Sci. Rep.* **8** 8419

[48] Lee D, Mohr A, Kwon J S and Wu H-J 2018 Kinetic Monte Carlo modeling of multivalent binding of CTB proteins with GM1 receptors *Comput. Chem. Eng.* **118** 283–95

[49] Lee D, Jayaraman A and Kwon J S 2020 Identification of cell-to-cell heterogeneity through systems engineering approaches *AIChE J.* **66** e16925

[50] Lee D, Jayaraman A and Kwon J S 2020 Development of a hybrid model for a partially known intracellular signaling pathway through correction term estimation and neural network modeling *PLoS Comput. Biol.* **16** e1008472

[51] Choi H-K and Kwon J S 2019 Multiscale modeling and control of kappa number and porosity in a batch-type pulp digester *AIChE J.* **65** e16589

[52] Choi H-K and Kwon J S 2020 Multiscale modeling and multiobjective control of wood fiber morphology in batch pulp digester *AIChE J.* **66** e16972

[53] Choi H-K and Kwon J S 2019 Modeling and control of cell wall thickness in batch delignification *Comput. Chem. Eng.* **128** 512–23

[54] Son S H, Choi H-K and Kwon J S 2021 Application of offset-free Koopman-based model predictive control to a batch pulp digester *AIChE J.* **67** e17301

[55] Son S H, Choi H-K and Kwon J S 2020 Multiscale modeling and control of pulp digester under fiber-to-fiber heterogeneity *Comput. Chem. Eng.* **143** 107117

[56] Shah P, Choi H-K and Kwon J S 2023 Achieving optimal paper properties: a layered multiscale kMC and LSTM-ANN-based control approach for kraft pulping *Processes* **11** 809

[57] Kim J, Pahari S, Ryu J, Zhang M, Yang Q, Yoo C G and Kwon J S 2024 Advancing biomass fractionation with real-time prediction of lignin content and MWD: a kMC-based multiscale model for optimized lignin extraction *Chem. Eng. J.* **479** 147226

[58] Pahari S, Kim J, Choi H-K, Zhang M, Ji A, Yoo C G and Kwon J S 2023 Multiscale kinetic modeling of biomass fractionation in an experiment: understanding individual reaction mechanisms and cellulose degradation *Chem. Eng. J.* **467** 143021

[59] Lee C H, Kim J, Ryu J, Won W, Yoo C G and Kwon J S 2024 Lignin structure dynamics: advanced real-time molecular sensing strategies *Chem. Eng. J.* **487** 150680

[60] Lee C H, Pahari S, Sitapure N, Barteau M A and Kwon J S 2022 DFT-kMC analysis for identifying novel bimetallic electrocatalysts for enhanced NRR performance by suppressing her at ambient conditions via active-site separation *ACS Catal.* **12** 15609–17

[61] Lee C H, Pahari S, Sitapure N, Barteau M A and Kwon J S 2023 Investigating high-performance non-precious transition metal oxide catalysts for nitrogen reduction reaction: a multifaceted DFT-kMC-LSTM approach *ACS Catal.* **13** 8336–46

[62] Stamatakis M, Chen Y and Vlachos D G 2011 First-principles-based kinetic Monte Carlo simulation of the structure sensitivity of the water–gas shift reaction on platinum surfaces *J. Phys. Chem.* C **115** 24750–62

[63] Kimaev G, Chaffart D and Ricardez-Sandoval L A 2020 Multilevel Monte Carlo applied for uncertainty quantification in stochastic multiscale systems *AIChE J.* **66** e16262

[64] Salciccioli M, Stamatakis M, Caratzoulas S and Vlachos D G 2011 A review of multiscale modeling of metal-catalyzed reactions: mechanism development for complexity and emergent behavior *Chem. Eng. Sci.* **66** 4319–55

[65] Li J, Croiset E and Ricardez-Sandoval L 2015 Carbon nanotube growth: first-principles-based kinetic Monte Carlo model *J. Catal.* **326** 15–25

[66] Chaffart D and Ricardez-Sandoval L A 2018 Robust optimization of a multiscale heterogeneous catalytic reactor system with spatially-varying uncertainty descriptions using polynomial chaos expansions *Can. J. Chem. Eng.* **96** 113–31

[67] Chaffart D and Ricardez-Sandoval L A 2017 Robust dynamic optimization in heterogeneous multiscale catalytic flow reactors using polynomial chaos expansion *J. Process Control* **60** 128–40

[68] Li J, Liu G, Ren B, Croiset E, Zhang Y and Ricardez-Sandoval L 2019 Mechanistic study of site blocking catalytic deactivation through accelerated kinetic Monte Carlo *J. Catal.* **378** 176–83

[69] Chaffart D, Shi S, Ma C, Lv C and Ricardez-Sandoval L A 2022 A moving front kinetic Monte Carlo algorithm for moving interface systems *J. Phys. Chem.* B **126** 2040–59

[70] Chaffart D, Shi S, Ma C, Lv C and Ricardez-Sandoval L A 2023 A semi-empirical force balance-based model to capture sessile droplet spread on smooth surfaces: a moving front kinetic Monte Carlo study *Phys. of Fluids* **35** 032109

[71] Sitapure N, Lee H, Ospina-Acevedo F, Balbuena P B, Hwang S and Kwon J S 2021 A computational approach to characterize formation of a passivation layer in lithium metal anodes *AIChE J.* **67** e17073

[72] Lee H, Sitapure N, Hwang S and Kwon J S 2021 Multiscale modeling of dendrite formation in lithium-ion batteries *Comput. Chem. Eng.* **153** 107415

[73] Hwang G, Sitapure N, Moon J, Lee H, Hwang S and Kwon J S 2022 Model predictive control of lithium-ion batteries: development of optimal charging profile for reduced intracycle capacity fade using an enhanced single particle model (SPM) with first-principled chemical/mechanical degradation mechanisms *Chem. Eng. J.* **435** 134768

[74] Shah P, Pahari S, Bhavsar R and Kwon J S 2024 Hybrid modeling of first-principles and machine learning: A step-by-step tutorial review for practical implementation *Comput. Chem. Eng.* **194** 108926

[75] Kwon J S 2024 Adding big data into the equation *Nat. Chem. Eng.* **1** 724

[76] Raissi M, Perdikaris P and Karniadakis G E 2019 Physics-informed neural networks: a deep learning framework for solving forward and inverse problems involving nonlinear partial differential equations *J. Comput. Phys.* **378** 686–707

[77] Mohammed S F, Bangi and Kwon J S 2020 Deep hybrid modeling of chemical process: application to hydraulic fracturing *Comput. Chem. Eng.* **134** 106696

[78] Pahari S, Shah P and Kwon J S 2024 Achieving robustness in hybrid models: a physics-informed regularization approach for spatiotemporal parameter estimation in PDEs *Chem. Eng. Res. Des.* **204** 292–302

[79] Pahari S, Shah P and Kwon J S 2024 Unveiling latent chemical mechanisms: hybrid modeling for estimating spatiotemporally varying parameters in moving boundary problems *Ind. Eng. Chem. Res.* **63** 1501–14

[80] Hassanpour H, Mhaskar P, House J M and Salsbury T I 2020 A hybrid modeling approach integrating first-principles knowledge with statistical methods for fault detection in HVAC systems *Comput. Chem. Eng.* **142** 107022

[81] Bangi M S F and Kwon J S 2022 Deep hybrid model-based predictive control with guarantees on domain of applicability *AIChE J.* **69** e18012

[82] Chaffart D and Ricardez-Sandoval L A 2018 Optimization and control of a thin film growth process: a hybrid first principles/artificial neural network based multiscale modelling approach *Comput. Chem. Eng.* **119** 465–79

[83] Ghosh D, Hermonat E, Mhaskar P, Snowling S and Goel R 2019 Hybrid modeling approach integrating first-principles models with subspace identification *Ind. Eng. Chem. Res.* **58** 13533–43

[84] Shah P, Ziyan Sheriff M, Bangi M S F, Kravaris C, Kwon J S, Botre C and Hirota J 2022 Deep neural network-based hybrid modeling and experimental validation for an industry-scale fermentation process: identification of time-varying dependencies among parameters *Chem. Eng. J.* **441** 135643

[85] Shah P, Ziyan Sheriff M, Bangi M S F, Kravaris C, Kwon J S, Botre C and Hirota J 2023 Multi-rate observer design and optimal control to maximize productivity of an industry-scale fermentation process *AIChE J.* **69** e17946

[86] Bangi M S F, Kao K and Kwon J S 2022 Physics-informed neural networks for hybrid modeling of lab-scale batch fermentation for β-carotene production using *Saccharomyces cerevisiae Chem. Eng. Res. Des.* **179** 415–23

[87] Zheng Y, Hu C, Wang X and Wu Z 2023 Physics-informed recurrent neural network modeling for predictive control of nonlinear processes *J. Process Control* **128** 103005

[88] Wu G, Yion W T G, Le K, Dang N Q and Wu Z 2023 Physics-informed machine learning for MPC: application to a batch crystallization process *Chem. Eng. Res. Des.* **192** 556–69

[89] Xiao M and Wu Z 2023 Modeling and control of a chemical process network using physics-informed transfer learning *Ind. Eng. Chem. Res.* **62** 17216–27

[90] Sitapure N and Kwon J S 2023 Introducing hybrid modeling with time-series-transformers: a comparative study of series and parallel approach in batch crystallization *Ind. Eng. Chem. Res.* **62** 21278–91

[91] Ainslie J, Ontanon S, Alberti C, Cvicek V, Fisher Z, Pham P, Ravula A, Sanghai S, Wang Q and Yang L 2020 ETC: encoding long and structured inputs in transformers arXiv: 2004.08483

[92] Sitapure N, Epps R, Abolhasani M and Kwon J S 2020 Multiscale modeling and optimal operation of millifluidic synthesis of perovskite quantum dots: towards size-controlled continuous manufacturing *Chem. Eng. J.* **413** 127905

[93] Sitapure N, Epps R W, Abolhasani M and Kwon J S 2021 CFD-based computational studies of quantum dot size control in slug flow crystallizers: handling slug-to-slug variation *Ind. Eng. Chem. Res.* **60** 4930–41

[94] Sitapure N and Kwon J S 2022 Neural network-based model predictive control for thin-film chemical deposition of quantum dots using data from a multiscale simulation *Chem. Eng. Res. Des.* **183** 595–607

[95] Sitapure N, Qiao T, Son D and Kwon J S 2020 Kinetic Monte Carlo modeling of the equilibrium-based size control of $CsPbBr_3$ perovskite quantum dots in strongly confined regime *Comput. Chem. Eng.* **139** 106872

[96] Kwon J S, Nayhouse M, Christofides P D and Orkoulas G 2014 Modeling and control of crystal shape in continuous protein crystallization *Chem. Eng. Sci.* **107** 47–57

[97] Narasingam A and Kwon J S 2017 Development of local dynamic mode decomposition with control: application to model predictive control of hydraulic fracturing *Comput. Chem. Eng.* **106** 501–11

[98] Narasingam A, Siddhamshetty P and Kwon J S 2018 Handling spatial heterogeneity in reservoir parameters using proper orthogonal decomposition based ensemble Kalman filter for model-based feedback control of hydraulic fracturing *Ind. Eng. Chem. Res.* **57** 3977–89

[99] Narasingam A, Siddhamshetty P and Kwon J S 2017 Temporal clustering for order reduction of nonlinear parabolic PDE systems with time-dependent spatial domains: application to a hydraulic fracturing process *AIChE J.* **63** 3818–31

[100] Narasingam A and Kwon J S 2018 Data-driven identification of interpretable reduced-order models using sparse regression *Comput. Chem. Eng.* **119** 101–11

[101] Brunton S L, Proctor J L and Kutz J N 2016 Discovering governing equations from data by sparse identification of nonlinear dynamical systems *Proc. Natl Acad. Sci.* **113** 3932–7
[102] Bhadriraju B, Narasingam A and Kwon J S 2019 Machine learning-based adaptive model identification of systems: application to a chemical process *Chem. Eng. Res. Des.* **152** 372–83
[103] Bhadriraju B, Bangi M S F, Narasingam A and Kwon J S 2020 Operable adaptive sparse identification of systems: application to chemical processes *AIChE J.* **66** e16980
[104] Bhadriraju B, Kwon J S and Khan F 2021 Risk-based fault prediction of chemical processes using operable adaptive sparse identification of systems (OASIS) *Comput. Chem. Eng.* **152** 107378
[105] Bhadriraju B, Kwon J S and Khan F 2023 An adaptive data-driven approach for two-timescale dynamics prediction and remaining useful life estimation of Li-ion batteries *Comput. Chem. Eng.* **175** 108275
[106] Bhadriraju B, Kwon J S and Khan F 2024 A data-driven framework integrating Lyapunov-based MPC and OASIS-based observer for control beyond training domains *J. Process Control* **138** 103224
[107] Narasingam A and Kwon J S 2019 Koopman Lyapunov-based model predictive control of nonlinear chemical process systems *AIChE J.* **65** e16743
[108] Narasingam A and Kwon J S 2020 Application of Koopman operator for model-based control of fracture propagation and proppant transport in hydraulic fracturing operation *J. Process Control* **91** 25–36
[109] Narasingam A and Kwon J S 2020 Closed-loop stabilization of nonlinear systems using Koopman Lyapunov-based model predictive control *2020 59th IEEE Conf. on Decision and Control (CDC)* (Piscataway, NJ: IEEE) pp 704–9
[110] Narasingam A, Son S H and Kwon J S 2023 Data-driven feedback stabilisation of nonlinear systems: Koopman-based model predictive control *Int. J. Control* **96** 770–81
[111] Son S H, Choi H-K, Moon J and Kwon J S 2022 Hybrid Koopman model predictive control of nonlinear systems using multiple EDMD models: an application to a batch pulp digester with feed fluctuation *Control Eng. Pract.* **118** 104956
[112] Son S H, Narasingam A and Kwon J S 2022 Development of offset-free Koopman Lyapunov-based model predictive control and mathematical analysis for zero steady-state offset condition considering influence of Lyapunov constraints on equilibrium point *J. Process Control* **118** 26–36
[113] Xiao M, Vellayappan K, Pravin P S, Gudena K and Wu Z 2024 Optimization-based multi-source transfer learning for modeling of nonlinear processes *Chem. Eng. Sci.* **295** 120117
[114] Mou T, Liu J, Zou Y, Li S and Xibilia M G 2024 Enhanced industrial process modeling with transfer-incremental-learning: a parallel sae approach and its application to a sulfur recovery unit *Control Eng. Pract.* **148** 105955
[115] Setianto F, Tsani E, Sadiq F, Domalis G, Tsakalidis D and Kostakos P 2021 GPT-2c: a parser for honeypot logs using large pre-trained language models *Proc. of the 2021 IEEE/ACM Int. Conf. on Advances in Social Networks Analysis and Mining* 649–53
[116] Kit Y and Mokji M M 2022 Sentiment analysis using pre-trained language model with no fine-tuning and less resource *IEEE Access* **10** 107056–65
[117] Christofides P D, Liu J and De La Pena D M 2011 *Networked and Distributed Predictive Control: Methods and Nonlinear Process Network Applications* (Berlin: Springer Science)
[118] OpenAI 2023 *GPT-4 technical report* arXiv: 2303.08774

[119] Gu J, Luo J, Li M, Huang C and Heng Y 2020 Modeling of pressure drop in reverse osmosis feed channels using multilayer artificial neural networks *Chem. Eng. Res. Des.* **159** 146–56
[120] Luo J, Li M, Hoek E M V and Heng Y 2023 Supercomputing and machine learning-aided optimal design of high permeability seawater reverse osmosis membrane systems *Sci. Bull.* **68** 397–407

Chapter 6

Optimization-based algorithms for solving inverse problems of parabolic PDEs

Yi Heng, Chen Wang, Qingqing Yang and Junxuan Deng

Heat transfer problems often exhibit causal behavior and are modeled using partial differential equations. When direct measurements are impractical, inverse solution techniques can estimate unknown values efficiently. This chapter discusses optimization-based algorithms for solving heat transfer inverse problems, including conjugate gradient, stochastic gradient, and Bayesian optimization methods. Their applications in heat flux estimation for pool boiling and chip heat dissipation are highlighted, demonstrating their effectiveness in addressing real-world thermal challenges.

6.1 Introduction

The inverse heat transfer problem (IHTP) represents a pivotal research domain within the heat transfer discipline, concerned with inferring the internal thermal characteristics and boundary conditions of a system given specified boundary conditions and incomplete observational data. The exploration of IHTP endeavors to tackle practical engineering hurdles, notably the determination of elusive thermal parameters of materials (comprising thermal conductivity, thermal diffusivity, etc), heat sources, and surface thermal boundary conditions that elude direct measurements. These parameters are important for material design, industrial process refinement, and evaluative analysis of thermal energy systems. The intricate nature of IHTP emanates from several facets. First, heat transfer processes are frequently beset by nonlinearities and spatial disparities, impeding precise thermal property delineation. Second, empirical data may be afflicted by noise and stochasticity,

doi:10.1088/978-0-7503-6174-3ch6

further complicating the analytical landscape. Moreover, owing to the presence of multi-physics couplings, such as heat conduction interwoven with fluid dynamics, heat transfer inverse problems frequently intertwine with diverse physical phenomena, thereby augmenting solution intricacy.

In the past few decades, the IHTP has garnered considerable attention within the realm of research, primarily due to the computational complexities involved and their practical implications. Diverse categories of IHTP, boundary conditions identification [1, 2], thermal properties estimation [3–5], initial condition identification [6, 7] or source term recognition [8–11], among others, have been subject to extensive investigation.

This chapter focuses on a specific type of IHTP, namely the reconstruction of unknown heat flux distributions on the inaccessible boundaries based on temperature observations obtained from the accessible boundary. Such problems find their applications across various branches of engineering. Following seminal works [12–14], a plethora of application-oriented IHTP studies have emerged. These endeavors predominantly rely on solution methodologies such as Tikhonov regularization [15], space marching [16], function specification [17], iterative regularization [18–20], and Bayesian inference strategies employing the maximum likelihood estimator [21]. Additionally, alternative methods such as the utilization of a stochastic algorithm such as quantum-behaved particle swarm optimization [22] or the decentralized fuzzy inference approach [23] have demonstrated applicability in resolving IHTP scenarios of limited scale.

The numerical solution of IHTP typically poses considerable challenges, attributed to both the pronounced ill-posed nature of the problem and the elevated computational burden, especially when employing three-dimensional (3D) computational domains. Many prior IHTP investigations have been confined to one or two spatial dimensions, albeit at the expense of solution accuracy due to the reduction in geometric complexity. Notably, recent endeavors focusing on 3D IHTP are summarized below. One study introduced a methodology employing dynamic state observers to tackle multi-dimensional IHTP scenarios [24], another work explored specialized techniques for addressing anisotropic 3D IHTP, notably applied to examine the impact of heat transfer in falling films through wave characteristics [25]. Furthermore, an iterative regularization scheme leveraging the CG method to efficiently resolve 3D IHTP associated with pool boiling processes was proposed [26]. Additionally, an analytical solution based on Green's function was introduced to address the scenario of a 3D IHTP [27].

Of the existing methodologies, Tikhonov regularization and iterative regularization stand out as commonly adopted techniques for IHTP solutions, with iterative regularization being particularly esteemed for its efficiency [18]. One study introduced an iterative regularization approach employing gradient methods for reconstructing space–time heat flux distribution. Furthermore, a method resembling CG techniques was devised for the iterative solution of transient boundary IHTP [28]. Such approaches have many successful applications in various practical contexts, including the estimation of transient heat flux on the inaccessible side of a film in falling film experiments [29], the reconstruction of instantaneous local boiling heat flux distributions on boiling surfaces [30], and the recovery of high-intensity periodic laser heat flux on the front surface of target objects [31].

6.1.1 Definition of the forward problem

Parts of this section have been reproduced with permission from [39]. Copyright 2012 Elsevier.

To streamline the forthcoming discourse and elucidate the rationale behind this study, we offer a succinct introduction to the transient heat conduction equation under consideration, representing the forward problem:

$$\begin{aligned}
&\rho c_p \frac{\partial \Theta}{\partial t} = \nabla \cdot \left(\lambda \nabla \Theta\right), \quad \text{in } \boldsymbol{\Omega} \times \left(0, t_f\right);\\
&\Theta(\cdot, 0) = \Theta_0(\cdot), \quad \text{on } \boldsymbol{\Omega};\\
&\lambda \frac{\partial \Theta}{\partial \mathbf{n}} = q_1, \quad \text{on } \Gamma_1 \times \left(0, t_f\right);\\
&\lambda \frac{\partial \Theta}{\partial \mathbf{n}} = q_2, \quad \text{on } \Gamma_2 \times \left(0, t_f\right);\\
&\lambda \frac{\partial \Theta}{\partial \mathbf{n}} = q, \quad \text{on } \Gamma_3 \times \left(0, t_f\right).
\end{aligned} \tag{6.1}$$

In the provided context, Θ represents the temperature defined over an observation time interval $[0, t_f]$ and across a 3D spatial domain $\boldsymbol{\Omega}$. The symbols q_1, q_2, and q denote the time- and space-dependent heat fluxes through different conductor boundaries Γ_1, Γ_2, and Γ_3, respectively. Additionally, ρ, c_p, and λ denote the density, heat capacity, and thermal conductivity, respectively. It is well-established that equations (6.1) constitute a well-posed parabolic PDE in the sense of Hadamard [32]. The existence, uniqueness, and stability of solutions to parabolic PDEs have been widely discussed in the literature, as shown in Evans' study [33].

6.1.2 Definition of the inverse problem

Parts of this section have been reproduced with permission from [39]. Copyright 2012 Elsevier.

The 3D IHTP investigated in this chapter emerges from various industrial applications, involving the reconstruction of the unknown heat flux q along boundary Γ_3 utilizing temperatures Θ_m^δ observed on the boundary Γ_1. The precision of the measured temperatures, denoted as δ, is typically unknown or, at least, challenging to accurately estimate in practical scenarios. Let X and Y denote appropriate function spaces for the heat flux q and the temperature data Θ_m^δ, respectively. The 3D IHTP, formulated in operator notation, reads as follows.

Find $q \in X$ such that

$$F(q) = \Theta_m^\delta, \tag{6.2}$$

where the function operator F: $X \to Y$ is implicitly defined by equation (6.1).

For addressing the solution of the IHTP described by equation (6.2), one can employ either the CG-based iterative regularization method or Tikhonov's regularization method. Below, we outline their primary limitations and subsequently introduce our efficient solution strategy aimed at mitigating these challenges.

The CG-based iterative regularization approach aims to minimize the least-squares error between the temperature measurements and the model predictions:

$$J(q) := \|F(q) - \Theta_m^\delta\|^2, \tag{6.3}$$

with $\|\cdot\|$ denoting the norm in the set of L^2 space. To address the ill-posed nature of the problem, an additional stopping rule, such as the discrepancy principle [34], must be implemented to properly terminate the iterative solution process. Specifically, the minimization procedure halts when the temperature prediction error approximately matches the *a priori* known noise level. Alternatively, other stopping rules such as the L-curve criterion [35] can be employed if the measurement noise level can be described. The utilization of the L-curve criterion usually requires a significant number of iterations, often surpassing 100–1000 iterations in certain instances, to generate the complete L-curve for further analysis. Furthermore, while this approach performs well across various problems, it lacks rigorous theoretical justification. Indeed, the CG-based iterative regularization approach utilizing the L-curve rule as a stopping criterion may not be mathematically stable.

The Tikhonov regularization [15] aims to minimize an objective functional that encompasses both the data fitting error and a penalty term representing the solution norm. Specifically,

$$J(q) := \|F(q) - \Theta_m^\delta\|^2 + \alpha\|q\|^2, \ \alpha > 0. \tag{6.4}$$

While Tikhonov's regularization method is mathematically rigorous, its efficacy heavily relies on the selection of the regularization parameter α. Determining the optimal regularization parameter typically entails solving numerous minimization problems across a wide range of α values, resulting in a notable increase in computational complexity.

6.2 Fast and robust 3D IHTP solution strategies

This section introduces two fast and robust 3D IHTP solution strategies, namely the improved Tikhonov regularization method (see section 6.2.1) and the fast Bayesian solving strategy (see section 6.2.2). Both of them aims to obtain the inverse solution by solving the optimization problem defined by equation (6.4).

6.2.1 Optimization-based conventional Tikhonov method for IHTP

Parts of this section have been reproduced with permission from [39]. Copyright 2012 Elsevier.

In this subsection it aims to minimize the objective functional defined by equation (6.4) and leverages the model function approach [36] to diminish the necessity for extensive α testing. Moreover, we propose an efficient CG method to tackle each resulting minimization problem for a fixed α value. This strategy significantly reduces the overall computational burden, rendering our approach notably faster than conventional Tikhonov regularization-based solutions. Furthermore, compared to previous CG-based iterative regularization approaches, our method exhibits enhanced robustness and does not require *a priori* knowledge of measurement noise, rendering it applicable across a broad spectrum of practical problems.

6.2.1.1 *Enhanced Tikhonov regularization*

The solution to the 3D IHTP defined by equation (6.2) employing Tikhonov regularization is achieved through the minimization of the objective functional defined by equation (6.4). The Tikhonov cost functional equation (6.4), by definition, accounts for the balance between the data fitting error, represented by the term $\|F(q) - \Theta_m^\delta\|$, and the regularization term, which measures the variation of the heat flux estimate $\|q\|^2$. The parameter α associated with the penalty term $\|q\|^2$ serves as the Tikhonov regularization parameter, regulating the quality of reconstruction, particularly when temperature measurements are subject to noise. It can be demonstrated that, for a suitably chosen α, the minimizer of the Tikhonov cost functional defined by equation (6.4) exhibits stability concerning perturbations in the data Θ_m^δ. However, since the error bound δ is typically unknown in practical scenarios, the conventional method for determining α involves conducting numerous tests to ascertain the optimal regularization parameter. Such a procedure incurs significant computational expense.

Here, we explore the adoption of an efficient technique introduced in [36] to expedite the solution process utilizing Tikhonov regularization. The fundamental concept entails adjusting α by minimizing an alternative objective functional [37], referred to as the modified L-curve method, where F is a linear operator,

$$\Psi_\mu(\alpha) = \|F(q_\alpha) - \Theta_m^\delta\|^2 \|q_\alpha\|^{2\mu}, \ \mu > 0. \tag{6.5}$$

In the studies by Regińska [37], the author has established that if the regularization parameter α_* selected by the standard L-curve method, which exhibits maximum curvature along the entire L-curve defined by $(\log\|F(q_\alpha) - \Theta_m^\delta\|^2, \ \log\|q_\alpha\|^2)$ possesses a slope of $-\frac{1}{\mu}$, then the minimizer of the functional equation (6.5) aligns with α_*. However, given the absence of an explicit method for computing the minimizer in the study [37], one still needs to explore a broad range of α values and plot the functional $\Psi_\mu(\alpha)$ to identify the minimum point. The rapid algorithm introduced in the study [36] endeavors to substitute the data fitting error and the solution norm with suitable model functions, thereby substantially reducing the computational burden associated with identifying the minimizer of equation (6.5).

For thoroughness, let us briefly revisit the concept of the model function approach. By introducing $q{:} =q(\alpha)$, and examining the Euler equation derived from equation (6.4), we can rephrase the original Tikhonov cost functional equation (6.4) as follows:

$$\begin{aligned} J(\alpha) &= \|F(q(\alpha)) - \Theta_m^\delta\|^2 + \alpha\|q(\alpha)\|^2 \\ &= \|\Theta_m^\delta\|^2 - \|F(q(\alpha))\|^2 - \alpha\|q(\alpha)\|^2. \end{aligned} \tag{6.6}$$

Note that the first derivative of $J(\alpha)$ with respect to α is $J'(\alpha) = \|q(\alpha)\|^2$, as stated in the study of Kunisch and Zou [38]. In our approach, we locally approximate the term $\|F(q(\alpha))\|^2$ in (6.6) by $T\|q(\alpha)\|^2$, where the constant T is iteratively determined within the algorithm. Subsequently, equation (6.6) takes the form

$$J(\alpha) + (\alpha + T)J'(\alpha) \approx \|\Theta_m^\delta\|^2. \tag{6.7}$$

The model function $m(\alpha)$, designed to emulate the behavior of $J(\alpha)$, should satisfy the following differential equation:

$$m(\alpha) + (\alpha + T)m'(\alpha) = \|\Theta_m^\delta\|^2. \tag{6.8}$$

Thus, $m(\alpha)$ can be represented in the form of a straightforward family of functions.

$$m(\alpha) = \|\Theta_m^\delta\|^2 + \frac{C}{\alpha + T}, \tag{6.9}$$

where the constants C and T need to be determined locally. We utilize the first derivative of the model function $m(\alpha)$ to approximate the first derivative of $J(\alpha)$, denoted as $J'(\alpha) \approx m'(\alpha)$. It is evident that the original function $\Psi_\mu(\alpha)$ can be locally approximated by another straightforward function, denoted as $\Psi_{C,\,T,\,\mu}(\alpha)$

$$\begin{aligned} \Psi_{C,\,T,\,\mu}(\alpha) &= (m(\alpha) - \alpha m'(\alpha))(m'(\alpha))^\mu \\ &= \left(\|\Theta_m^\delta\|^2 + \frac{C}{\alpha + T} + \frac{\alpha C}{(\alpha + T)^2}\right)\left(\frac{-C}{(\alpha + T)^2}\right)^\mu \end{aligned} \tag{6.10}$$

by introducing the model function $m(\alpha)$, the minimizer of which can be explicitly calculated.

A comprehensive mathematical analysis of this approach can be found in our study [36]. Here, we provide the model function algorithm 6.1 in the modified *L*-curve method. In this algorithm, the constants C and T are updated by

$$C_k = -\frac{(\|F(q(\alpha_k))\|^2 + \alpha_k\|q(\alpha_k)\|^2)^2}{\|q(\alpha_k)\|^2}, \ T_k = \frac{\|F(q(\alpha_k))\|^2}{\|q(\alpha_k)\|^2}. \tag{6.11}$$

We mention that step 5 in algorithm 6.1 might be restarted with an iterative α_{k+1} as

$$\alpha_{k+1} = \omega\left(-T_k + \frac{(2 + 2\mu)C_k T_k}{2C_k + 4\mu C_k} - \frac{2\mu\|\Theta_m^\delta\|^2(T_k + \alpha)^2}{2C_k + 4\mu C_k}\right), \tag{6.12}$$

with a contraction factor $\omega \in (0,1)$. A theoretical justification for this choice can be found in remark 3.8 of [36].

Algorithm 6.1. Model function approach in a modified L-curve [39]. (Reproduced with permission from [39]. Copyright 2012 Elsevier.)

Input: $\epsilon > 0$, Θ_m^δ, $\mu > 0$.
1: Choose initial guess $\alpha_1 > \alpha_*$, and set $k = 1$.
2: Do
3: Solve the minimization problem to find $q = q(\alpha_k)$.
4: Update C_k and T_k according to equation (6.11) and construct the corresponding model function
$m_k(\alpha) = \|\Theta_m^\delta\|^2 + \frac{C_k}{\alpha + T_k}$.
5: Insert $m_k(\alpha)$ into equation (6.10), update α_{k+1} as the minimizer of $\Psi_{C_k,\, T_k,\, \mu}(\alpha)$, set $k: = k + 1$.
6: While $|\frac{\alpha_{k+1} - \alpha_k}{\alpha_k}| \leqslant \epsilon$.
7: If the stopping rule is fulfilled return q, α_k.

6.2.1.2 *CG iteration*

Step 3 in algorithm 6.1 is executed using the efficient CG method. The CG iteration computes a heat flux estimate $\hat{q}^{k+1}$ of q on the boundary Γ_3 as follows:

$$\hat{q}^{k+1}(\mathbf{x}, t) = \hat{q}^k(\mathbf{x}, t) - \mu^k p^k(\mathbf{x}, t), \text{ for } k = 0, 1, 2, \ldots, \tag{6.13}$$

where $\hat{q}^0 = 0$, $p^0 = \nabla J^0 |_{\Gamma_3}$. Here, μ^k represents the search step length in each iteration k. The conjugate search direction $p^k(\mathbf{x}, t)$ is updated as follows:

$$p^k(\mathbf{x}, t) = \nabla J^k(\mathbf{x}, t)|_{\Gamma_3} + \gamma^k p^{k-1}(\mathbf{x}, t), \tag{6.14}$$

where the gradient is determined by

$$\nabla J^k(\mathbf{x}, t)|_{\Gamma_3} = \psi^k(\mathbf{x}, t)|_{\Gamma_3} + \alpha\hat{q}^k. \tag{6.15}$$

The conjugate coefficient γ^k in (6.15) is determined from

$$\gamma^0 = 0, \gamma^k = \frac{\int_0^{t_f}\int_{\Gamma_3}[\nabla J^k]^2 \mathrm{d}\mathbf{x}\mathrm{d}t}{\int_0^{t_f}\int_{\Gamma_3}[\nabla J^{k-1}]^2 \mathrm{d}\mathbf{x}\mathrm{d}t}. \tag{6.16}$$

In each iteration, μ^k and ψ^k are determined by solving the sensitivity problem and the adjoint problem. The sensitivity problem is formulated as

$$\begin{cases} \rho c_p \dfrac{\partial S^k}{\partial t} = \nabla \cdot (\lambda \nabla S^k), \text{ in } \boldsymbol{\Omega} \times (0, t_f); \\ S^k(\cdot, 0) = 0, \quad \text{on } \boldsymbol{\Omega}; \\ \lambda \dfrac{\partial S^k}{\partial \mathbf{n}} = 0, \quad \text{on } \Gamma_1 \cup \Gamma_2 \times (0, t_f); \\ \lambda \dfrac{\partial S^k}{\partial \mathbf{n}} = p^k, \quad \text{on } \Gamma_3 \times (0, t_f), \end{cases} \tag{6.17}$$

and the search step length is determined by

$$\mu^k = \frac{\int_0^{t_f}\int_{\Gamma_1}[\boldsymbol{\Theta}(\mathbf{x},\, t;\, \hat{q}^k) - \boldsymbol{\Theta}_m^\delta(\mathbf{x},\, t)]S^k(\mathbf{x},\, t)\mathrm{d}\mathbf{x}\mathrm{d}t + \alpha\int_0^{t_f}\int_{\Gamma_3}\hat{q}^k p^k\mathrm{d}\mathbf{x}\mathrm{d}t}{\int_0^{t_f}\int_{\Gamma_1}[S^k(\mathbf{x},\, t)]^2\mathrm{d}\mathbf{x}\mathrm{d}t + \alpha\int_0^{t_f}\int_{\Gamma_3}[p^k]^2\mathrm{d}\mathbf{x}\mathrm{d}t}, \tag{6.18}$$

where $\boldsymbol{\Theta}(\mathbf{x},\, t; \hat{q}^k)$ represents the temperature solution of the forward problem (6.1) for $q = \hat{q}^k$. The resulting adjoint problem corresponds to

$$\begin{cases} \rho c_p \dfrac{\partial \psi^k}{\partial t} = -\nabla\cdot(\lambda\nabla\psi^k), & \text{in } \boldsymbol{\Omega}\times(0,\, t_f); \\ \psi^k(\cdot,\, t_f) = 0, & \text{on } \boldsymbol{\Omega}; \\ \lambda\dfrac{\partial\psi^k}{\partial\mathbf{n}} = 2(\boldsymbol{\Theta}_m^\delta - \boldsymbol{\Theta}), & \text{on } \Gamma_1\times(0,\, t_f); \\ \lambda\dfrac{\partial\psi^k}{\partial\mathbf{n}} = 0, & \text{on } (\Gamma_2\cup\Gamma_3)\times(0,\, t_f). \end{cases} \tag{6.19}$$

The final-time value problem (6.19) can be converted into a standard initial value problem by introducing a new time variable $t' = t_f - t$.

It is worth noting that if α is set to zero, the CG iteration described above becomes a specific case, as discussed in equation (6.19). Consequently, the results presented in this section extend the applicability of the CG-based solution method, enabling it to address the Tikhonov cost functional (6.6) as well.

6.2.2 Bayesian optimization-based method

Parts of this section have been reproduced with permission from [60]. Copyright 2024 Elsevier.

Bayesian inverse problems (BIPs) pertain to the inference of model parameters or latent variables' distributions within the Bayesian inference framework, grounded on observed data. Addressing BIPs often entails dealing with uncertainties in model parameters, the influence of noise, and constraints imposed by the observed data, all of which necessitate modeling and solution through Bayesian methods. In BIPs, prior distributions are typically assigned to parameters, and posterior distributions are updated based on the observed data. This process is encapsulated by Bayes' theorem as

$$\mathcal{P}(q|\boldsymbol{\Theta}_m^\delta) = \frac{\mathcal{P}(\boldsymbol{\Theta}_m^\delta|q)\mathcal{P}(q)}{\mathcal{P}(\boldsymbol{\Theta}_m^\delta)}, \tag{6.20}$$

where $\mathcal{P}(q \mid \Theta_m^\delta)$ represents the posterior probability density function (PPDF) of the parameter q given data $\boldsymbol{\Theta}_m^\delta$, $\mathcal{P}(\boldsymbol{\Theta}_m^\delta \mid q)$ is the likelihood probability density of the data given parameter q, $\mathcal{P}(q)$ is the prior probability density of the parameter, and $\mathcal{P}(\boldsymbol{\Theta}_m^\delta)$ is the marginal probability of the data.

In the 1950s, statistical physicists, such as Metropolis, integrated Bayesian theory into complex physical problems by employing the Markov chain Monte Carlo

(MCMC) sampling method for parameter inference. Bayesian theory derives the PPDF of unknown parameters using prior information and likelihood functions. Prior information can be obtained through experience or assumptions, while the likelihood function adjusts this prior information to account for errors, thereby yielding the PPDF.

Numerous deterministic optimization theories and algorithms have been extensively applied in solving inverse problems. With the rapid advancements in computational capabilities and the increasing demands for robustness and reliability, optimization under uncertainty has emerged as a significant focus in contemporary engineering research related to identification, design, and control. Solving BIPs facilitates a better understanding of the uncertainties in model parameters, optimizes model fitting and predictive capabilities, and enables the inference and prediction of system behavior. In practical applications, Bayesian methods offer a robust tool for a comprehensive analysis and interpretation of complex system behaviors.

Recent research has increasingly focused on the inversion and identification of one-dimensional (1D) or two-dimensional (2D) boundary heat flux [40–44], thermal property parameters [45], and heat source terms [9]. A prominent study in the field of IHTP is the Bayesian solution framework proposed by Wang *et al* [21, 46]. This research establishes posterior probability models for cases with known and unknown errors under sparse solving grids, employs Gibbs sampling for conditional distribution inference, and validates the method for boundary heat flux and thermal parameter identification in both 1D and 2D IHTP scenarios.

Addressing the limitations of traditional MCMC in high-dimensional posterior sampling, this study incorporates a more efficient probabilistic sampler, the Hamiltonian Monte Carlo (HMC) method. Initially proposed in 1987 by Duane, Kennedy, Pendleton, and Roweth, HMC draws inspiration from Hamiltonian dynamics in physics. It transforms the sampling problem into the simulation of Hamiltonian dynamics on a potential energy surface, generating samples by traversing trajectories on this energy surface [47]. By introducing momentum variables and leveraging gradient information, HMC more efficiently explores the probability distribution's structure during the sampling process. Neal *et al* [48] have further developed and popularized HMC, establishing it as a widely used sampling method. This method shows unique advantages in solving high-dimensional problems by proposing states based on Hamiltonian dynamics and implementing the leapfrog method. Proposed states may significantly differ from the current state but are highly likely to be accepted, thus avoiding random walk behaviors. Compared to traditional Gibbs samplers, HMC exhibits substantially higher sampling efficiency for high-dimensional distributions.

6.2.2.1 ***Bayesian formulation***

In this section, within the framework of the classic Bayesian model described by equation (6.20), specific configurations have been established for the prior model and likelihood function relevant to the problem at hand. The Markov random field (MRF), a type of probabilistic graphical model, is frequently employed in the fields of computer vision and pattern recognition, including applications such as image

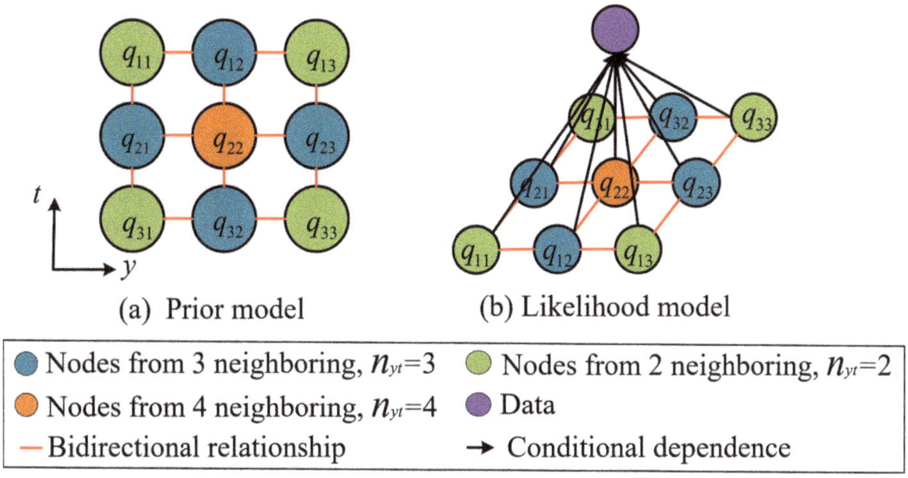

Figure 6.1. The relationship diagram between the Markov prior model and the likelihood probability model. (a) The probability graph of the prior variables and (b) the probability graph of the likelihood. (Reproduced with permission from [60]. Copyright 2024 Elsevier.)

segmentation, object recognition, and image denoising. This study utilizes the MRF for prior modeling, representing the MRF model of discrete random variables as a graph structure on a two-dimensional space–time plane. As depicted in figure 6.1(a), nodes correspond to image values, while edges denote relationships between nodes. By leveraging the Markov property [49], the dependencies among nodes are characterized, allowing for globally consistent modeling and inference of images. Figure 6.1(b) illustrates the probabilistic graphical relationship inherent to the likelihood. The discrete heat flux nodes q_{ij}, which we aim to determine, are systematically arranged on the two-dimensional space–time plane, with adjacency defined by nodes connected through a red undirected line segment. The assumption that each heat flux node interacts solely with its neighboring nodes, and not with non-adjacent nodes, encapsulates the Markov property. In Gaussian prior models, the covariance parameter settings can be configured to reflect this Markov property.

Assuming that the covariance matrix is denoted by $\mathbf{R} = \{r_{ij}\}_{M\times M}$, then

$$r_{ij} = \begin{cases} n_{yt}, & \text{if } i = j; \\ -1, & \text{if } q_i \text{ and } q_j \text{ are adjacent;} \\ 0, & \text{else;} \end{cases} \tag{6.21}$$

where n_{yt} represents the total number of adjacent nodes for each heat flux node on a 2D space–time plane. As illustrated in figure 6.1(a), positions on the 2D space–time plane are categorized into three types: corners, edges, and interiors. The corresponding values of n_{yt} are 2, 3, and 4, respectively. This enables us to present the specific expression of the prior model:

$$\mathcal{P}_0(q) \propto \lambda^{M/2} \exp\left(-\frac{\lambda}{2}\mathbf{q}^{\mathrm{T}}\mathbf{R}\mathbf{q}\right), \tag{6.22}$$

where λ controls the scale of the random vector distribution, where taking $\alpha = \lambda/2$ represents the regularization parameter. M is the length of the vector $\mathbf{q}$, which essentially corresponds to the finite element discretization scale.

The construction of a prior model is a critical step in Bayesian inference, profoundly influencing the model's accuracy. Prior information typically derives from domain-specific expertise and experience, and its value may be comparable to that of the data itself. The selection and incorporation of prior information to mitigate errors arising from data modeling require careful evaluation. If the prior information provided by domain experts is deemed more reliable, a greater proportion of it can be integrated into the dataset construction. In research emphasizing theoretical analysis, simple distribution forms, such as zero-mean Gaussian distributions, are often chosen as priors. However, for probabilistic models with Gaussian distribution forms, a zero-mean parameter implies the absence of a location parameter, thereby heavily relying on the data model for the inference of posterior parameters, specifically the likelihood probability model. To quantitatively assess the varying degrees of influence of prior information on inversion results, this study considers two prior scenarios: a zero-mean prior model and a scenario involving the construction of a prior dataset that serves as the mean parameter based on prior knowledge provided by domain experts.

In numerous engineering case studies, Gaussian white noise is commonly used to simulate the data noise of sensor measurements, so the specific form of $P_l(\cdot)$ is defined as

$$\mathcal{P}_l(\Theta_m^\delta|\mathbf{q}) \propto \exp\left\{\frac{-1}{2\sigma_y}\|\Theta_m^\delta - F(\mathbf{q})\|_2^2\right\}. \tag{6.23}$$

In fact, the likelihood function $\mathcal{P}_l(\cdot)$ essentially establishes a many-to-one probabilistic relationship from unknown quantities to data, as shown in figure 6.1(b).

By substituting the prior probability expression (6.22) and the likelihood probability (6.23) into the Bayes formula (6.20), the posterior distribution $\mathcal{P}_b(\cdot)$ can be obtained, which is specifically represented as

$$\mathcal{P}_b(\mathbf{q}|\Theta_m^\delta) \propto \exp\left[\frac{-1}{2v_\sigma}\|\Theta_m^\delta - F(\mathbf{q})\|_2^2 - \alpha\mathbf{q}^{\mathrm{T}}\mathbf{R}\mathbf{q}\right], \tag{6.24}$$

the expression $\|\cdot\|_2$ represents the Euclidean norm in the noise data space. It is evident that the posterior probability (6.24) is normally distributed.

6.2.2.2 Comparative analysis of three types of MCMC algorithms

The core concept of the MCMC algorithm is to iteratively update the current state, enabling the Markov chain to gradually converge to a smooth distribution. Common MCMC algorithms include the Metropolis–Hastings (MH) algorithm, the Gibbs sampling algorithm, and the HMC algorithm. These algorithms facilitate the sampling process under a smooth distribution through a series of state transition rules. The advantage of MCMC algorithms lies in their ability to handle various forms of probability distributions, making them particularly suitable for high-dimensional and complex problems. However, MCMC algorithms also face challenges such as slow convergence and difficult parameter tuning. As mentioned in the introduction, MCMC algorithms have been widely applied to solving IHTP [49–55].

In this section, we conduct posterior parameter inference based on three typical MCMC algorithms—MH, Gibbs, and HMC—for the established IHTP. We then quantify the computational efficiency of these three samplers using the effective sample size (ESS) method.

The MH algorithm is a classical MCMC method, originally proposed by Metropolis *et al* in 1953 and later refined and extended by Hastings in 1970. This refinement established the MH algorithm as a widely used MCMC sampling technique. The fundamental concept of the MH algorithm is to construct a Markov chain whose equilibrium distribution matches the target probability distribution. Based on the posterior probability density described in this chapter, the applicable MH computational procedure is outlined in algorithm 6.2. The MH algorithm is extensively used in Bayesian inference and statistical modeling, particularly for handling complex posterior distributions. For instance, Wang *et al* proposed a framework for parameter estimation of thermal objects in 2D IHTP using MH inference in 2015. However, the efficiency and convergence of the MH algorithm are influenced by the choice of the proposal distribution, necessitating careful tuning of the parameters to achieve optimal sampling results.

Algorithm 6.2. The MHg sampler for IHTP. (Reproduced with permission from [60]. Copyright 2024 Elsevier.)

Input: $\mathcal{P}_b(\cdot)$, $\mathcal{P}_0(\cdot)$, I, α, v_σ.
1: Set $\mathcal{P}_b(\cdot)$ as target density.
2: Initialize $\mathbf{q}_n^{(0)}$.
3: For $i = 1$: I do
4: $\mathbf{q}_{\text{proposal}} \sim \mathcal{P}\left(\cdot \mid \mathbf{q}_n^{(i-1)}\right)$.
5: Compute $\mathcal{P}_{\text{accept}}$.
6: If $\mathcal{P}_{\text{accept}} > u \sim \boldsymbol{U}(0, 1)$.
7: $\mathbf{q}_n^{(i)}$: $=\mathbf{q}_{\text{proposal}}$.
8: Else $\mathbf{q}_n^{(i)}$: $=\mathbf{q}_n^{(i-1)}$.
9: End for
Output: $\mathbf{q}_n^{(1:I)}$.

The Gibbs algorithm is an MCMC variant designed for sampling from joint probability distributions. Proposed by American physicist Josiah Willard Gibbs in 1984, it is a specialized MH algorithm for multi-dimensional probability distributions with known conditional distributions. The core idea of the Gibbs sampling method is to update each random variable iteratively using its full conditional probability density, thereby achieving sampling of the overall joint probability density. Wang *et al* [21] provided an inference method for the Gibbs full conditional distribution for IHTP based on a Bayesian approach to solving Newman's boundary-conditional heat conduction equations, demonstrating effective inverse validation on 1D and 2D sparse spatiotemporal solution grids. The computational flow is detailed in algorithm 6.3. Building on this research, this chapter develops a parallel solution strategy for 3D IHTP, which is shown in algorithm 6.7. The primary aim is comparative analysis, thus details are omitted in this section. The advantage of the Gibbs method lies in its sampling process, which does not require a proposal distribution but only necessitates the full conditional probability distribution for sampling. Additionally, Gibbs sampling methods typically exhibit fast convergence and straightforward implementation. These methods are widely applied in Bayesian statistics and machine learning, particularly for high-dimensional data and complex posterior distributions. However, Gibbs sampling also faces limitations, such as the difficulty of obtaining and computing conditional probability distributions and potential convergence issues.

Algorithm 6.3. The Gibbs sampler for IHTP. (Reproduced with permission from [60]. Copyright 2024 Elsevier.)

Input: $\mathcal{P}_b(\cdot)$, $\mathcal{P}_0(\cdot)$, I, α, v_σ.
1: Infer the full conditional density of $P_b(\cdot)$.
2: Initialize $\mathbf{q}_n^{(0)}$.
3: For $i = 1{:}\ I$ do
4: $q_1^{(i+1)} \sim \mathcal{P}\left(q_1 \mid q_2^{(i)}, q_3^{(i)}, \cdots, q_m^{(i)}\right)$.
5: $q_2^{(i+1)} \sim \mathcal{P}\left(q_1 \mid q_1^{(i+1)}, q_3^{(i)}, \cdots, q_m^{(i)}\right)$.
6: $\cdots$
7: $q_m^{(i+1)} \sim \mathcal{P}\left(q_1 \mid q_1^{(i+1)}, q_2^{(i+1)}, \cdots, q_{m-1}^{(i+1)}\right)$.
8: End for
9: Output: $\mathbf{q}_n^{(1:I)}$.

The HMC algorithm is a statistical sampling method based on Hamiltonian dynamics. It performs efficient sampling in the parameter space by simulating the laws of particle motion in a Hamiltonian dynamical system and introducing auxiliary variables. Specifically, the unknown quantity q is treated as the 'position' in the Hamiltonian system, while 'momentum' variables, controlling kinetic energy, are introduced as auxiliary variables. The proposed state is generated by solving the Hamiltonian dynamical equations, and the acceptance decision for this new state follows the MH criterion. The samples are obtained by iterating these steps. For the target probability density $\mathcal{P}_b(\cdot)$ established by equation (6.24), the introduced 'momentum' p is typically assumed to follow an independent multivariate

Gaussian distribution with zero mean. The diagonal elements of the covariance matrix are $v_1, \cdots v_m$. In this chapter, we denote its density function by $\mathcal{P}_k(\mathbf{p})$. Then the Hamiltonian joint probability density function $\mathcal{P}_{\mathrm{H}}(\cdot)$ can be expressed as

$$\mathcal{P}_{\mathrm{H}}(\mathbf{q}, \mathbf{p}) = \mathcal{P}_b(\mathbf{q}|\cdot) \cdot \mathcal{P}_k(\mathbf{p}). \tag{6.25}$$

It should be noted that $\mathcal{P}_b(\cdot)$ and $\mathcal{P}_k(\cdot)$ are also known as the canonical distributions of the states q and p. Let $\mathcal{U}$ and $\mathcal{K}$ denote the potential energy functiona and kinetic energy function in the Hamiltonian system, respectively. Then they can be obtained by taking the logarithm of their respective canonical distributions, i.e.

$$\mathcal{U}(\mathbf{q}) = -\log[\mathcal{P}_b(\cdot)],\ \mathcal{K}(\mathbf{p}) = -\log[\mathcal{P}_{\mathrm{k}}(\cdot)]. \tag{6.26}$$

In fact, from the point of view of the law of motion of the object, setting $\mathcal{P}_k(\cdot)$ as an independent Gaussian distribution with zero mean, one can make the kinetic energy function $\mathcal{K}(\cdot)$ have the following specific expression:

$$\mathcal{K}(\mathbf{p}) = \sum_{i=1}^{M} \frac{p_i^2}{2v_i}, \tag{6.27}$$

which corresponds to the quadratic kinetic energy function that describes the motion of a mass in classical mechanics.

In the implementation of HMC, q is sampled from $\mathcal{P}_{\mathrm{H}}(\cdot)$ via Metropolis updates. The invariance of the Hamiltonian energy constant H, needs to be guaranteed during each update of q and p. The general computational framework of the HMC algorithm of the IHTP is as described in algorithm 6.4.

Algorithm 6.4. The Hamiltonian Monte Carlo sampler for IHTP. (Reproduced with permission from [60]. Copyright 2024 Elsevier.)

Input: ϵ, L, I, $\mathcal{U}_n$, $\mathcal{K}$.
1: Initialize $\mathbf{q}_n^{(0)}$, $\mathbf{p}_n^{(0)}$.
2: For $i = 1$: I do
3: $\mathbf{q}_n = \mathbf{q}_n^{(i-1)}$; $\mathbf{p}_n = \mathbf{p}_n^{(i-1)}$.
4: $\mathbf{p}_n = \mathbf{p}_n - (\epsilon/2) \cdot \nabla_{\mathbf{p}}\ \mathcal{K}(\mathbf{p}_n)$.
5: For $j = 1$: L do
6: $\mathbf{q}_n = \mathbf{q}_n + \epsilon \cdot \nabla_{\mathbf{p}}\ \mathcal{K}(\mathbf{p}_n)$.
7: If $j \neq L$ Then $\mathbf{p}_n = \mathbf{p}_n - \epsilon \cdot \nabla_q\ U_n(\mathbf{q}_n)$.
8: End for
9: $\mathbf{p}_n = -\mathbf{p}_n + (\epsilon/2) \cdot \nabla_q\ \mathcal{U}_n(\mathbf{q}_n)$.
10: $\mathbf{P}^{(i-1)} = \mathcal{U}_n(\mathbf{q}_n^{(i-1)})$; $\mathbf{P}^{(i)} = \mathcal{U}_n(\mathbf{q}_n)$.
11: $\mathbf{K}^{(i-1)} = \mathcal{K}(\mathbf{p}_n^{(i-1)})$; $\mathbf{K}^{(i)} = \mathcal{K}(\mathbf{p}_n)$.
12: If $\exp(\mathbf{P}^{(i-1)} - \mathbf{P}^{(i)} + \mathbf{K}^{(i-1)} - \mathbf{K}^{(i)}) > u \sim \boldsymbol{U}(0, 1)$.
13: $\mathbf{q}_n^{(i)} = \mathbf{q}_n$.

14: Else
15: $\mathbf{q}_n^{(i)} = \mathbf{q}_n^{(i-1)}$.
16: End for
17: Output: $\mathbf{q}_n^{(1:I)}$.

The resolution of the parameter space of the 3D IHTP often involves a large number of unknowns. However, due to the uncertainty in the number of samples required for MCMC to converge, it is common to preset a relatively large sample value I to ensure the convergence of the sampling path. In solving such problems of interest in this section, the sample size I generally needs to be set with at least three orders of magnitude. At such a sampling scale, the total sample size required to reconstruct the heat flux in the 3D spatiotemporal domain will increase exponentially with an increase in unknowns. The HMC algorithm considered in this study directly updates $\mathbf{q}_n$ with Metropolis updates in each iteration, and the number of gradient computations determined by ϵ and L in these updates will have a decisive impact on the speed of HMC. However, the Gibbs sampler samples $q_1 \cdots q_m$ sequentially, and the computational cost of this serial computation is significantly higher than the gradient computation in each iteration of HMC [56].

ESS can be thought of as the number of independent samples required to obtain an estimate of the Monte Carlo mean with variance equal to the variance of the MCMC mean estimate. In Bayesian statistics, MCMC usually makes the samples non-independent. Therefore, the ESS is usually used instead of the actual sample size when determining whether the MCMC model has converged.

Let $\mathbf{q}^{1:I}$ denote I samples drawn from an approximate distribution of the target distribution P, and the ESS statistic is defined as the ratio of the true sample size to the autocorrelation time (AT), i.e.

$$\mathrm{ESS}_{\mathcal{P}}(\mathbf{q}^{1:I}) = \frac{I}{\tau}. \tag{6.28}$$

Autocorrelation time is used to quantify the convergence rate of means of Markov chains constrained by static, ergodic, and variance constraints. Various methods have been developed to calculate $\hat{\tau}$ from the data itself. This section adopts four reliable methods to comprehensively and quantitatively compare the sampling efficiency of HMC, Gibbs, and MH: batch mean [57], spectral fitting [58], initial convex sequence (ICS) [56], and AR modeling [59].

To quantitatively illustrate the selectivity of HMC in solving IHTP, a comparative analysis was conducted using a typical 1D IHTP boundary heat flux estimation case [61] with HMC, Gibbs, and MH. The convergence of sample paths generated by HMC, Gibbs, and MH at different sampling scales was evaluated using the four methods with $\hat{\tau}$, as shown in figure 6.2. From the graph, it is evident that for samples generated by Gibbs, all methods except batch mean converge in $\hat{\tau}$, whereas for samples generated by HMC, all four methods indicate convergence. In contrast, the stability of $\hat{\tau}$ calculated by the four methods for MH is significantly lower than for HMC. In terms of ESS estimates, HMC surpasses Gibbs and MH. Additionally,

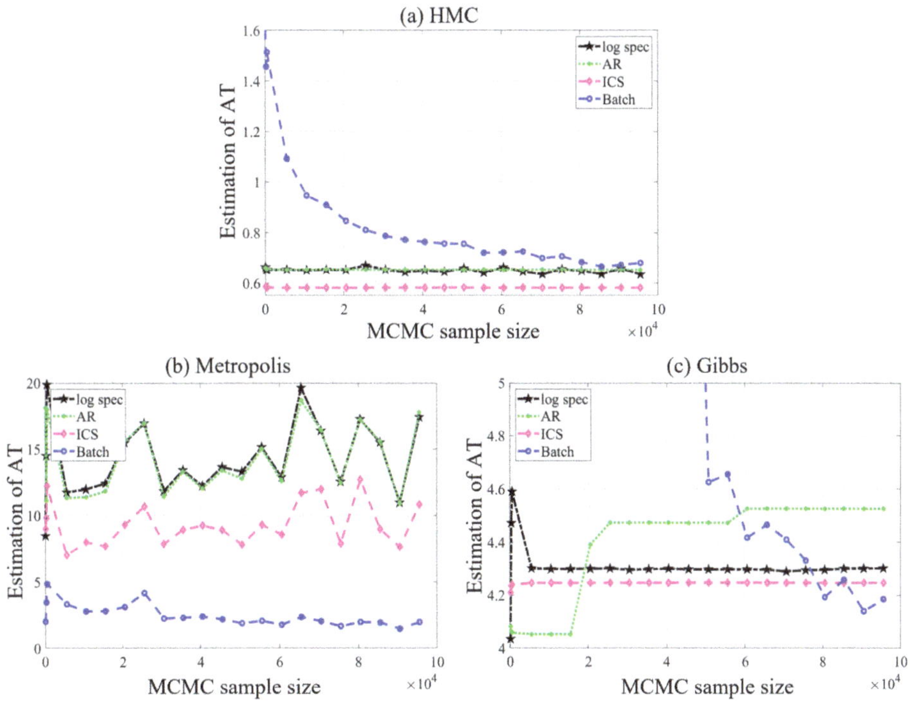

Figure 6.2. Sample convergence analysis for HMC, MH, and Gibbs. Panels (a)–(c) show the convergence analysis results for HMC, MH, and Gibbs, respectively. (Reproduced with permission from [60]. Copyright 2024 Elsevier.)

MH has the highest $\hat{\tau}$ among the three samplers, suggesting strong autocorrelation in the generated samples. Theoretically, under the same sample size setting, the samples from HMC exhibit a larger effective independent sample size compared to Gibbs and MH. Hence, in parameter inference for this type of IHTP, the HMC algorithm is considered more efficient compared to the MH and Gibbs algorithms.

6.2.2.3 ***Fast Bayesian parallel sampling***

The research focuses on the 3D IHTP solving domain $\Omega_s = (0, X) \times (0, Y) \times [0, t_f]$ discretized using the Galerkin method, where finite element interpolation includes the time dimension, making Bayesian calculations of unknown parameters extremely difficult, especially on high-density grids. To address this computational challenge, a fast Bayesian parallel sampling strategy (FBPS) is proposed, with its computational process outlined in algorithm 6.5. This solving strategy's model construction utilizes an MRF prior assumption and Gaussian white noise likelihood function-based posterior model, as described by equation (6.24). For sampler selection, as discussed in section 6.2.2.2, a more efficient HMC algorithm is employed for parameter inference in the corresponding PPDF subtasks in each parallel process.

In order to speed up the solution, we slice the solution region to reduce the dimensionality, i.e. the 3D problem is transformed into a number of 2D problems to be solved in parallel. Let n_{cut} denote the number of slices, and $\mathbf{U}(.)$ and $\mathbf{V}(.)$ denote the mean and variance parameters of the corresponding 2D posterior probability densities, respectively.

Algorithm 6.5. FBPS. (Reproduced with permission from [60]. Copyright 2024 Elsevier.)

Input: $\mathcal{P}_b(\cdot)$, $\mathcal{P}_0(\cdot)$, ϵ, L, I, $\mathcal{K}$, α, v_σ.
1: S1: Start n_{cut} parallel processes.
2: S2: Parallel compute $\mathbf{U}_1, \cdots, \mathbf{U}_{n_{\text{cut}}}; \mathbf{V}_1, \cdots, \mathbf{V}_{n_{\text{cut}}}; \mathcal{U}_1, \cdots, \mathcal{U}_{n_{\text{cut}}}$.
3: S3: Parallel execute algorithm 6.4 and generate $\mathbf{q}_1^{(1:I)}, \cdots, \mathbf{q}_{n_{\text{cut}}}^{(1:I)}$.
4: Calculate the posterior estimate value $\mathbf{q}_1^{\text{MAP}}, \cdots, \mathbf{q}_{n_{\text{cut}}}^{\text{MAP}}, \mathbf{q}_1^{\text{mean}}, \cdots, \mathbf{q}_{n_{\text{cut}}}^{\text{mean}}$.
5: S4: Merge $\mathbf{q}_1^{\text{MAP}}, \cdots, \mathbf{q}_{n_{\text{cut}}}^{\text{MAP}}$ obtain $\mathbf{q}^{\text{MAP}}$; merge $\mathbf{q}_1^{\text{mean}}, \cdots, \mathbf{q}_{n_{\text{cut}}}^{\text{mean}}$ obtain $\mathbf{q}^{\text{mean}}$.
6: Output: $\mathbf{q}^{\text{MAP}}, \mathbf{q}^{\text{mean}}$.

Algorithm 6.6. MH-based high-throughput computing framework. (Reproduced with permission from [60]. Copyright 2024 Elsevier.)

Input: $\mathcal{P}_b(\cdot)$, $\mathcal{P}_0(\cdot)$, I, α, v_σ.
1: S1: Start n_{cut} parallel processes.
2: S2: Parallel compute $\mathbf{U}_1, \cdots, \mathbf{U}_{n_{\text{cut}}}; \mathbf{V}_1, \cdots, \mathbf{V}_{n_{\text{cut}}}$.
3: S3: Parallel execute algorithm 6.2 and generate $\mathbf{q}_1^{(1:I)}, \cdots, \mathbf{q}_{n_{\text{cut}}}^{(1:I)}$.
4: Calculate the posterior estimate value $\mathbf{q}_1^{\text{MAP}}, \cdots, \mathbf{q}_{n_{\text{cut}}}^{\text{MAP}}, \mathbf{q}_1^{\text{mean}}, \cdots, \mathbf{q}_{n_{\text{cut}}}^{\text{mean}}$.
5: S4: Merge $\mathbf{q}_1^{\text{MAP}}, \cdots, \mathbf{q}_{n_{\text{cut}}}^{\text{MAP}}$ obtain $\mathbf{q}^{\text{MAP}}$; merge $\mathbf{q}_1^{\text{mean}}, \cdots, \mathbf{q}_{n_{\text{cut}}}^{\text{mean}}$ obtain $\mathbf{q}^{\text{mean}}$.
6: Output: $\mathbf{q}^{MAP}, \mathbf{q}^{\text{mean}}$.

Algorithm 6.7. Gibbs-based high-throughput computing framework. (Reproduced with permission from [60]. Copyright 2024 Elsevier.)

Input: $\mathcal{P}_b(\cdot)$, $\mathcal{P}_0(\cdot)$, I, α, v_σ.
1: S1: Start n_{cut} parallel processes.
2: S2: Parallel execute algorithm 6.3 and generate $\mathbf{q}_1^{(1:I)}, \cdots, \mathbf{q}_{n_{\text{cut}}}^{(1:I)}$.
3: Calculate the posterior estimate value $\mathbf{q}_1^{\text{MAP}}, \cdots, \mathbf{q}_{n_{\text{cut}}}^{\text{MAP}}, \mathbf{q}_1^{\text{mean}}, \cdots, \mathbf{q}_{n_{\text{cut}}}^{\text{mean}}$.
4: S3: Merge $\mathbf{q}_1^{\text{MAP}}, \cdots, \mathbf{q}_{n_{\text{cut}}}^{\text{MAP}}$ obtain $\mathbf{q}^{\text{MAP}}$; merge $\mathbf{q}_1^{\text{mean}}, \cdots, \mathbf{q}_{n_{\text{cut}}}^{\text{mean}}$ obtain $\mathbf{q}^{\text{mean}}$.
5: Output: $\mathbf{q}^{\text{MAP}}, \mathbf{q}^{\text{mean}}$.

Furthermore, a high-throughput computational framework based on Gibbs and MH is developed to provide a fair comparison of the efficiency between both and

FBPS. The high-throughput frameworks for Gibbs and MH are illustrated in algorithms 6.7 and 6.6, respectively. Notably, the inherent serial nature of the Gibbs method necessitates a combination of domain splitting and high-throughput parallel techniques to make large-scale 3D IHTP solving feasible. Nevertheless, compared to the HMC algorithm, Gibbs exhibits lower efficiency, as evident from the results presented in table 6.1. The table shows the computational times of algorithms 6.5–6.7 across several IHTP cases, highlighting the varying efficiency across different grid densities.

From the results displayed in table 6.1, it can be observed that FBPS can achieve an estimation of unknown heat flux of dimensions $152 \times 63 \times 11$ on a high-density grid such as in case 3 within minutes. For sparse grids as in case 1, FBPS efficiently computes on a standard desktop machine with eight cores. Even on refined grids such as in case 2, FBPS maintains a calculation time within minutes. Maximizing the value of n_{cut} within the given range of measurement data can enhance result integrity but demands more computational power, i.e. additional parallel nodes and cores. As shown in case 3, with a roughly 2.4 times increase in n_{cut} compared to case 2 on the same grid scale, maintaining nearly the same computation time necessitates an about 4.2 times increase in total core count (with five parallel nodes). The computation durations of the three methods from table 6.1 highlight that FBPS is approximately 55 times faster than the Gibbs method and around five times faster than the MH method in solving 2D/3D IHTP.

To quantitatively compare the solving accuracy of the FBPS, Gibbs, and MH algorithms, a relative error analysis is conducted on a 3D IHTP case of chip heat transfer, with the results detailed in table 6.2. Here, the relative error formula is given by

$$e_r = |(\hat{q} - q^h)/q^h| \tag{6.29}$$

and values of e_r^{min}, e_r^{mean}, and e_r^{max} represent the minimum, mean, and maximum relative error at each time-stamp. Across the studied timeframe of 10 s simulating boundary heat flux density, with a spatial–temporal grid size of $61 \times 61 \times 21$, the

Table 6.1. The CPU computing time of FBPS (algorithm 6.5), MH (algorithm 6.6), and Gibbs (algorithm 6.7) of cases 1–3 for $I = 5 \times 10^3$. (Reproduced with permission from [60]. Copyright 2024 Elsevier.)

	Case 1 (2D)	Case 2 (3D)	Case 3 (3D)
Ω scale	1×0.05 mm^2	$1 \times 1 \times 0.05$ mm^3	$1 \times 1 \times 0.05$ mm^3
Number of processes	/	63	152
Slices scale n_{cut}	/	63	152
Grid scale	26×11	63×11	63×11
Core and node	8 cores/1 node	36 cores/1 node	150 cores/5 nodes
CPU time (FBPS: algorithm 6.5)	0.849 min	10.133 min	9.985 min
CPU time (MH: algorithm 6.6)	4.628 min	32.168 min	45.297 min
CPU time (Gibbs: algorithm 6.7)	46.855 min	561.901 min	520.071 min

Table 6.2. Comparison of the relative error quantification statistics for the estimated chip heat flux density results in the first 10 s of the chip heat transfer case using FBPS (algorithm 6.5), Gibbs (algorithm 6.7), and MH (algorithm 6.6). (Reproduced with permission from [60]. Copyright 2024 Elsevier.)

Time (s)		FBPS	Gibbs	MH
1 s	e_r^{mean}	0.57%	0.61%	0.58%
	e_r^{max}	5.04%	4.57%	5.11%
	e_r^{min}	6.00×10^{-5}%	1.90×10^{-4}%	2.76×10^{-5}%
2 s	e_r^{mean}	0.60%	0.62%	0.60%
	e_r^{max}	2.80%	3.08%	2.61%
	e_r^{min}	4.82×10^{-4}%	4.04×10^{-5}%	3.07×10^{-5}%
3 s	e_r^{mean}	1.08%	1.08%	1.08%
	e_r^{max}	4.96%	4.90%	4.81%
	e_r^{min}	2.80×10^{-4}%	5.63×10^{-4}%	1.54×10^{-3}%
4 s	e_r^{mean}	1.41%	1.41%	1.41%
	e_r^{max}	6.21%	6.22%	6.11%
	e_r^{min}	3.93×10^{-4}%	5.35×10^{-4}%	9.05×10^{-6}%
5 s	e_r^{mean}	1.67%	1.68%	1.68%
	e_r^{max}	7.12%	7.04%	7.05%
	e_r^{min}	8.73×10^{-5}%	2.05×10^{-4}%	1.10×10^{-3}%
6 s	e_r^{mean}	1.92%	1.92%	1.92%
	e_r^{max}	7.92%	7.99%	7.87%
	e_r^{min}	4.38×10^{-4}%	1.38×10^{-3}%	7.54×10^{-4}%
7 s	e_r^{mean}	2.15%	2.15%	2.15%
	e_r^{max}	8.59%	8.58%	8.56%
	e_r^{min}	2.48×10^{-3}%	8.33×10^{-4}%	1.15×10^{-3}%
8 s	e_r^{mean}	2.32%	2.32%	2.32%
	e_r^{max}	8.90%	8.93%	8.87%
	e_r^{min}	6.96×10^{-4}%	4.08×10^{-4}%	2.26×10^{-3}%
9 s	e_r^{mean}	2.39%	2.39%	2.39%
	e_r^{max}	8.27%	8.37%	8.25%
	e_r^{min}	2.11%	6.59×10^{-4}%	1.31×10^{-3}%
10 s	e_r^{mean}	2.40%	2.40%	2.40%
	e_r^{max}	8.88%	8.97%	8.88%
	e_r^{min}	5.46×10^{-4}%	4.39×10^{-4}%	1.40×10^{-4}%

statistical analysis of the relative errors indicates relatively comparable results for all three methods, with a maximum error of less than 9% over the 10 s simulation period. This underscores the reliability of the posterior model (6.24) despite the absence of positional parameters in the prior density, demonstrating its effectiveness in this linear case study.

When setting up parallel computing, understanding the available CPU or core count can be a beneficial, albeit a somewhat vague concept. Currently, most physical

CPUs come with two or more independent running cores that may share some cache and memory access. Nonetheless, some processors allow these cores to handle multiple tasks simultaneously, with operating systems such as Windows featuring the concept of logical CPUs, which may outnumber physical cores. Consequently, in configuring parallel node resources, the choice of node configuration is based on the size of slices in a given case—the larger the slice count, the more physical CPU cores are included in the computation node.

The FBPS strategy proposed in this research is applicable for ultra-high-dimensional parameter estimation in large-scale 3D IHTP scenarios. Through dimensionality reduction of the spatiotemporal problem domain, parameter calculations spanning five orders of magnitude can be achieved within minutes. Utilizing the HMC sampler has enabled more efficient estimation in a high-dimensional parameter space compared to traditional MCMC methods. Through simulation testing on an engineering case study, the superiority of FBPS in solving efficiency over traditional Bayesian strategies is demonstrated, with the potential for efficiency enhancement up to 55 times. The improvement in reliability and computational efficiency of FBPS is considerable.

6.3 Applications and analysis

This section primarily discusses the applications of the above inversion solution strategies (algorithms 6.1 and 6.5) in different types of engineering IHTPs. To assess the inversion effects of the strategies under varying levels of noise contamination, four different noise levels, ζ, were defined. These ζ levels were set as 0.1 K, 0.2 K, 0.3 K, and 0.4 K, where v_ζ and δ_ζ represent the noise variance and white noise random variable when the noise level is ζ, respectively. The simulated noisy data $\boldsymbol{\Theta}_m^\delta$ are defined accordingly,

$$\begin{aligned} \boldsymbol{\Theta}_m^\delta &:= \boldsymbol{\Theta}_m + \delta_\zeta,\ \ \delta_\zeta \sim \mathcal{N}(0, v_\zeta), \\ v_\zeta &= \zeta \cdot \left(\max \boldsymbol{\Theta}_m - \min \boldsymbol{\Theta}_m\right). \end{aligned} \tag{6.30}$$

6.3.1 Chip heat dissipation

The thermal management of microelectronic components is a pivotal issue directly affecting device performance and reliability. Accurately measuring the temperature of microstructures under real-world conditions has long posed significant challenges [62], with the assessment of temperature gradients being even more complex. Traditional temperature sensing technologies, including thermocouples, thermal resistors, infrared thermometers, and fiber optic temperature sensors, are inadequate for directly providing heat flux information. Consequently, the advancement of soft measurement techniques has introduced various solutions to these challenges. Soft measurement techniques based on IHTP have demonstrated substantial potential, markedly reducing experimental demands and yielding precise thermal physical quantities, especially in scenarios where direct measurement is impractical or prohibitively expensive. Figure 6.3(a) shows a physical photograph of the chip

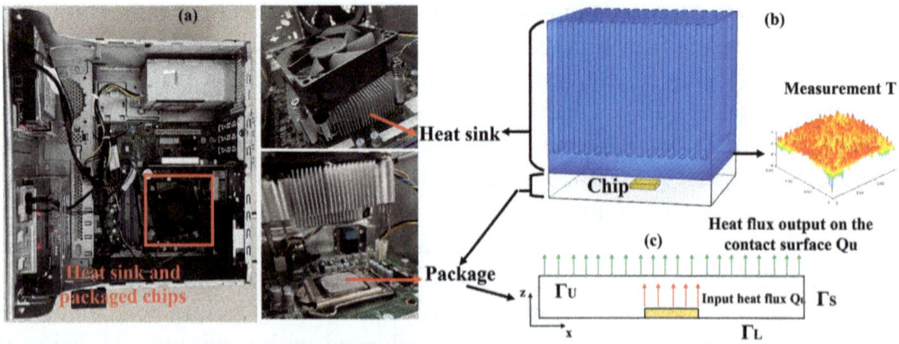

Figure 6.3. Schematic diagram of the structure and location of the semiconductor chip and heat sink assembly. (a) Physical images of the host chip and heat sink, (b) geometric diagram of the chip and heat sink, and (c) x–z cross-sectional diagram of the chip packaging structure. (Reproduced with permission from [60]. Copyright 2024 Elsevier.)

heat sink. The simulated structure of the chip, package and heat sink are shown in figure 6.3(b). The x–z-direction projection of the geometric domain Ω is shown in figure 6.3(c).

Due to the heat generated by the chip, which is first transferred to the heat sink through the contact surface, the heat flux on this surface is a crucial quantity for evaluating the heat dissipation effect. Additionally, it provides recommendations for optimizing heat sink structural design. The modeling of the chip and its packaging structure is initiated. The regular packaging structure is considered as a continuous 3D geometric domain $\Omega:=[0, l_x] \times [0, l_y] \times [0, l_z]$, with the x–z-plane projection of Ω shown in figure 6.3(c). Semiconductor chips generally sit on the internal bottom surface of the package and, during operation, the chip generates heat leading to an increase in temperature. In order to enhance heat dissipation and ensure the chip's normal lifespan, heat sink devices are typically installed on semiconductor components. Figure 6.3 illustrates the chip, packaging, and heat sink structure within a computer host system, with figure 6.3(a) showcasing the physical structure of the semiconductor component and the heat sink device.

Assume that the heat flux generated by the semiconductor chip enters Ω from the boundary Γ_1 from bottom to top, and then diffuses into the heat sink through the boundary Γ_3, represented by the blue sheet-like structure in figure 6.3(b). The distribution of heat flux on the Γ_3 surface quantifies the heat transfer efficiency during the chip's heat transfer process, providing valuable insights for heat sink structural design and material selection. Therefore, if high-resolution temperature measurement data can be obtained on the Γ_3 surface in experiments, the goal is to reconstruct the heat flux density field on the Γ_3 surface at a scale comparable to the measurement data.

The geometrical parameters of the forward model and the thermophysical parameters of the material for this example are shown in table 6.3. In order to systematically study the heat flux density in this example, the analytical expression for the theoretical assumptions of the heat flux density is given in this study based on

Table 6.3. Geometry and thermophysical parameters of the solution domain. (Reproduced with permission from [60]. Copyright 2024 Elsevier.)

	Parameter	Symbol	Unit	Value
Geometry	Length of pack	l_x	m	0.03
	Width of pack	l_y	m	0.03
	Height of pack	l_z	m	0.01
	Length of chip	$l_{x_2} - l_{x_1}$	m	0.005
	Width of chip	$l_{y_2} - l_{y_1}$	m	0.005
Thermophysical	Thermal conductivity	k	$W \cdot (m \cdot K)^{-1}$	148
Properties	Thermal diffusivity	a	$m^2 \cdot s^{-1}$	8.92×10^{-5}

the literature [62]. Let the upper surface heat flux $Q_3(x, y, t)$ be analytically expressed over the region $[0, l_x] \times [0, l_y] \times [0, l_z]$,

$$Q_3(x, y, t) := \omega(x, y) \cdot \tau(t), \tag{6.31}$$

where the spatial distribution function of the heat flux $\omega(x, y)$ is

$$\omega(x, y) = (\cos^2(\pi x/l_x)\sin^2(\pi y/l_y) + \cos^2(\pi y/l_y)\sin^2(\pi x/l_x)) + 0.5. \tag{6.32}$$

The peak value of the heat flux as a function of time is

$$\tau(t) = (1 - e^{-0.1t}). \tag{6.33}$$

It is assumed that the heat flux Q_1 input from the lower bottom surface is generated by the chip of uncalculated thickness with constant input to the interior of the region. Therefore, considering only the constant heat flux input within the area of the chip, Q_1 is defined over the region $[0, l_x] \times [0, l_y] \times [0, l_z]$ as

$$Q_1\left(x, y\right) = 1 \times 10^5, \left(x, y\right) \in \left[l_{x_1}, l_{y_1}\right] \times \left[l_{x_2}, l_{y_2}\right]. \tag{6.34}$$

The distribution of the above heat flux function $Q_1(x, y, t)$ for the first 10 s is shown in figure 6.4.

In the implementation presented, a thermal flux distribution on the upper surface over a total time of 10 s was reconstructed, with $t_f = 10$ s being set. For the forward simulation generating measurement data, a time interval of 0.5 s was chosen, resulting in a total of 21 frames of simulated measurement data. A square solving grid with a step size of 5×10^{-4} m was utilized spatially. Under the aforementioned grid settings, the values at all grid points on the upper surface were taken as measurement data, with a total scale in the x, y, and t directions of $61 \times 61 \times 21$. Figure 6.5 displays the temperature distribution simulated by equation (6.1), showcasing selected time frames at 2, 4, 6, 8 and 10 s, with the temperature distributions at these moments under four noise levels also illustrated.

The inversion grid was obtained using a spatiotemporal finite element method that discretizes the time dimension as a spatial dimension, as seen in the following :

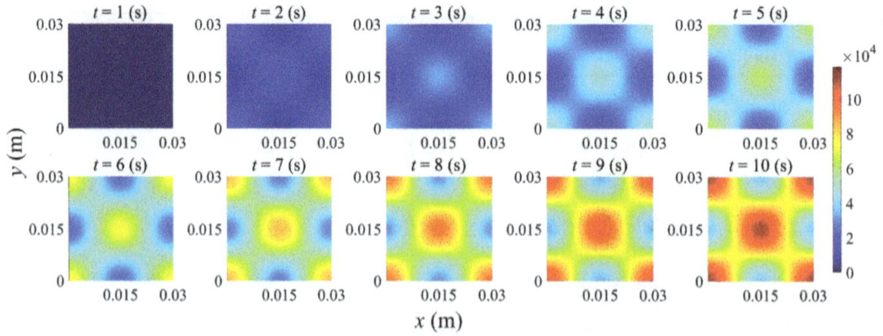

Figure 6.4. Theoretically assumed distribution of heat flux on the upper surface of the chip package (heat flux unit: W·m^{-2}). (Reproduced with permission from [60]. Copyright 2024 Elsevier.)

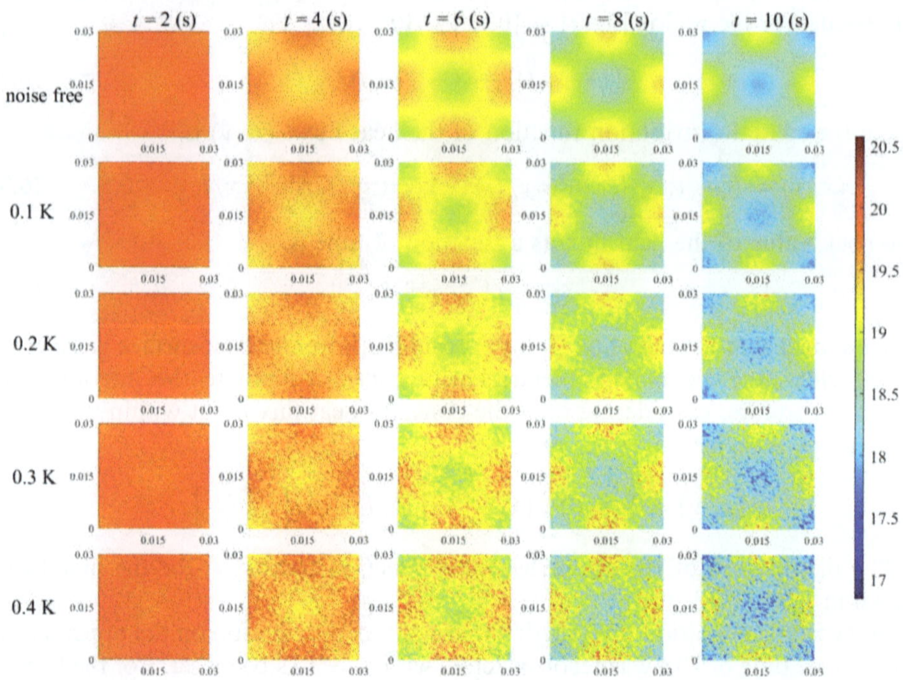

Figure 6.5. Measured temperature distribution simulated on the top surface of the chip package (temperature unit: °C). (Reproduced with permission from [60]. Copyright 2024 Elsevier.)

$$q(y, t) = \sum_{i-1}^{M} q_i B_i(y, t), \quad n \in \{1, \cdots, n_{\text{cut}}\}. \tag{6.35}$$

To achieve accurate thermal flux reconstruction with high grid density, the inversion grid size matched that of the forward simulation. Similarly, the number of slices depended on the forward simulation grid size, aligning the slice count with the scale in the x-direction of the measurement data. High-performance computing employed a host with 36 CPU cores per node, with 61 parallel processes set up.

Table 6.4 provides the optimal regularization parameters and input parameters to the algorithm for four noise levels. The MAP estimate of the surface thermal flux solved using FBPS is depicted in figure 6.6 for the four noise levels. In practical

Table 6.4. Algorithmic input parameters for the chip heat transfer calculus. (Reproduced with permission from [60]. Copyright 2024 Elsevier.)

Parameter	Symbol	Value
Time steps in leapfrog	ϵ	0.01
Total time of leapfrog	L	45
MCMC iterations	I	5×10^{-3}
Noise variance	v_σ	0.05
Regularization parameter	α	0.1 K: 8×10^{-11}
		0.2 K: 2×10^{-10}
		0.3 K: 7×10^{-10}
		0.4 K: 1×10^{-9}

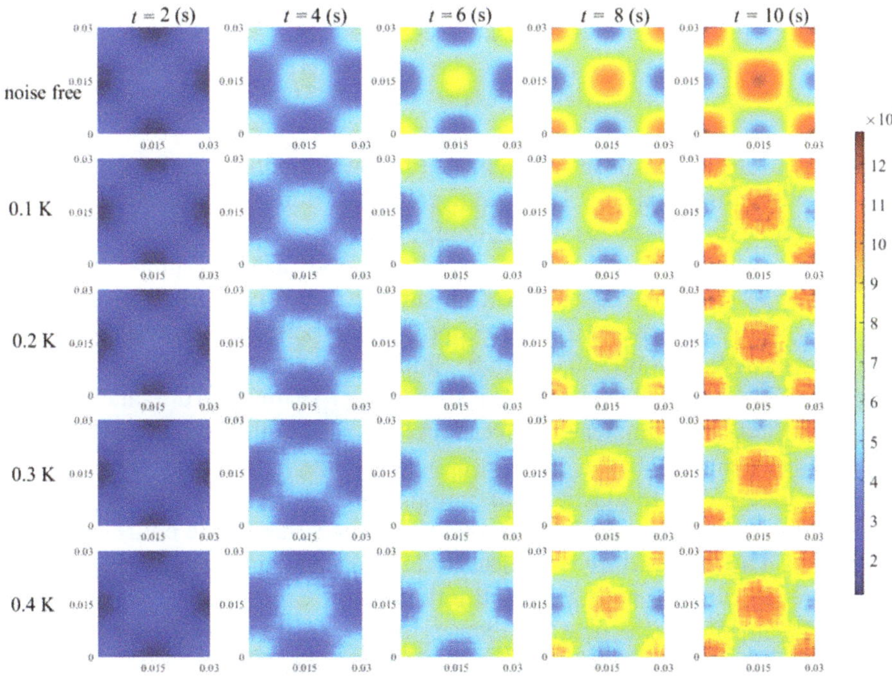

Figure 6.6. Heat flux results for the upper surface of a chip package reconstructed by the FBPS at four different noise levels (heat flux units: W·m^{-2}). (Reproduced with permission from [60]. Copyright 2024 Elsevier.)

engineering measurements, noise levels around 0.1 K are typically considered normal for sensor measurements. Overall, FBPS demonstrates good estimation performance for measurement data with normal noise levels in the absence of prior data. However, at higher noise levels, there is a slight decrease in estimation performance. Generally, selecting too small a value for α can lead to oscillations in the thermal flux density results, while too large a value can cause excessive smoothing. From the inversion results in table 6.4, it is evident that higher noise data necessitates larger regularization parameters for smoothing.

6.3.2 Pool boiling

In a typical pool boiling process, a constant heat flux is applied to the heating surface, while unknown heat flux on the boiling surface undergoes instantaneous changes. Estimating the difficult-to-measure boiling heat flux involves solving the IHTP by measuring temperature data on the heating surface. Solving this transient 3D IHTP is vital to accurately estimate the transient local boiling heat flux density distribution on the heater surface, providing physically meaningful boundary conditions required for simulating two-phase movement in boiling, essential for precise modeling of the entire boiling process and conducting heat transfer mechanism analysis.

This section will focus on the practical issues encountered in the single bubble nucleate boiling experiments conducted by Wagner *et al* [63, 64]. Two types of thin stainless-steel heating foils with thicknesses of 50 and 25 μm were used in the experiments. The boiling fluid chosen was refrigerant HFC 7100. Here we employed the material thermal properties from the study by Heng *et al* [39] for inverse calculations.

In the first set of experiments, a heating foil with dimensions of $2.416 \times 2.288 \times 0.05\ \mathrm{mm}^3$ was considered as a 3D regular geometric domain. The spatial–temporal discretization scale chosen for FBPS computation matched the resolution of the measurement system (pixel scale of 152×144, unit pixel size of 16 μm × 16 μm, and a frame rate of 987 Hz). A constant bottom surface heat flux $Q_1 = 5356\ \mathrm{W\,m^{-2}}$ was used in this set of experiments. In the second set of experiments, the heating foil had dimensions of $5.104 \times 4.08 \times 0.025\ \mathrm{mm}^3$, and the measurement system acquired a total pixel scale of 320×256, with a unit pixel size of 16 μm × 16 μm, and frame rate of 987 Hz. A relatively higher input heat flux $Q_1 = 12\,900\ \mathrm{W\,m^{-2}}$ was used in this group of experiments.

In both cases, due to the lack of initial temperature information across the domain, it was assumed that the initial temperature value of the entire heating foil was uniform. The temperature field measured in the first frame was used as the initial temperature distribution across the entire domain.

Figure 6.7 illustrates the heat flux density on the surface of a 25 μm heating foil estimated through the inverse estimation of FBPS (algorithm 6.5). We selected all instances of heat flux occurrence, namely from 0.002 s to 0.006 s, and from 0.014 s to 0.018 s, totaling ten frames. It is evident that the peak of heat flux occurs around 0.004 s. Figure 6.8 displays the heat flux density on the surface of a 50 μm heating

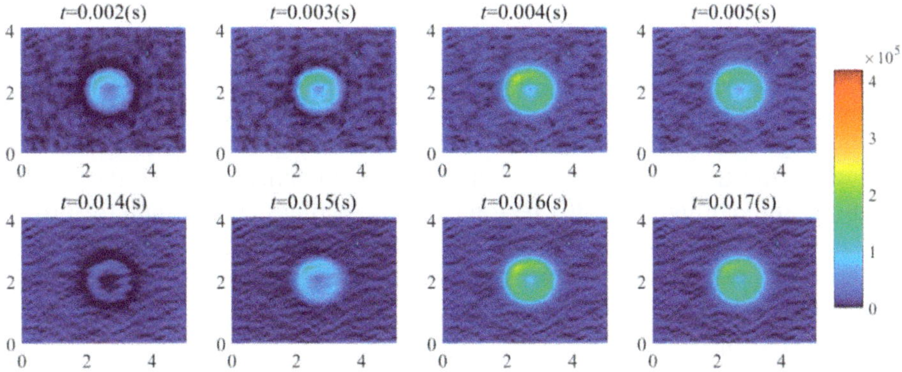

Figure 6.7. Boiling surface heat flux reconstructed by the FBPS (algorithm 6.5) for the 25 μm arithmetic case (first row from 0.002 to 0.006 s, second row from 0.014 to 0.018 s, in W·m^{-2})

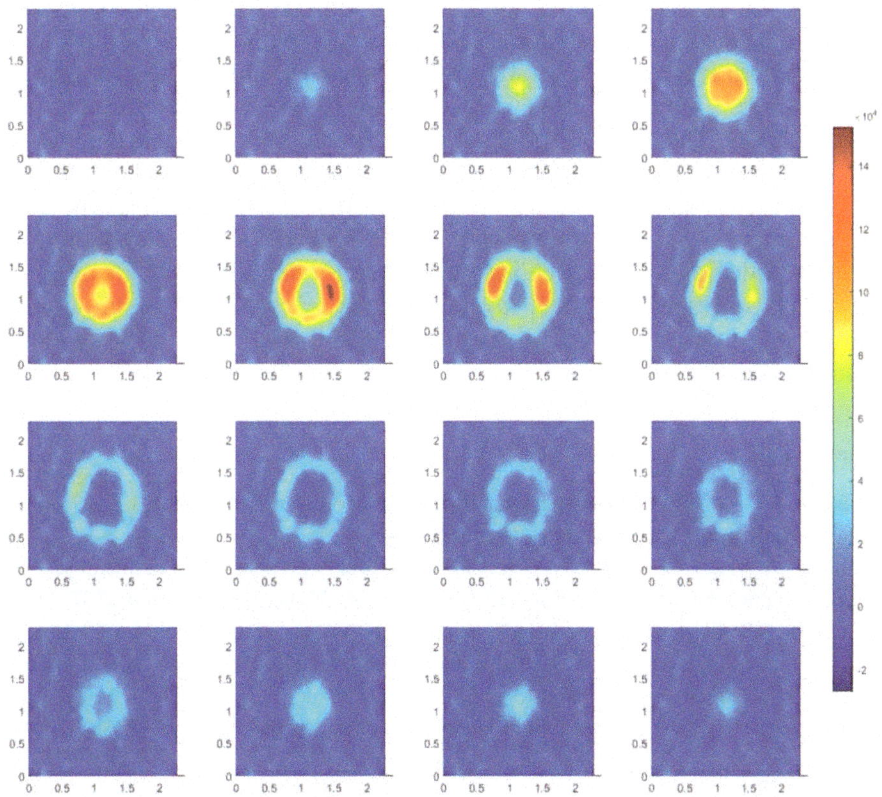

Figure 6.8. Boiling surface heat flux reconstructed by the improved Tikhonov method (algorithm 6.1) for the 50 μm arithmetic case (from 0.020 s to 0.036 s, in W·m^{-2}). (Reproduced with permission from [39]. Copyright 2012 Elsevier.)

foil estimated using an improved Tikhonov method (algorithm 6.1). The estimation results reveal significant variations in transient boiling heat flux over a single bubble cycle. During time frames 22, 23, and 24, bubbles form and continuously expand, resulting in increasingly larger Gaussian-type high heat flux distributions. Subsequently, the bubbles continue to expand and rise, eventually detaching completely from the contact surface. During the detachment process, the estimated peak of the annular high heat flux region is nearly 30 times greater than the provided input heat flux. These estimated heat flux values, which cannot be directly measured in experiments, validate micro-layer theory [65]. The theory posits that a significant portion of heat during the boiling process is transferred through evaporation in the micro-region at the three-phase contact line.

6.4 Conclusions

This study addresses the technical challenges of solving large-scale Bayesian inverse problems by introducing a robust and efficient FBPS solver tailored for 3D IHTP applications in regular geometric domains, including pool boiling and chip cooling. By mitigating the high computational costs associated with high-dimensional inverse problems within the Bayesian Monte Carlo framework, the study offers a comprehensive parallel framework that demonstrates its effectiveness through a chip packaging case. The FBPS strategy optimizes the speed and accuracy of the HMC sampler, attracting attention from researchers interested in enhancing HMC parameters such as ε and L, similar to the NUTS (No-U-Turn Sampler) implementation in *Stan*[1]. Future research directions include developing automatic parameter tuning methods for enhanced computational efficiency, creating a non-linear IHTP solver for complex inverse parameter estimation in unstable PDE scenarios, and applying FBPS to irregular regional challenges, such as biometric heat source identification.

References

[1] Sladek J *et al* 2012 Inverse heat conduction problems in three-dimensional anisotropic functionally graded solids *J. Eng. Math.* **75** 157–71

[2] Zhang C *et al* 2015 A model of regularization parameter determination in low-dose *x*-ray CT reconstruction based on dictionary learning *Comput. Math. Methods Med.* **2015** 831790

[3] Czél B, Woodbury K A and Gróf G 2014 Simultaneous estimation of temperature-dependent volumetric heat capacity and thermal conductivity functions via neural networks *Int. J. Heat Mass Transfer* **68** 1–13

[4] Zhang Q *et al* 2016 Determination of temperature dependent thermophysical properties using an inverse method and an infrared line camera *Int. J. Heat Mass Transfer* **96** 242–8

[5] Cui M, Zhu Q and Gao X 2014 A modified conjugate gradient method for transient nonlinear inverse heat conduction problems: a case study for identifying temperature-dependent thermal conductivities *J. Heat Transfer* **136** 091301

[1] Stan: Software for Bayesian Data Analysis https://mc-stan.org/.

[6] Bourquin F and Nassiopoulos A 2011 Inverse reconstruction of initial and boundary conditions of a heat transfer problem with accurate final state *Int. J. Heat Mass Transfer* **54** 3749–60
[7] Liu C-S 2011 A self-adaptive LGSM to recover initial condition or heat source of one-dimensional heat conduction equation by using only minimal boundary thermal data *Int. J. Heat Mass Transfer* **54** 1305–12
[8] Qiu S, Zhang W and Peng J 2018 Simultaneous determination of the space-dependent source and the initial distribution in a heat equation by regularizing Fourier coefficients of the given measurements *Adv. Math Phys.* **2018** 8247584
[9] Hasanov A 2012 Identification of spacewise and time dependent source terms in 1D heat conduction equation from temperature measurement at a final time *Int. J. Heat Mass Transfer* **55** 2069–80
[10] Lee K H 2019 Application of repulsive particle swarm optimization for inverse heat conduction problem—parameter estimations of unknown plane heat source *Int. J. Heat Mass Transfer* **137** 268–79
[11] Yang F and Fu C-L 2010 The method of simplified Tikhonov regularization for dealing with the inverse time-dependent heat source problem *Comput. Math. Appl.* **60** 1228–36
[12] Stolz G 1960 Numerical solutions to an inverse problem of heat conduction for simple shapes *J. Heat Transfer* **82** 20–5
[13] Beck J V 1962 Calculation of surface heat flux from an internal temperature history *ASME Paper* 62-HT-46
[14] Shumakov N *et al* 1957 A method for the experimental study of the process of heating a solid body *Sov. Phys. Tech. Phys.* **1957** 771–80
[15] Tikhonov A N and Arsenin V Y 1977 *Solutions of Ill-Posed Problems* (Washington, DC: Winston)
[16] Raynaud M and Bransier J 1986 A new finite-difference method for the nonlinear inverse heat conduction problem *Numer. Heat Transfer* **9** 27–42
[17] Beck J V, Blackwell B and Clair C R S 1985 *Inverse Heat Conduction: Ill-posed Problems* (Hoboken, NJ: Wiley)
[18] Kaltenbacher B, Neubauer A and Scherzer O 2008 *Iterative Regularization Methods for Nonlinear Ill-posed Problems* (Berlin: Walter de Gruyter)
[19] Alifanov O M 2012 *Inverse Heat Transfer Problems* (Berlin: Springer Science)
[20] Jarny Y, Ozisik M and Bardon J 1991 A general optimization method using adjoint equation for solving multidimensional inverse heat conduction *Int. J. Heat Mass Transfer* **34** 2911–9
[21] Wang J and Zabaras N 2004 A Bayesian inference approach to the inverse heat conduction problem *Int. J. Heat Mass Transfer* **47** 3927–41
[22] Tian N *et al* 2011 Estimation of unknown heat source function in inverse heat conduction problems using quantum-behaved particle swarm optimization *Int. J. Heat Mass Transfer* **54** 4110–6
[23] Wang G, Zhu L and Chen H 2011 A decentralized fuzzy inference method for solving the two-dimensional steady inverse heat conduction problem of estimating boundary condition *Int. J. Heat Mass Transfer* **54** 2782–8
[24] Sousa P F, Carvalho S R and Guimarães G 2008 Dynamic observers based on Green's functions applied to 3D inverse thermal models *Inverse Prob. Sci. Eng.* **16** 743–61
[25] Soemers M 2008 Numerical methods for the solution of a three-dimensional anisotropic inverse heat conduction problem *Dissertation* Techniche Hochschule, Aachen

[26] Egger H *et al* 2009 Efficient solution of a three-dimensional inverse heat conduction problem in pool boiling *Inverse Prob.* **25** 095006

[27] Fernandes A P *et al* 2010 Use of 3D-transient analytical solution based on Green's function to reduce computational time in inverse heat conduction problems *Appl. Math. Modell.* **34** 4040–9

[28] Huang C-H and Wang S-P 1999 A three-dimensional inverse heat conduction problem in estimating surface heat flux by conjugate gradient method *Int. J. Heat Mass Transfer* **42** 3387–403

[29] Groß S *et al* 2005 Identification of boundary heat fluxes in a falling film experiment using high resolution temperature measurements *Int. J. Heat Mass Transfer* **48** 5549–62

[30] Heng Y *et al* 2008 Reconstruction of local heat fluxes in pool boiling experiments along the entire boiling curve from high resolution transient temperature measurements *Int. J. Heat Mass Transfer* **51** 5072–87

[31] Zhou J *et al* 2010 Inverse estimation of surface heating condition in a three-dimensional object using conjugate gradient method *Int. J. Heat Mass Transfer* **53** 2643–54

[32] Hadamard J 1923 *Lectures on Cauchy's Problem in Linear PDEs* (New Haven, CT: Yale University Press)

[33] Evans L C 2022 *PDEs* (Providence, RI: American Mathematical Society) p 19

[34] Engl H W, Hanke M and Neubauer A 1996 *Regularization of Inverse Problems* (Berlin: Springer Science) p 375

[35] Hansen P C 1998 *Rank-deficient and Discrete Ill-posed Problems: Numerical Aspects of Linear Inversion* (Philadelphia, PA: SIAM)

[36] Heng Y *et al* 2010 Model functions in the modified *L*-curve method—case study: the heat flux reconstruction in pool boiling *Inverse Prob.* **26** 055006

[37] Regińska T 1996 A regularization parameter in discrete ill-posed problems *SIAM J. Sci. Comput.* **17** 740–9

[38] Kunisch K and Zou J 1998 Iterative choices of regularization parameters in linear inverse problems *Inverse Prob.* **14** 1247

[39] Lu S, Heng Y and Mhamdi A 2012 A robust and fast algorithm for three-dimensional transient inverse heat conduction problems *Int. J. Heat Mass Transfer* **55** 7865–72

[40] Silveira J V T *et al* 2023 Estimation of heat generation in semiconductors by inverse heat transfer analysis *IEEE Trans. Compon., Packag. Manuf. Technol.* **13** 315–22

[41] Chen S *et al* 2023 Reconstruction of the heat flux input of coated gun barrel with the interfacial thermal resistance *Case Stud. Therm. Eng.* **49** 103242

[42] Han J *et al* 2021 Online estimation of the heat flux during turning using long short-term memory based encoder–decoder *Case Stud. Therm. Eng.* **26** 101002

[43] Cui M *et al* 2016 A modified Levenberg–Marquardt algorithm for simultaneous estimation of multi-parameters of boundary heat flux by solving transient nonlinear inverse heat conduction problems *Int. J. Heat Mass Transfer* **97** 908–16

[44] Zhu F *et al* 2022 A deep learning method for estimating thermal boundary condition parameters in transient inverse heat transfer problem *Int. J. Heat Mass Transfer* **194** 123089

[45] Ademane V, Kadoli R and Hindasageri V 2023 Simultaneous estimation of reference temperature and heat transfer coefficient in transient film cooling problems *J. Therm. Eng.* **9** 702–17

[46] Wang J and Zabaras N 2005 Hierarchical Bayesian models for inverse problems in heat conduction *Inverse Prob.* **21** 183–206

[47] Duane S *et al* 1987 Hybrid Monte Carlo *Phys. Lett.* B **195** 216–22
[48] Neal R M *et al* 2011 MCMC using Hamiltonian dynamics *Handbook of MCMC* (London: Chapman and Hall/CRC Press) vol 2 p 2
[49] Andreevich M A 2006 An example of statistical investigation of the text *Eugene Onegin* concerning the connection of samples in chains *Sci. Context* **19** 591–600
[50] Ramos N P *et al* 2022 Simultaneous Bayesian estimation of the temperature-dependent thermal properties of a metal slab using a three-dimensional transient experimental approach *Int. J. Therm. Sci.* **179** 107671
[51] Berger J, Orlande H R B and Mendes N 2017 Proper generalized decomposition model reduction in the Bayesian framework for solving inverse heat transfer problems *Inverse Prob. Sci. Eng.* **25** 260–78
[52] Deng S and Hwang Y 2007 Solution of inverse heat conduction problems using Kalman filter-enhanced Bayesian back propagation neural network data fusion *Int. J. Heat Mass Transfer* **50** 2089–100
[53] Deng S and Hwang Y 2006 Applying neural networks to the solution of forward and inverse heat conduction problems *Int. J. Heat Mass Transfer* **49** 4732–50
[54] Yu Z, Wei-qi Q and Yuan-pei S 2020 Application of an adaptive MCMC method for the heat flux estimation *Inverse Prob. Sci. Eng.* **28** 859–76
[55] Yan L, Yang F and Fu C 2011 A new numerical method for the inverse source problem from a Bayesian perspective *Int. J. Numer. Methods Eng.* **85** 1460–74
[56] Hoffman M D and Gelman A 2014 The No-U-Turn sampler: adaptively setting path lengths in Hamiltonian Monte Carlo *J. Mach. Learn. Res.* **15** 1593–623
[57] Neal R M 1993 Probabilistic inference using MCMC methods *PhD thesis* University of Toronto
[58] Heidelberger P and Welch P D 1981 A spectral method for confidence interval generation and run length control in simulations *Commun. ACM* **24** 233–45
[59] Plummer M *et al* 2006 CODA: convergence diagnosis and output analysis for MCMC *R News* **6** 7–11
[60] Wang C *et al* 2024 A fast Bayesian parallel solution framework for large-scale parameter estimation of 3D inverse heat transfer problems *Int. Commun. Heat Mass Transfer* **155** 107409
[61] Luo J *et al* 2019 A novel formulation and sequential solution strategy with time-space adaptive mesh refinement for efficient reconstruction of local boundary heat flux *Int. J. Heat Mass Transfer* **141** 1288–300
[62] Chen W-L and Yang Y-C 2008 Estimation of the transient heat transfer rate at the boundary of an electronic chip packaging *Numer. Heat Transf.* A **54** 945–61
[63] Wagner E and Stephan P 2009 High-resolution measurements at nucleate boiling of pure FC-84 and FC-3284 and its binary mixtures *J. Heat Transfer* **131** 121008
[64] Wagner E *et al* 2007 Nucleate boiling at single artificial cavities: bubble dynamics and local temperature measurements *6th Int. Conf. on Multiphase Flow (Leipzig, Germany)* pp 9–13
[65] Stephan P and Hammer J 1994 A new model for nucleate boiling heat transfer *Heat Mass Transfer* **30** 119–25

Chapter 7

Deep learning-based approach for solving forward and inverse partial differential equation problems

Yi Heng, Jianghang Gu, Guohong Xie and Jia Yi

Forward and inverse problems of partial differential equations (PDEs) are fundamental in various multidisciplinary applications. While traditional methods for solving forward PDEs are well-established, they have limitations, particularly for inverse problems that require solving numerous direct problems iteratively. This chapter explores both classical and emerging deep learning-based approaches for solving forward and inverse PDE problems, demonstrating how machine learning techniques improve computational efficiency and solution accuracy in complex systems.

7.1 Introduction

7.1.1 Forward problems

We formally define a general PDE problem with a partial differential operator L: $U \rightarrow U^*$ of order $p > 0$ from the solution function space U to its dual input function space U^* with support in a compact set $D \subset \mathbb{R}^n$. Let U be the Sobolev space of the same order of p. The boundary condition B specifies the solution function constraints on the domain boundary ∂D. Combining these two factors leads to

$$\begin{cases} \mathcal{L}u(\mathbf{r}) = f(\mathbf{r}), \ \forall \ \mathbf{r} \in D, \\ \mathcal{B}_r u(\mathbf{r})|_{r \in \partial D} = h(\mathbf{r}), \end{cases} \tag{7.1}$$

where the input function $f \in U^*$: $D \rightarrow R$ is usually given and the solution function $u \in U$: $D \rightarrow R$ is to be solved.

doi:10.1088/978-0-7503-6174-3ch7

Many established PDE solvers can be chosen to solve the generic problem defined in equation (7.1) based on the characteristics of the specific problem. Traditional methods include Fourier and Laplace transform methods, FEM and FDM, etc. The idea of using neural networks to solve PDEs in an efficient manner, to the best of our knowledge, dates back to the 1990s. However, the lack of computational power and auto-differentiation capabilities severely limited the efficacy of neural networks in this domain.

7.1.2 Inverse problems

Inverse problems aim to reconstruct the unknown information from the observed data, as follows: the unknown quantity q^* is reconstructed from the measured data Θ containing noise ε, $\mathcal{F}$ is usually the nonlinear operator of the forward problem system. In particular, many inverse problems in biomedical imaging, geophysics, engineering science, or other fields can be written in such a form that if $\mathcal{F}$ is known, the inverse problem is called a no-blind inverse problem, and if $\mathcal{F}$ is partially known or unknown, the inverse problem is a called a semi-bind or blind inverse problem:

$$\Theta = \mathcal{F}(q^*) + \varepsilon. \tag{7.2}$$

Unlike the forward problems, most inverse problems are ill-posed, that is, the solution is unstable, and a small noise ε may cause the solution to oscillate, or the solution is not unique or even non-existent. Traditional methods for solving inverse problems usually involve iterative process of solving forward problems. Deep learning methods are promising for fast inverse solution strategies.

7.2 Deep-learning-based methods

7.2.1 Deep-learning-based methods for forward problems

Deep learning algorithms can usually be broadly divided into three categories: operator learning methods, physics-informed neural network (PINN) methods, and kernel extracting methods. These are summarized as below:

(a) *Operator learning method*

The operator learning method is a type of deep learning method that uses a data-driven method directly and approximates the operator equation between the input and output by constructing end-to-end networks. Different to function regression, operator learning aims to map an infinite dimensional function (the input) to an infinite dimensional function (output). From the computational point of view, the new paradigm of operator learning allows one to simulate the dynamics of complex nonlinear systems.

For example, Guo *et al* [1] proposed a fast method for solving Navier–Stokes equations based on end-to-end self-coding-decoding convolutional neural networks and distance functions. Similarly, Bhatnagar *et al* [2] also used an approximate model based on a convolutional neural network to predict the flow field, which can effectively estimate the velocity and pressure field orders of magnitude faster than traditional solvers, so as to study the influence of wing shape and operating conditions on aerodynamic forces and near-real-time flow fields. A DeepONet method based on the general operator approximation theorem of single-layer neural networks proposed by Chen *et al* [3] was proposed in the literature [4]. The DeepONet was composed of a deep neural network for encoding discrete input function space and another deep neural network for encoding the output function domain. It has been shown in [4] that DeepONet can learn various explicit operators, such as Laplace operators of integrals and fractions, as well as implicit operators representing deterministic and random differential equations. Another benefit of DeepONet is that it can be applied to simulation data, experimental data, or a mixture of both, and that the experimental data may span multiple orders of magnitude on a spatiotemporal scale, allowing scientists to better estimate dynamics by pooling existing data. Anandkumar *et al* [5] used the message passing mechanism of graph neural networks to construct a fast algorithm for solving partial differential equations based on graph neural networks, and represented infinite dimensional maps by combining nonlinear activation functions and a class of integral operators. The kernel integral was calculated by message passing on graph networks, because it becomes unstable as the number of hidden layers increases. In addition, Anandkumar also proposed a Fourier neural operator (FNO) [6], which directly parameterized the integral kernel in Fourier space to approximate the abstract operator, so that the network has stronger nonlinear fitting ability. For relatively simple settings, including Burgers equations, Darcy flows, and Navier–Stokes equations, FNO demonstrates excellent accuracy and efficiency. Although the operator learning method has high solving efficiency, the model interpretability is poor, the network needs to be retrained when the geometric domain changes, and the model generalization is difficult to evaluate, which limits the application of this method in industry.

(b) *PINN methods*

PINN methods are neural networks that encode the PDE as a component of the neural network itself, and approximate the solution of the PDE by training the neural network to minimize the loss function. PINNs consider the underlying PDE information, rather than solving the PDE solely based on a data-driven mechanism [7], and Raissi *et al* [7] introduce and illustrate PINN methods for solving nonlinear PDEs, such as the Schrödinger, Burgers, and Allen–Cahn equations. In fact, the concept of PINNs is not entirely new, that by Dissanayake and Phan–Thien [8] can be considered one of the earliest PINNs. If the predictive power of a discrete-time stepping scheme neural networks is effectively utilized, the framework can be used directly by inserting it into any differential problem.

Similar to PINNs, there is also the deep Ritz method (DRM) [9], in which the loss is defined as the energy of the solution to the problem. The forward problem of PDE is reduced to the weak form by the variational method, and the neural network is used as trial function in FEM. The equations are solved during the training of the neural network. In [10], the convergence of the DRM in solving high-dimensional elliptic PDE is analysed, and it is proved that the generalization error of the DRM is independent of PDE dimension. The literature [11] estimates the error of the DRM in solving linear elliptic PDEs with Dirichlet boundary conditions, and proves that the DRM inherits the approximation ability of neural networks to smooth functions. In [12], the DRM is applied to Laplace and Stokes equations, and a posterior error estimator is proposed as the iteration stop standard of neural networks to ensure an upper limit of solving errors.

In addition, there is a deep learning framework based on the Galerkin method or Petrov–Galerkin method, called the deep Galerkin method (DGM), which is essentially similar to the Galerkin method. The weak solution is approximated by a neural network rather than a linear combination of basis functions. The loss function of the neural network is to multiply the residual by the test function, train the deep neural network to satisfy the differential operator, initial conditions, and boundary conditions, and use random gradient descent at random sampling space points. In [13], an improvement to the DGM was proposed, and an adaptive sampling framework was developed, which optimized the training point selection method of the method. This technology improved the accuracy and efficiency of the basic numerical method by better solving the computation solution of the time-domain part. In [14], the DGM method is used to solve a class of Stokes equations and the convergence of the DGM method for exact solutions is proved.

(c) Kernel extracting methods

The kernel function method is an organic combination of the advantages and disadvantages of the above two methods. The kernel function of the equation is approximated by the deep learning method, and the constraint of the partial differential equation is substituted into the loss function, so that the model has interpretability and good generalization performance. Representative works include GF-Net, proposed by Teng *et al* [15], and the DeepGreen method, proposed by Craig *et al* [16] based on the Koopman operator principle. In [15], Teng *et al* proposed a novel neural network method, GF-Net, to learn the Green's function of classical linear reaction–diffusion equations in an unsupervised manner. The proposed method overcomes the challenge of finding the Green's function of the equation on any domain by using physical knowledge neural network method and domain decomposition. The effectiveness of the method in square, ring, and L-shaped domain experiments has been demonstrated in [15]. In [16], Craig *et al* proposed the DeepGreen method based on the Koopman operator principle, which can be used to solve nonlinear boundary value problems. This method can solve nonlinear boundary value problems several orders of magnitude faster than traditional methods in nonlinear Helmholtz and Sturm–Liouville problems, nonlinear elasticity, and two-dimensional nonlinear Poisson equations. An embedded neural network model based on Green's function is proposed in [17], where the input consists of a far-field displacement field and the output is the expected but initially unknown pulling and separating interface. Specifically, the Green function is embedded as a loss function term along with other terms based on the mean square error and field equations related to the loaded elastomer, and the method is first verified by the analytical solution of a simply supported beam affected by the end moment, and an uneven tractive force is applied to part of one boundary of the beam. In [18], a machine learning method for regression of Green's function from boundary conditions is proposed, and a multilevel Green's function is developed to replace the traditional Poisson solver.

7.2.2 Deep learning-based methods for inverse problems

The traditional inverse problem-solving method relies on the iterative regularization method, which needs to solve a large number of forward problems iteratively, which requires a large amount of computation and thus has low computational efficiency. However, the analytical methods in traditional inverse problems are only aimed at specific problems and are difficult to generalize. Solution methods based on deep learning for inverse problems have attracted much attention in recent years. Deep learning methods can directly calculate a large number of reconstructed inverse problems by using a large amount of training data, which provides an opportunity to reduce the computational burden caused by numerous iterations of solving forward problems in the traditional methods. Fast inversion solution methods based on deep learning can be divided into direct mapping methods and prior knowledge embedded deep learning frameworks. The relevant research progress is summarized as follows:

(a) *Direct mapping method*

The direct mapping method refers to directly approaching the unknown $\mathcal{F}^{-1}$ by training the deep learning network, and optimizing the objective function by adjusting the parameters of the deep neural network NN($\cdot$). It is an end-to-end black box model with data-driven core:

$$\Theta = \mathcal{F}(q^*) + \varepsilon. \tag{7.3}$$

For example, a novel end-to-end tomography image reconstruction technology based on a deep convolutional encode–decoder network was proposed in the literature [19]. This technique takes the sinusoidal tomography data as the input and outputs high-quality and quantitative tomography images directly and quickly. A direct and efficient reconstruction algorithm based on deep learning was developed in [20], which used a filtered back projection algorithm for the first layer and a U-net architecture for the remaining layers. In [21], a convolutional neural network architecture based on multi-resolution decomposition and residual learning was constructed for sparse view reconstruction of parallel beam x-ray computed tomography. The proposed network was superior to the full-variant regularization iterative reconstruction, and the reconstruction time of 512 × 512-pixels images on GPU was less than 1 s. The authors of [22] propose a method that incorporates deep learning into the iterative image reconstruction framework to reconstruct images from severely incomplete measurement data, and uses a convolutional neural network as a quasi-projection operator in the least-squares minimization process. The structure of the network is inspired by the near-end gradient descent method, in which the near-end gradient operator is replaced by a deep convolutional neural network. The gradient descent step is performed by using a linear reconstruction operator.

(b) *Physics-informed deep learning framework*

Direct mapping methods are usually based on convolutional neural networks, which require a large number of parameters to be trained and a large amount of training data, and are highly dependent on the problem itself. When the forward model of the problem changes, the deep learning network model needs to be retrained. In order to solve the training problem, generalization problem, and interpretability problem caused by direct mapping, more and more research is focusing on developing a deep learning framework embedded with prior knowledge. In addition to solving the forward problem, PINNs [7] can be used to solve the parameter inversion problem by embedding PDEs into the loss function of the neural network and solving the approximate unknown parameters by reducing the mean square error of the loss function. Further, a new deep learning framework for solving PDE inverse problems was proposed in [23], which introduced a more general discretization method and embedded a numerical solver into the deep learning architecture, and could be applied to any static or nonlinear time-dependent PDE inverse problem. The authors of [24] proposed an idea and deep learning method based on classical regularization theory, using the prior information of the inverse problem encoded in the forward operator, noise model, and regularization function for training and learning at the same time. This method generated an iterative scheme similar to the gradient method, where the 'gradient' component was learned using the convolutional network. The authors of [25] proposed a method to solve electrical impedance tomography using deep neural networks. Based on numerical low-rank characteristics and linear perturbation analysis, 2D and 3D inverse mapping methods of electrical impedance tomography were constructed to achieve efficient inverse calculation of electrical conductivity.

7.3 Applications and analysis

7.3.1 Predicting the transport process in reverse osmosis desalination

In this section, we introduce a data-driven deep learning method to predict the transport process in reverse osmosis (RO) desalination. RO is one of the most advanced desalination technologies, accounting for more than 50% of the desalination market [26]. The spiral-wound membrane module is composed of a sequence of membrane sheet–feed spacer–membrane sheet–permeate spacer wrapped around a central collection tube. The primary role of the feed spacer is to support the membrane sheets on both sides, facilitating fluid turbulence and enhancing the mass transfer process. The mesoscale structure of a typical commercial spiral-wound membrane module is illustrated in figure 7.1.

Under typical engineering RO operational conditions, the flow type is laminar. The governing equations for fluid flow are the Navier–Stokes equations and the continuity equation, while the salt component transfer process conforms to the convection–diffusion equation:

$$\begin{cases} \rho(\mathbf{u} \cdot \nabla)\mathbf{u} = \nabla \cdot [-\boldsymbol{PI} + \mu(\nabla \mathbf{u} + (\nabla \mathbf{u})^T)], \\ \rho \nabla \cdot \mathbf{u} = 0, \\ \nabla \cdot (D_s \nabla c) = \mathbf{u} \cdot \nabla c, \end{cases} \tag{7.4}$$

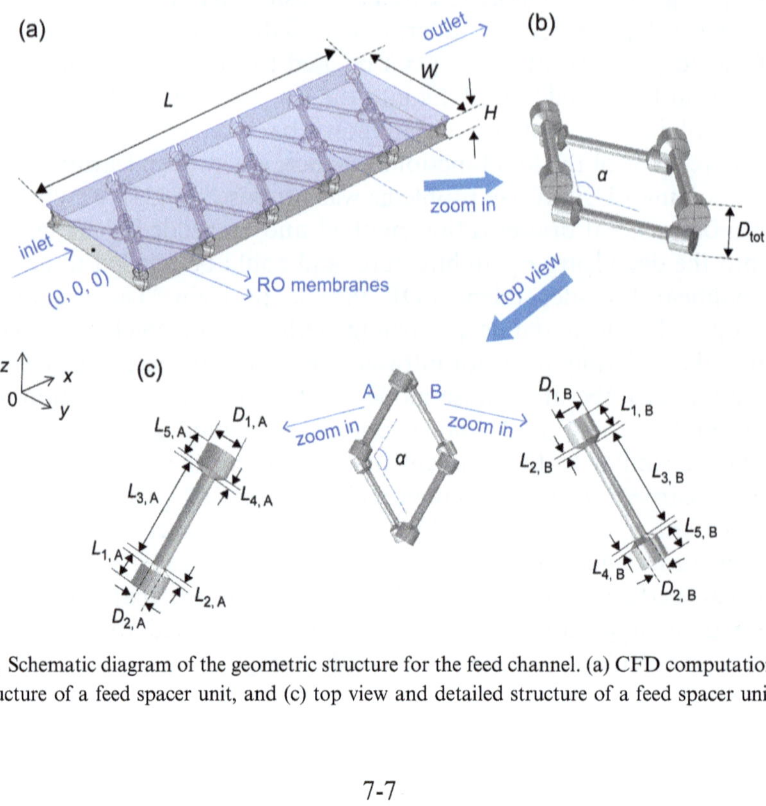

Figure 7.1. Schematic diagram of the geometric structure for the feed channel. (a) CFD computational region, (b) 3D structure of a feed spacer unit, and (c) top view and detailed structure of a feed spacer unit.

where the density $\rho(c)$ (kg m^{-3}) and viscosity $\mu(c)$ (Pa s) are related to the local molar concentration c (mol m^{-3}). **u** and P (Pa) represent the velocity vector and pressure, respectively, while D_s denotes the diffusion coefficient.

The mathematical model for the RO process can be described as a multi-scale, strongly nonlinear PDE problem that couples multiple physical fields such as fluid flow and salt component convection–diffusion. The optimization design of the membrane module requires serial iterative solutions to numerous complex PDE problems, which is inefficient and time-consuming.

We chose the multilayer perceptron (MLP) as a universal model to approximate the high-dimensional nonlinear relationship between the geometric parameters of the spiral-wound membrane module's feed spacer, the RO system's inlet flow, and the internal channel's velocity u, pressure P, and concentration c. The conversion formula between the inlet flow and the average inlet speed in the computational fluid dynamic (CFD) model can be found in our previous study [27], and the u solved by this chapter for MLP represents the numerical magnitude of velocity. This chapter uses the 3D coordinate values of the channel node data obtained from the CFD simulation and the channel's design parameters combined as the input end of the MLP, denoted as X, $X = (D_{1,\mathrm{A}}, D_{2,\mathrm{A}}, D_{1,\mathrm{B}}, D_{2,\mathrm{B}}, D_{\mathrm{tot}}, \alpha, Q_{\mathrm{tot}}, x, y, z)$. The feed channel's velocity, pressure, and concentration under different design parameter conditions are selected as the output end of the MLP, denoted as Y. The three networks are trained and predicted independently.

$$\delta_{l,i} = f\left(\sum\nolimits_{j=1}^{n_{l-1}} w_{l,i}\ \delta_{l-1,j} + b_{l,i}\right), \tag{7.5}$$

where $\delta_{l,\,i}$ represent the output of the ith neuron in layer l. $w_{l,\,i}$ denotes the weight coefficient of the ith neuron in layer l, while $b_{l,\,i}$ represents the bias coefficient of the ith neuron in the same layer. n_{l-1} ($l = 1, 2, \ldots, L$) is the number of neurons in layer l–1. L signifies the total number of layers in the MLP model, excluding the input layer. In this study, the MLP structure adopted consists of four layers. The number of neurons in the hidden layers and the output layer are sequentially 128, 64, 32, and 1, as illustrated in figure 7.2.

First, we established a 'multiphysics fully-coupled high-fidelity three-dimensional model' to describe the fluid flow and salt component transfer in the inlet channel of the reverse osmosis membrane module. Leveraging the design parameter space of commercial membrane components, Latin hypercube sampling techniques were employed, resulting in 726 different operational CFD models. Subsequently, these data were partitioned into a ratio of 650:66:10 for MLP training, validation, and testing, respectively.

Then we proceeded to train the MLP and the training process of the MLP model can be mathematically represented as

$$\theta^* = \underset{\theta}{\operatorname{argmin}} \frac{1}{N}\sum\nolimits_{k=1}^{N} \mathcal{L}(Y_k, f(X_k; \theta)) + \lambda\ \|w\|_2^2, \tag{7.6}$$

where $\mathcal{L}(\cdot)$, $f(\cdot)$, N, θ, and λ, respectively, represent the loss function, the MLP model, the number of samples in the training set, the hyperparameters of the neural

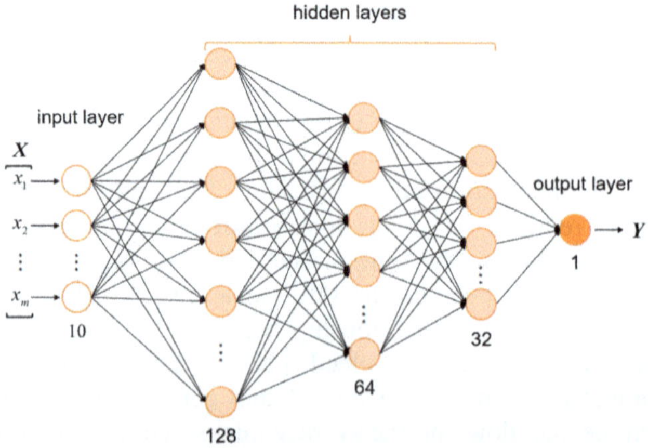

Figure 7.2. Architecture of the MLP network.

network, and the regularization coefficient. We employ the error backpropagation algorithm [28, 29] to sequentially compute the gradient of the loss function with respect to the parameters of each hidden layer neuron. We update the neuron parameters using the adaptive moment estimation (Adam) optimizer [30]. Finally, a random search is performed to identify a superior combination of hyperparameters, culminating in the training of the MLP model.

After finishing training the MLP, we selected two representative structures from the test set with the same inlet velocity and detailed comparative analysis was carried out on the pressure distribution (as shown in figures 7.3 and 7.4), velocity distribution (as shown in figure 7.5), and concentration distribution (as shown in figure 7.6) of these two characteristic inlet spacer structures, utilizing both the CFD and MLP approaches.

In the two representative structural models, the local pressure distribution along the cross-sectional line below the membrane surface (denoted as $z = -0.5\,H$, $y = 0.25\,L_y$, $0 \leqslant x \leqslant L$) estimated by both the CFD and MLP models is presented in figure 7.4. The results indicate that the derived proxy MLP model can provide relatively accurate predictions for local pressures. The different inlet spacer structures play a significant role in affecting the flow resistance within the channels.

The local velocity distribution predicted by both the MLP and CFD methods is depicted in figure 7.5. While there are slight deviations between the local velocity distribution predicted by the MLP method and the results of the CFD simulation, the overall predicted outcomes are generally consistent. The simulation for structure 2 exhibits two prominent 'high-velocity zones' along with minor vortices, leading to increased energy dissipation. In contrast, the velocity distribution in structure 1 is relatively more uniform. This further elucidates why the flow resistance in the representative structure 2 is considerably higher than in structure 1. The local concentration distribution simulated by the MLP, reflecting concentration polarization phenomena on the RO membrane surface (as shown in figure 7.6), can be

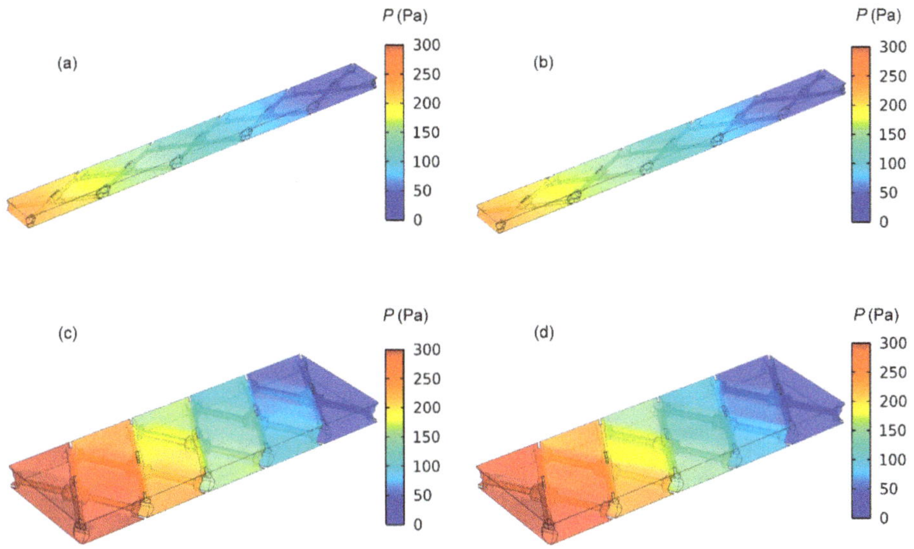

Figure 7.3. Distribution of pressure (P, Pa) for the CFD and MLP methods. (a), (b) Simulation results referring to the typical structure 1 and (c), (d) simulation results referring to the typical structure 2.

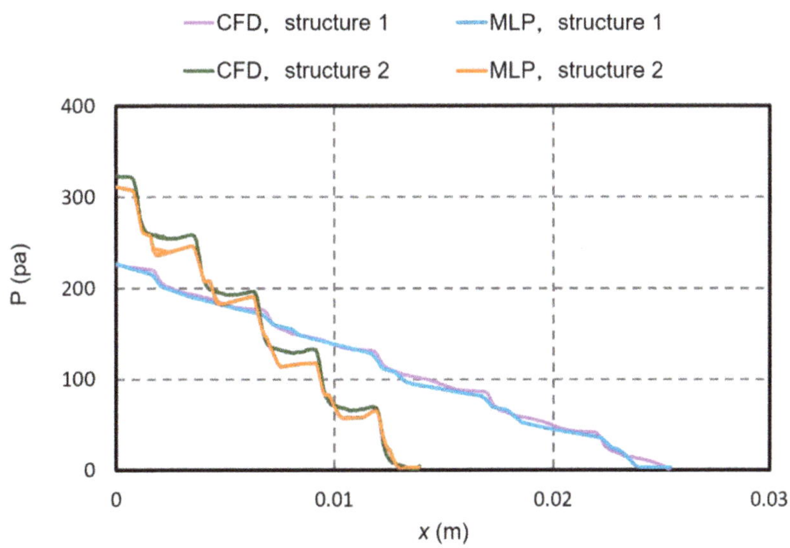

Figure 7.4. The value of pressure (P, Pa) along the surface line ($z = -0.5H$, $y = 0.25L_y$, $0 \leqslant x \leqslant L$) for the CFD and MLP methods.

employed to analyse intricate mass transfer characteristics in channels with varying inlet spacer structures.

Based on the above discussion, the MLP transfer model established in this chapter provides a relatively precise overall prediction of the flow field in RO feed channels under various conditions. In terms of computational efficiency, similar

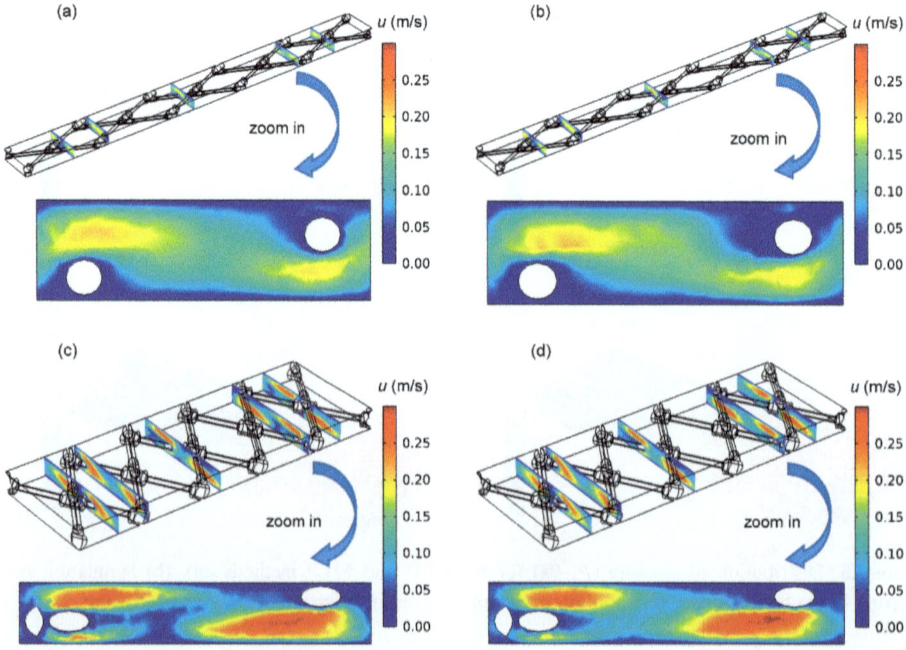

Figure 7.5. Distribution of velocity (u, m s^{-1}) for the CFD and MLP methods. (a), (b) Simulation results referring to the typical structure 1 and (c), (d) simulation results referring to the typical structure 2.

problems require approximately 40 min for computation using CFD models combined with supercomputing (utilizing five computing nodes, each with 24 cores), whereas the method based on MLP only requires 3 min on a single machine. This proxy transfer model can serve as an efficient solver and can be applied to rapid inference tasks in large-scale multi-scale optimization of RO systems. It offers valuable guidance for the design of optimization schemes for membrane components.

Overall, the highly scalable parallel calculation method proposed in this chapter is suitable for the massive screening and optimal design of multi-type feed spacers. This study provides a computable model and data support to reveal the complex mechanism of the flow and transport process in the RO membrane module, and has critical significance for the factory-scale application of high permeability RO membranes in the future.

7.3.2 Identification of highly transient surface heat flux

Parts of this section have been reproduced with permission from [34]. Copyright 2023 Elsevier.

In this section, we study a specific kind of inverse problem arising from the applications of pool boiling. The governing PDE system of the forward problem can be mathematically illustrated as follows:

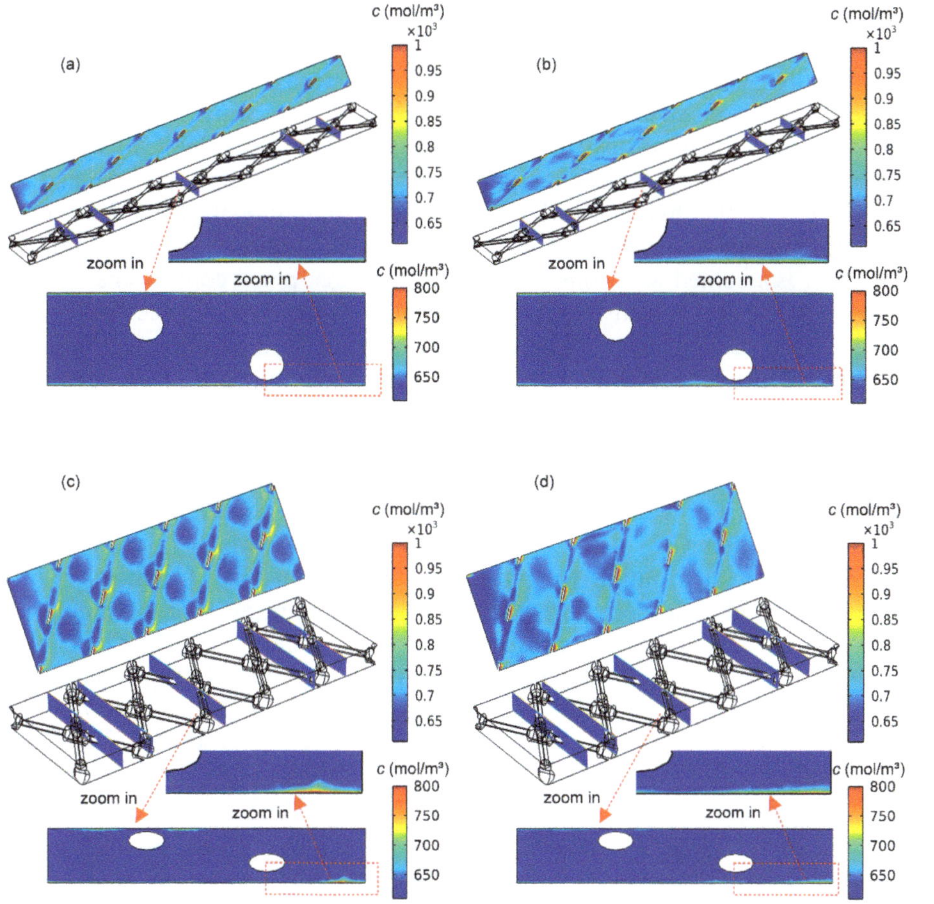

Figure 7.6. Distribution of concentration (c, mol m^{-3}) for the CFD and MLP methods: (a), (b) refer to the typical structure 1 and (c), (d) refer to the typical structure 2.

$$\begin{cases} \rho(\Theta)c_p(\Theta)\dfrac{\partial \Theta}{\partial t} = \nabla \cdot (k(\Theta)\ \nabla\ \Theta), \ \text{ in } \Omega \times (0,\, t_f) \\ \Theta(\cdot,\, 0) = \Theta_0(\cdot), \quad \text{ in } \Omega \\ -\,k(\Theta)\dfrac{\partial \Theta}{\partial n} = q_d, \quad \text{on } \Gamma_d \times (0,\, t_f) \\ -\,k(\Theta)\dfrac{\partial \Theta}{\partial n} = q_r, \ \text{ on } \Gamma_r \times (0,\, t_f) \\ -\,k(\Theta)\dfrac{\partial \Theta}{\partial n} = q_u, \ \text{ on } \Gamma_u \times (0,\, t_f) \end{cases}, \tag{7.7}$$

where q_d, q_u, and q_r denote the heat flux through the bottom surface Γ_d, the upper surface Γ_u, and the remaining boundary Γ_r of the 3D computational domain Ω, corresponding to the Neumann boundary conditions, and$\rho(\Theta)$, $c_p(\Theta)$, and$k(\Theta)$

denote the temperature-dependent density, heat capacity, and thermal conductivity, which leads to a nonlinear PDE system.

Specifically, we study the enhancement of heat transfer performance with the modified micro-nano structures by electrodeposition. Heat transfer problems in the computational domain reconstructed from computed tomography (CT) scanning of micro-nano porous structures prepared in our experiments are studied (see figure 7.7). For the hypothetical forward simulation models guided by micro-boundary layer theory in [31, 32] and our previous studies [33], the boiling heat flux q_u on Γ_u is assumed with considerations of the three-phase contact lines (TPCLs) since most of the heat is transferred by evaporation in the region of the TPCLs in the boiling heat transfer process. The evaporation process is mainly accelerated by abundant independent bubbles, each bubble corresponding to a TPCL region, and thus numerous ring-shaped local transient heat fluxes on the TPCLs of boiling surface Γ_u are suggested for the simulation according to microlayer theory. Specifically, the transient heat flux of TPCLs is composed of different segments $\mathrm{TPCL}_k\,|_{k=1,\ldots,N}$. For each segment TPCL_k, the corresponding $q_u^{t_k}$ (W cm^{-2}) is defined as equation (7.8) with stochastic number t_k (s), which is drawn from a standard normal distribution to mimic the random and asynchronous bubbles' generation, growth, and detachment in the boiling heat transfer process. $q_{\max}$ is an empirical number and set as a constant 1010 with reference to [33]. The constant heating flux is set as $q_d = 50$ W cm^{-2} and the remaining boundary q_r is assumed as the convection boundary condition in equations (7.8) and (7.9), where $h = 1000$ W (m$^2 \cdot$ K) $-$ 1, and $\Theta_{\text{amb}} = 373.15$ K. The initial condition is set as 373.15 K. The

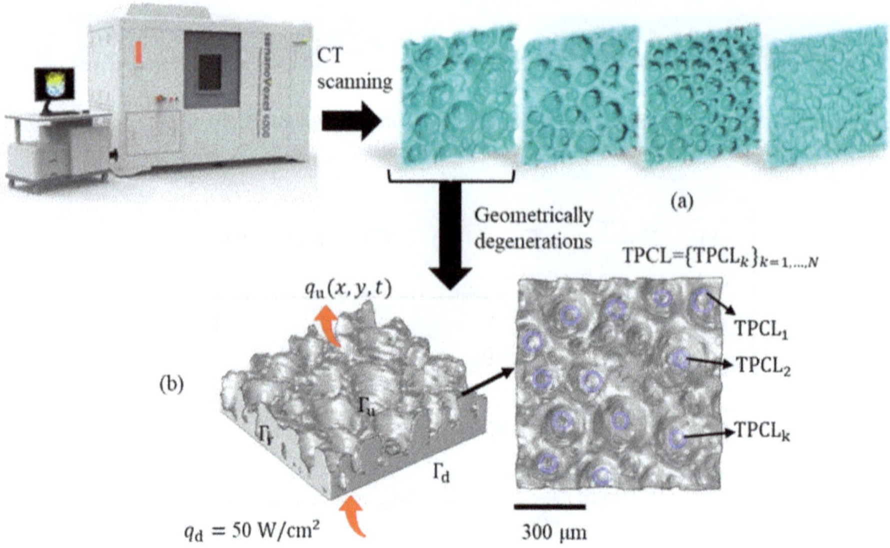

Figure 7.7. (a) CT scanning views of the four samples of micro-nano porous structures prepared in different current densities. (b) The simulation geometric model of sample M4 (left-hand side), and TPCL (marked in blue) on the boiling boundary (right-hand side). (Reproduced with permission from [34]. Copyright 2023 Elsevier.)

thermal conductivity of copper materials is temperature dependent. Values of 393 $W\,(m \cdot K)^{-1}$ at 400 K and 386 $W\,(m \cdot K)^{-1}$ at 500 K are used to provide interpolation values of thermal conductivity for the other nearby temperatures. The density and specific heat of the copper material obtained by electrodeposition in the current case are assumed to be constant since the properties of the copper material do not experience significant change under small temperature variations. The desired IHTP is to reconstruct the boiling heat flux q_u on the complex porous surface from simulated temperature data on Γ_d:

$$q_u^{t_k}(x, y, t) = \begin{cases} q_{\max} \cdot \sin(100\pi(t - 0.02)) + q_{\max}, \; t \in [t_k, t_k + 0.02], \; (x, y) \in \mathrm{TPCL}_k \\ 0, \; t \in [0, t_r) \cup (t_r + 0.02, 0.05], \; (x, y) \in \mathrm{TPCL}_k \\ h(\Theta - \Theta_{\mathrm{amb}}), \; t \in [0, 0.05], \; (x, y) \in \Gamma_u \backslash \mathrm{TPCL}_k \Big|_{k=1,\ldots,N} \end{cases} \tag{7.8}$$

$$q_r = h(\Theta - \Theta_{\mathrm{amb}}), \; \Gamma_r \times (0, t_f). \tag{7.9}$$

Assume that Θ_m^δ is the measured temperature with an error upper bound δ, and $\mathcal{F}$: $q_u \rightarrow \Theta_m^\delta$ is the function operator implicitly given by the forward heat transfer PDE system. To reconstruct the missing quantity q_u on the upper surface Γ_u, a classical solution approach is to solve the minimization problem in the form of the Tikhonov functional, which consists of a least-squares term to account for the difference between temperature measurements and predictions, and a parameterized penalty term that reflects the fluctuation of estimated heat flux:

$$J(q_u) := \left\| \mathcal{F}(q_u) - \Theta_m^\delta \right\|^2 + \alpha \left\| q_u \right\|^2 \tag{7.10}$$

$$\mathcal{F}^\dagger(\Theta_m^\delta) = q_u. \tag{7.11}$$

Different to the aforementioned regularization-based computational approaches, the purpose of this chapter is to find the implicit inverse operator $\mathcal{F}^\dagger$ straightforwardly, which circumvents the difficulty of handling packs of nonlinear PDE-constrained optimization problems when using conventional regularization-based methods to solve inverse problems. Based on our numerous simulated experiments, there exists a relationship between the temperature difference in adjacent time frames and boiling heat flux. For violent temperature variation in adjacent time frames, there will be a high heat flux. For trivial temperature variation in adjacent time frames, there will be a low heat flux. Additionally, more time steps of temperature frames could not guarantee better estimation performance due to the intrinsic uninterpretable nature of deep learning approaches. From our experiments, three-time-steps-input is an appropriate trade-off between accuracy and efficiency, and we chose $\Theta_m(\mathbf{x}, t_i)$, $\Theta_m(\mathbf{x}, t_{i-1})$, $\Theta_m(\mathbf{x}, t_{i-2})$ as input data.

As is demonstrated by figure 7.8, the FluxNet [35] is composed of a discrete wavelet transform (DWT) layer and a 'generator' together with a 'discriminator'. The DWT layer is utilized to extract detailed information whilst the 'generator' is employed to approximate the inverse operator. The minimax strategy is adopted

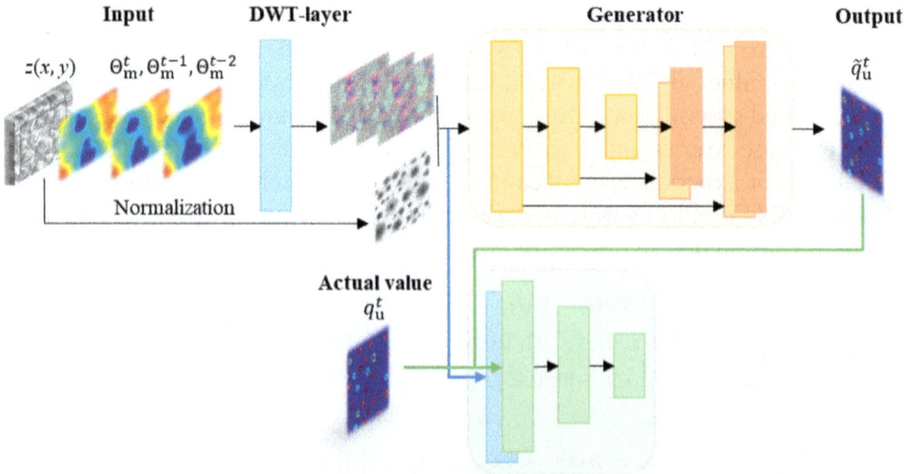

Figure 7.8. Network structure of FluxNet: input layer, DWT layer, 'generator', 'discriminator', and output layer. (Reproduced with permission from [34]. Copyright 2023 Elsevier.)

through adversarial training between the 'generator' and 'discriminator' for the purpose of encouraging a deeper understanding of data distribution.

The DWT is the wavelet transform for which the wavelets are discretely sampled. In this study, the coefficients of low-pass filter and high-pass filter are chosen with reference to Daubechies wavelet basis [34]. Detailed horizontal variations, vertical variations, and diagonal variations are acquired after a single-level wavelet decomposition. Temperatures in adjacent time frames $\Theta_m(\boldsymbol{x}, t_i)$, $\Theta_m(\boldsymbol{x}, t_{i-1})$, $\Theta_m(\boldsymbol{x}, t_{i-2})$ are fed as the input data of the DWT layer. Note that $\Theta_m(\boldsymbol{x}, t_i)$, $\Theta_m(\boldsymbol{x}, t_{i-1})$, $\Theta_m(\boldsymbol{x}, t_{i-2})$ are all considered as distributions dependent on the dimensional variables $\mathbf{x} \in \Gamma_d$ and temporal variable $t \in [0, t_f]$. The DWT layer is constructed as a pre-processing module to extract detailed information of the temperature distribution.

To approximate the nonlinear inverse operator, an encoder–decoder multilayer convolution neural network is employed and denoted as the 'generator' for convenience. The 'generator' maps $u = [\text{DWT}(\Theta_m(\boldsymbol{x}, t_i), \Theta_m(\boldsymbol{x}, t_{i-1}), \Theta_m(\boldsymbol{x}, t_{i-2}), z]$ to pseudo heat flux $\tilde{q}_u(\mathbf{x}, t_i)$, namely the estimated heat flux distribution on Γ_u (here $\mathbf{x} \in \Gamma_u$) at time instant t_i. In the 'generator', each hidden layer receives an input u_l from previous layer and transforms it to

$$u_l \to u_{l+1} = \text{sigmoid}(\kappa_{l+1,\,l}{}^* u_l + o_{l+1}). \tag{7.12}$$

The network is thus a composition of a sequence of nonlinear functions:

$$u = [\text{DWT}(\Theta_m(x, t_i), \Theta_m(x, t_{i-1}), \Theta_m(x, t_{i-2})), z] \tag{7.13}$$

$$\text{Gen}(u;\ \theta_g) = (l_{N_l} {}^\circ l_{N_l-1} {}^\circ \cdots {}^\circ l_2 {}^\circ l_1)u = \tilde{q}_u(x, t_i), \tag{7.14}$$

where the operator $\circ$ denotes the composition and θ_g represents trainable parameters in the network.

For the purpose of encouraging deeper understanding of data distribution, a 'discriminator' is introduced to compete with the 'generator' through adversarial training and optimization based on the minimax strategy. The 'discriminator' is also a multilayer convolution neural network, whose input is the concatenation of the feature maps $(u, \tilde{q}_u)$ or (u, q_u). When the 'discriminator' takes in data of q_u, it outputs the probability $\mathrm{Dis}(q_u; \theta_d)$. When the 'discriminator' takes in generated data of $\tilde{q}_u$, it will output the probability $\mathrm{Dis}(\tilde{q}_u; \theta_d)$. Through maximizing the probability of $1 - \mathrm{Dis}(\tilde{q}_u;\theta_d)$ and $\mathrm{Dis}(q_u;\theta_d)$, the 'discriminator' learns to distinguish between the generated $\tilde{q}_u$ and the real q_u. Simultaneously, the 'generator' learns to deceive the 'discriminator' by narrowing errors between $\tilde{q}_u$ and q_u and minimizing the probability $1 - \mathrm{Dis}(\tilde{q}_u; \theta_d)$. By repeating such a process, the 'generator' can successively approximate the solution of the IHTP. The $L1$ term is introduced as regulation with a positive parameter λ to force low-frequency correctness in the 'generator'. The value of λ is chosen through comparisons between a series of FluxNet models, where the only difference between these models is the value of λ. In this work, λ is set as an empirical value 50 based on this knowledge:

$$\mathcal{L}_{L1} = E\left[\left| q_u^i(x, t) - \tilde{q}_u^i(x, t)\right|\right] \tag{7.15}$$

$$\mathrm{Gen}^* = \mathrm{argmin}_{\mathrm{Gen}}\mathrm{max}_{\mathrm{Dis}}E\left[\ln\left(\mathrm{Dis}\left(q_u^i(x, t)\right)\right)\right] + E\left[\ln\left(1 - \mathrm{Dis}\left(\tilde{q}_u^i(x, t)\right)\right)\right] + \lambda\mathcal{L}_{L1}. \tag{7.16}$$

The loss function of the 'generator' and the 'discriminator' can be calculated as follows:

$$\frac{1}{N_s}\sum_{i=1}^{N_s}\left(E\left[\ln\left(\mathrm{Dis}\left(q_u^i(x, t)\right)\right)\right] + \lambda E\left[\left| q_u^i(x, t) - \tilde{q}_u^i(x, t)\right|\right]\right) \tag{7.17}$$

$$\frac{1}{N_s}\sum_{i=1}^{N_s}\left(E\left[\ln\left(\mathrm{Dis}\left(q_u^i(x, t)\right)\right)\right] + E\left[\ln\left(1 - \mathrm{Dis}\left(\tilde{q}_u^i(x, t)\right)\right)\right]\right), \tag{7.18}$$

where $i = 1, \ldots, N_s$, and N_s is the total number of training samples.

Automatic differentiation [36] is utilized to compute derivatives for optimization and the adaptive moment estimation (Adam) optimization algorithm is applied to update network parameters.

In this section, our recent model-based multi-bubble pool boiling experiments [33] with micro-nano porous structures are further studied. For the first time, CT scanned 3D geometrical structures of micro-nano porous copper samples prepared by us are analysed. Particularly, four types of porous copper samples (denoted as M1, M2, M3, M4) with different porosity are considered in establishing the training set (see figure 7.9). In the data preparation stage, 240 forward models are simulated by COMSOL Multiphysics on the Tianhe-2 supercomputer. A 8000 core hour cost was used in total, and parallel computational scale for the training process can reach up to 5560 cores. The temperature profile calculated by COMSOL Multiphysics was validated by macroscopic averaged wall superheats obtained from the experiments in [33]. The dataset was then divided into 80% training set and 20% testing set. The

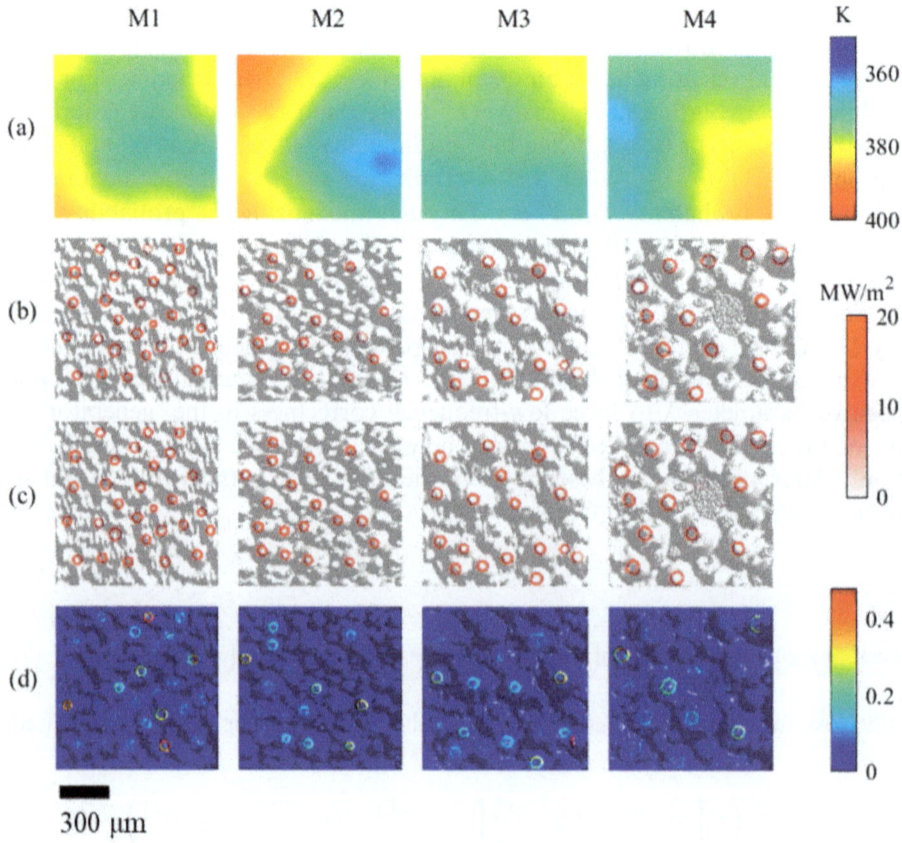

Figure 7.9. (a) The simulated temperature on the heating surface for samples M1, M2, M3, and M4 at the representative time instant $t = 0.028$ s. (b) The expected heat flux patterns for samples M1, M2, M3, and M4 at the representative time instant $t = 0.028$ s based on the realistic simulation models. (c) Estimated heat flux by FluxNet. (d) RMD error between the expected heat flux and estimated heat flux. (Reproduced with permission from [34]. Copyright 2023 Elsevier.)

time step Δt was 0.001 s in [33]. The corresponding estimation results are analysed for the aspects of accuracy and stability below.

First, the estimated heat flux by the FluxNet with noise-free data at the representative time instant of 0.028 s for M1, M2, M3, and M4 are shown in figure 7.9. The average estimation error by FluxNet in the whole observation time 0–0.05 s for four samples is about 0.0247 in terms of RMD error.

To verify the algorithm stability, we conducted five levels of noise tests ($\delta = 0.025$ K, $\delta = 0.05$ K, $\delta = 0.1$ K, $\delta = 0.2$ K, and $\delta = 0.5$ K) with uniformly distributed noise. In this case, the maximum temperature difference at $t = 0.025$ s is 12 K, which leads to the relative noise levels of 0.21%, 0.42%, 0.83%, 1.67%, and 4.17%, respectively. The average estimation errors of the FluxNet acquired by Θ_m^δ at $t = 0.025$ s are 0.0129, 0.0147, 0.0193, 0.0228, and 0.0795 under noise levels of $\delta = 0.025$ K, $\delta = 0.05$ K, $\delta = 0.1$ K, $\delta = 0.2$ K, and $\delta = 0.5$ K and accuracy with noise-free data is 0.0126.

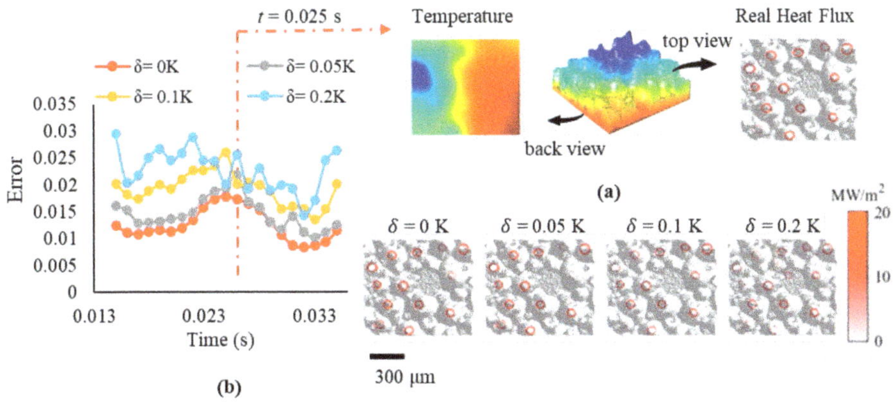

Figure 7.10. Stability tests: (a) heat flux estimated by FluxNet and (b) a performance study by FluxNet under noise levels of $\delta = 0$ K, $\delta = 0.05$ K, $\delta = 0.1$ K, $\delta = 0.2$ K in the main observation time of 0.015 s—0.035 s. (Reproduced with permission from [34]. Copyright 2023 Elsevier.)

A comparison of performance at the representative time instance $t = 0.025$ s under various noise levels is shown in figure 7.10. As mentioned above, the error bound of measured temperature in the studied pool boiling process is about 0.1 K. Hence, the algorithm stability is sufficient for real engineering applications.

7.4 Conclusions

In this chapter, we demonstrated both forward solution approach and inverse solution approach for two specific cases using deep learning-based methods, which tactfully circumvents the difficulty of handling packs of PDE problems. Our future work will be devoted to theoretically and systematically studying the deep learning-based methods towards solving both large-scale forward and inverse PDE problems, and tackling complex heat and mass transfer problems coupled with continuous fluid flow in real-world scenarios.

References

[1] Guo X, Li W and Iorio F 2016 Convolutional neural networks for steady flow approximation *Proc. 22nd ACM SIGKDD Int. Conf. on Knowledge Discovery and Data Mining*

[2] Bhatnagar S *et al* 2019 Prediction of aerodynamic flow fields using convolutional neural networks *Comput. Mech.* **64** 525–45

[3] Chen T and Chen H 1995 Universal approximation to nonlinear operators by neural networks with arbitrary activation functions and its application to dynamical systems *IEEE Trans. Neural Netw.* **6** 911–7

[4] Lu L *et al* 2021 Learning nonlinear operators via DeepONet based on the universal approximation theorem of operators *Nat. Mac. Intell.* **3** 218–29

[5] Li Z *et al* 2020 Neural operator: graph kernel network for partial differential equations arXiv: 2003.03485

[6] Li Z *et al* 2020 Fourier neural operator for parametric partial differential equations arXiv: 2010.08895

[7] Raissi M, Perdikaris P and Karniadakis G E 2019 Physics-informed neural networks: a deep learning framework for solving forward and inverse problems involving nonlinear partial differential equations *J. Comput. Phys.* **378** 686–707

[8] Dissanayake M and Phan-Thien N 1994 Neural-network-based approximations for solving partial differential equations *Commun. Numer. Methods Eng.* **10** 195–201

[9] Yu B 2018 The deep Ritz method: a deep learning-based numerical algorithm for solving variational problems *Commun. Math. Stat.* **6** 1–12

[10] Lu Y, Lu J and Wang M 2021 *A priori* generalization analysis of the deep Ritz method for solving high dimensional elliptic partial differential equations *Proc. Mach. Learn. Res.* **134** 1–46

[11] Müller J and Zeinhofer M 2022 Error estimates for the deep Ritz method with boundary penalty *Proc. Mach. Learn. Res.* **145** 1–20

[12] Minakowski P and Richter T 2023 *A priori* and *a posteriori* error estimates for the deep Ritz method applied to the Laplace and Stokes problem *J. Comput. Appl. Math.* **421** 114845

[13] Aristotelous A C, Mitchell E C and Maroulas V 2023 ADLGM: an efficient adaptive sampling deep learning Galerkin method *J. Comput. Phys.* **447** 111944

[14] Li J *et al* 2022 The deep learning Galerkin method for the general stokes equations *J. Sci. Comput.* **93** 5

[15] Teng Y *et al* 2022 Learning Green's functions of linear reaction-diffusion equations with application to fast numerical solver *Proc. Mach. Learn. Res.* **180** 1–16

[16] Gin C R *et al* 2021 DeepGreen: deep learning of Green's functions for nonlinear boundary value problems *Sci. Rep.* **11** 1–14

[17] Wei C *et al* 2022 Deep-Green inversion to extract traction-separation relations at material interfaces *Int. J. Solids Struct.* **250** 111698

[18] Tang J *et al* 2022 Neural Green's function for Laplacian systems *Comput. Graph.* **107** 186–96

[19] Häggström I *et al* 2019 DeepPET: a deep encoder–decoder network for directly solving the PET image reconstruction inverse problem *Med. Image Anal.* **54** 253–62

[20] Antholzer S, Haltmeier M and Schwab J 2019 Deep learning for photoacoustic tomography from sparse data *Inverse Prob. Sci. Eng.* **27** 987–1005

[21] Jin K H *et al* 2017 Deep convolutional neural network for inverse problems in imaging *IEEE Trans. Image Process.* **26** 4509–22

[22] Kelly B, Matthews T P and Anastasio M A 2017 Deep learning-guided image reconstruction from incomplete data arXiv: 1709.00584

[23] Pakravan S *et al* 2021 Solving inverse-PDE problems with physics-aware neural networks *J. Comput. Phys.* **440** 110414

[24] Adler J and Öktem O 2017 Solving ill-posed inverse problems using iterative deep neural networks *Inverse Prob.* **33** 124007

[25] Fan Y and Ying L 2020 Solving electrical impedance tomography with deep learning *J. Comput. Phys.* **404** 109119

[26] Goh P *et al* 2018 Membrane fouling in desalination and its mitigation strategies *Desalination* **425** 130–55

[27] Luo J, Li M and Heng Y 2020 A hybrid modeling approach for optimal design of non-woven membrane channels in brackish water reverse osmosis process with high-throughput computation *Desalination* **489** 114463

[28] Werbos P 1974 Beyond regression: new tools for prediction and analysis in the behavioral sciences *PhD thesis* Harvard University, Cambridge, MA

[29] Rumelhart D E, Hinton G E and Williams R J 1986 Learning representations by back-propagating errors *Nature* **323** 533–6
[30] Kingma D P and Ba J 2014 Adam: a method for stochastic optimization arXiv: 1412.6980
[31] Van Stralen S 1966 The mechanism of nucleate boiling in pure liquids and in binary mixtures —part I *Int. J. Heat Mass Transfer* **9** 995–1020
[32] Stephan P and Hammer J 1994 A new model for nucleate boiling heat transfer *Heat Mass Transfer* **30** 119–25
[33] Hong M *et al* 2021 Model-based experimental analysis of enhanced boiling heat transfer by micro-nano porous surfaces *Appl. Therm. Eng.* **192** 116809
[34] Gu J *et al* 2023 A fast inversion approach for the identification of highly transient surface heat flux based on the generative adversarial network *Appl. Therm. Eng.* **220** 119765
[35] Vonesch C, Blu T and Unser M 2007 Generalized Daubechies wavelet families *IEEE Trans. Signal Process.* **55** 4415–29
[36] Baydin A G *et al* 2018 Automatic differentiation in machine learning: a survey *J. Mach. Learn. Res.* **18** 1–43

Chapter 8

An active subspace based swarm intelligence method with its application in optimal design problem

Jiu Luo, Ke Chen, Junzhi Chen, Yutong Lu and Yi Heng

Solving high-dimensional, constrained optimal design problems is computationally challenging due to increasing model complexity. This chapter introduces a dimensionality reduction method, the supervised active subspace approach, to enhance search efficiency for intelligent swarms. The methodology is demonstrated through an optimal design model for reverse osmosis desalination, showcasing its effectiveness. The active subspace-based swarm intelligence optimization framework is also applicable to other complex design challenges, including protein structure modeling and aircraft topology optimization.

8.1 Introduction

Parts of this section have been reproduced with permission from [25]. Copyright 2024 Springer Nature.

Multi-scale design, widely applied in many research areas, can usually be mathematically described as a partial differential equation (PDE) constrained optimization problem:

$$\begin{aligned} &\min_{\mathbf{x}} \ J(\mathbf{x}, \mathbf{m}) \\ &\text{s. t.} \\ &\mathbf{F}(\mathbf{x}, \mathbf{m}) = \mathbf{0} \\ &\mathbf{C}(\mathbf{x}, \mathbf{m}) \leqslant \mathbf{0}, \end{aligned} \tag{8.1}$$

doi:10.1088/978-0-7503-6174-3ch8

where **x** and **m** denote the design variable and state variable. J is the objective of interest. The constraints **F** and **C** are usually related to high-dimensional, multi-scale, multi-physics PDE systems subject to realistic initial and boundary conditions in practice. PDE-constrained optimization has broad and important applications in the engineering and science domains, including optimal design, optimal control, and inverse problems. Using traditional gradient-based approaches is usually expensive in terms of computational cost, particularly in high-dimensional cases. The primary focus of research is on the computation of feasible local optima. The rate of convergence may be slow due to the analytical complexity, computational intractability, and non-convex nature of the system.

PDE-constrained optimization has extensive and important applications in engineering and science. Current research mainly focuses on calculating feasible local optimal solutions. Traditional gradient-based optimization methods, such as the conjugate gradient method [1], Gaussian–Newton method [2], and Levenberg–Marquardt method [3], usually need to solve hundreds or more PDEs. These methods are very time-consuming and require a lot of computing resources. It is worth mentioning that mixed-integer PDE-constrained optimization (MIPDECO) combines the difficulties of integer programming and PDE processing. Traditional algorithms, such as the branch-bound method [4] and cut plane method [5], usually face dimension disasters, and require substantial storage space [6] when dealing with discrete variables. One of the main difficulties in solving the MIPDECO problem is how to solve the forward simulation problem accurately and effectively.

There are mainly two solution frameworks referring to gradient-based means: discretize-then-optimize and optimize-then-discretize. In the discretize-then-optimize framework, differential equations are discretized completely, reducing the infinite-dimensional optimization problem into a finite-dimensional optimization problem and solving it with efficient nonlinear programming solvers. This framework typically leads to a large and sparse optimization system which is then solved. Liu *et al* [7] utilize the discretize-then-optimize framework to solve elliptic PDE-constrained optimal control problems and introduce discretization criteria and regularization terms to enhance performance. Wilcox *et al* [8] adopt the discontinuous Galerkin method to discretize the linear hyperbolic system and discuss the computation of derivatives for PDE-constrained optimization problems under the discretization scheme. Inversely, the optimize-then-discretize framework derives the continuous optimality conditions analytically through the Lagrange multiplier method and Karush–Kuhn–Tucker conditions in general and then discretizes the adjoint differential equations with appropriate means, such as the finite difference method and finite element method. The optimize-then-discretize method can become more numerically complex and cause convergence problems due to inaccurate gradients [8]. Dontchev *et al* [9] utilize Runge–Kutta approximations to solve optimal control problems and demonstrate that in the adjoint system, the error of the approximating control can reach $O(h^2)$ where h is the mesh spacing. Becker *et al* [10] propose a stabilization scheme for finite element methods and solve convection–diffusion-equation based optimal control problems using both discretize-then-optimize and optimize-and-discretize frameworks. The results show that these two frameworks coincide with the suggested discretization scheme. There are also some numerical methods emerging from these two frameworks. Habeck *et al* [11] derive the lower and upper bounds of ordinary differential equations (ODEs) and utilize the regular mixed-integer nonlinear programming (MINLP) method branch-and-bound to solve the optimization problem, which is shown to be convergent. In engineering practice, the feasible region of optimal design or control problems is normally rugged with many local minima. However, gradient-based optimization relies greatly on the choice of initial values and is easily trapped in a local optimum.

In recent years, derivative-free optimization has aroused great interest. The lack of derivative information brings about efficiency for the optimization procedure. Derivative-free methods are normally divided into three classes: local search, global search, and stochastic search [12]. Local search requires a single initial point and proceeds by making small movements through various strategies. In contrast, global search explores the whole bounded region during each iteration. Stochastic methods, such as population-based evolutionary optimization algorithms, have become efficient and powerful techniques for solving the mixed-integer programming (MIP) problems, such as the genetic algorithm (GA) and particle swarm optimization (PSO). Population-based optimization methods are naturally capable of handling both continuous and discrete variables and are appropriate for parallel computing due to their characteristic of multi-agents. When dealing with multi-modal and highly constrained optimization problems, these approaches show robustness with less possibility of being trapped at a local minimum.

However, the convergence of the original population-based optimization method is not fast enough to handle complex problems. Several variants have been proposed to enhance the performance. An information-guided GA approach is provided by Young [13] to solve the MINLP optimization problem, which demonstrates a novel information-entropy-guided mutation and has been proved to be more efficient than the traditional GA. Tometzki *et al* [14] propose novel initialization techniques for a hybrid GA to solve a two-stage stochastic MIP problem, demonstrating faster convergence and better results compared to the regular GA. The GA is also used to solve discrete-time optimal control problems [15]. Abo-Hammour *et al* [16] propose a novel continuous GA to solve the ODE-constrained optimization problem, which outperforms other conventional methods. The approach randomly generates control smooth curves for each individual during the initialization stage, and the featured crossover and mutation operators are proposed accordingly. Sahoo *et al* [17] use the discretize-then-optimize framework and real-coded GA to solve the PDE-constrained optimization problem, which proved to be an efficient and stable approach.

8.2 Modeling and methods

PDE-constrained optimization problems usually involve high-dimensional design and control variables. If the traditional gradient-based optimization algorithm is used for such problems, the optimal solution can easily fall into the local optimum, and the optimization performance greatly depends on the initial solution. The swarm intelligent optimization algorithm without gradient often has problems such as slow convergence rates and large-scale calculational demand for objective functions. To solve this problem, this chapter introduces an optimization method based on supervised dimensional reduction learning of the AS, combined with the population intelligence optimization algorithm (particle swarm algorithm), to study the PDE-constrained optimization problem by identifying the direction of the fluctuation function in the original variable space.

8.2.1 Active subspace method

The AS method was initially proposed for sensitivity analysis of high-dimensional problems and for establishing surrogate models. When the model's input space is high-dimensional, achieving an even distribution of sample points requires numerous samplings. Each sample necessitates the computation of a complex system, and large-scale sampling is highly time-consuming, posing significant challenges for industrial applications. AS technology can identify and use the special mapping structure, transforming the high-dimensional problem into low-dimensional space. By analyzing the centered gradient covariance matrix, AS technology can monitor the input disturbance direction of the greatest influence on the output to accelerate the analysis process. This is particularly effective for high-dimensional problems (dozens to thousands of dimensions).

In the AS method, we define the center-free covariance matrix $\mathbf{C}$ of the gradient vector as follows:

$$\mathbf{C} = \mathbb{E}[(\nabla_{\mathbf{x}} f)(\nabla_{\mathbf{x}} f)^{\mathrm{T}}] = \int (\nabla_{\mathbf{x}} f)(\nabla_{\mathbf{x}} f)^{\mathrm{T}} \rho \mathrm{d}\mathbf{x}, \tag{8.2}$$

where $\mathbb{E}[\cdot]$ is the expectation operator, $\nabla_{\mathbf{x}}$ is the gradient of the objective function, and ρ represents the probability density. When the expected values cannot be calculated analytically, it can be approximated by sampling methods, such as Monte Carlo sampling and Latin hypercubic sampling [18]. By the sampling method, the matrix $\mathbf{C}$ can be approximated as

$$\mathbf{C} \approx \frac{1}{M}\sum_{i=1}^{M}(\nabla_{\mathbf{x}} f(\mathbf{x}_i))(\nabla_{\mathbf{x}} f(\mathbf{x}_i))^{\mathrm{T}}. \tag{8.3}$$

The subscript i represents the ith sample in the sampling set, and M is the sampling quantity. The matrix $\mathbf{C}$ defines the expectation of the outer product of the gradient vector, whose diagonal elements are the average of the partial derivative square. Since $\mathbf{C}$ is symmetric and positive semi-definite, its eigenvalue decomposition can be expressed as

$$\mathbf{C} = \mathbf{W}\Lambda\mathbf{W}^{\mathrm{T}},\ \Lambda = \begin{bmatrix} \Lambda_1 & \\ & \Lambda_2 \end{bmatrix},\ \mathbf{W} = [\mathbf{W}_1\mathbf{W}_2], \tag{8.4}$$

where $\mathbf{W}$ represents the $m \times m$ orthogonal matrix containing eigenvectors, also the rotation matrix of the original input space. $\boldsymbol{\Lambda}$ is the descending diagonal matrix of eigenvalues, that to some extent reflects the degree of change of the function value in the direction of the relevant eigenvector. The relationship between the eigenvalue decomposition of the matrix **C** and the functional gradient can be expressed by the following equation:

$$\int_{\Omega}((\nabla_{\mathbf{x}} f)^{\mathrm{T}}\mathbf{w}_i)^2\rho \mathrm{d}\mathbf{x} = \lambda_i, \quad i = 1, \cdots, m, \tag{8.5}$$

this is due to $\lambda_i = \mathbf{w}_i^{\mathrm{T}}\mathbf{C}\mathbf{w}_i = \mathbf{w}_i^{\mathrm{T}}\left(\int_{\Omega}(\nabla_{\mathbf{x}} f)(\nabla_{\mathbf{x}} f)^{\mathrm{T}}\rho \mathrm{d}\mathbf{x}\right)\mathbf{w}_i = \int_{\Omega}((\nabla_{\mathbf{x}} f)^{\mathrm{T}}\mathbf{w}_i)^2\rho \mathrm{d}\mathbf{x}$. Therefore, if an eigenvalue of matrix **C** is very small or close to 0, the mean inner product of the directional derivative and the eigenvector $\mathbf{w}_i$ is very small (orthogonal). If f is a continuous function, the inner product between the gradient of any point and the eigenvector will be zero. Through the energy decay of the eigenvalues in specific cases, we can compare the eigenvalues with the eigenvector matrix into active and inactive parts, Λ_1 and Λ_2, $\mathbf{W}_1$ and $\mathbf{W}_2$. Vectors in the inactive and the AS can be expressed as

$$\mathbf{y} = \mathbf{W}_1^{\mathrm{T}}\mathbf{x}, \ \eta = \mathbf{W}_2^{\mathrm{T}}\mathbf{x}. \tag{8.6}$$

Similarly, **x** can be reconstructed as

$$\mathbf{x} = \mathbf{W}\mathbf{W}^{\mathrm{T}}\mathbf{x} = [\mathbf{W}_1\mathbf{W}_2]\begin{bmatrix}\mathbf{W}_1^{\mathrm{T}} \\ \mathbf{W}_2^{\mathrm{T}}\end{bmatrix}\mathbf{x} = \mathbf{W}_1\mathbf{y} + \mathbf{W}_2\eta, \tag{8.7}$$

With the development of the linear AS, nonlinear types of AS methods have gradually been proposed, such as kernel-based AS [19] and active manifold subspace [20]. The basic idea of the kernel-based AS is to project the original variable space through feature learning into the nonlinear parameter space Φ, such as the random Fourier features, and then build the AS. This method is verified to be more accurate than linear ASs in constructing Gaussian response surfaces. Similarly, the active manifold approach projects the high-dimensional variable space into the one-dimensional variable function space, and reduces the dimension by defining the manifold curve y with that most consistent with the direction of the gradient,

$$\arg\max_{y} \int_0^1 \langle \nabla f(y(x)), y'(x)\rangle \mathrm{d}x, \tag{8.8}$$

where $\langle\cdot\rangle$ represents the inner product in the Euclidean distance space. When the function gradient is consistent with the derivative direction, equation (8.8) achieves the maximal integration of the inner product. This method demonstrates effective application in the fields of parameter sensitivity analysis and surrogate modeling.

Large-scale simulation-based optimization functions usually struggle to obtain an analytical gradient solution or find it difficult to calculate the gradient. Based on such problems, the construction of a gradient covariance matrix in the AS method can approximate [21, 22] using a local linear, global model, finite difference, or Gaussian process. How to sample the original space and the number of samples need to be considered. For the AS construction problem, the sampling number M can be expressed as [23]

$$M = \gamma n_{\mathrm{AS}} \log(n_{\mathrm{Var}}), \tag{8.9}$$

where γ is the oversampling factor with a range of 2–15, and n_{Var} and n_{AS} are the original and AS variable dimensions, respectively.

8.2.2 Particle swarm optimization algorithm

The PSO is an intelligent optimization method proposed by Kennedy and Eberhart [24], inspired by the social behavior of biological populations. The algorithm simulates the way individuals interact and information is shared, with each particle constantly changing the search pattern based on its own and the learning experience of others to find approximately optimal results. In the variant-based particle swarm algorithm, each particle i can be described by the m-dimensional position vector and the velocity vector. Initially, each particle is assigned a random position and velocity, keeping them within the feasible domain. Based on individual and global optima, they update individual speed and position (see figure 8.1), and finally perform variation operation. The above procedure is repeated until the stop condition is met. The basic mode of the speed and position update variation process is

$$v_i^{k+1} = \omega_k v_i^k + c_1 r_1\left(p_i^k - x_i^k\right) + c_2 r_2\left(p_g^k - x_i^k\right), \tag{8.10}$$

$$x_i^{k+1} = x_i^k + v_i^k, \tag{8.11}$$

$$x_{i,m}^{k+1} = x_i^{k+1}\,(1 + \mathrm{Gauss}(\theta)), \tag{8.12}$$

where v_i^k and x_i^k represent the velocity and position vector of particle i in the k iteration, v_i^{k+1}, x_i^{k+1}, and $x_{i,m}^{k+1}$ are the particle velocity, position, and post-mutation position in the $k+1$ iteration, and ω represents the inertia coefficient. c_1 and c_2 are individual and global acceleration factors, and r_1 and r_2 are random vectors from 0 to 1. p_i^k is the historical best position for particles, and p_g^k is the best position for all particles in the kth iteration. Gauss(θ) returns a random number that follows a Gaussian distribution, where θ is the standard deviation.

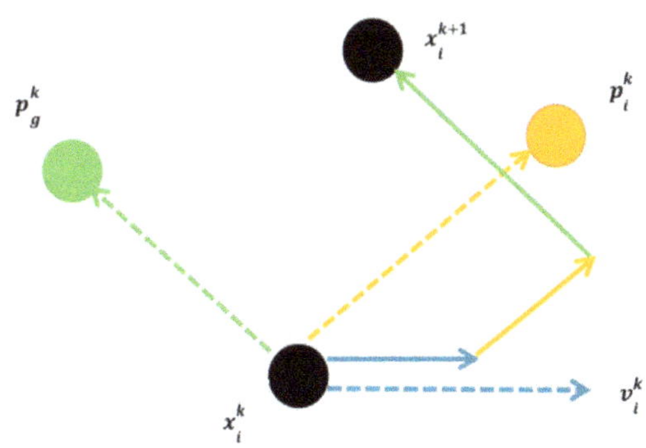

Figure 8.1. Conception of particle position update.

To enhance the optimization ability of each particle, the damping inertia coefficient is used in each iteration. The following damping strategy is an improvement of the original damping strategy with the aim of keeping the population enhancing the local search capability near the search endpoint:

$$\begin{cases} \omega^{k+1} = \omega^k - \dfrac{\omega^k k}{\text{Maxiter}}, & \text{if } k \leqslant \beta \cdot \text{Maxiter} \\ \omega^{k+1} = \omega^k, & \text{if } k > \beta \cdot \text{Maxiter}, \end{cases} \tag{8.13}$$

where Maxiter is the maximum number of iterations and the range of β is from 0 to 1.

8.2.3 Active subspace particle swarm optimization algorithm

This chapter combines the linear AS method with the particle swarm algorithm, proposing the active subspace particle swarm optimization (ASPSO) algorithm [25]. Although various nonlinear AS algorithms are primarily used in sensitivity analysis or agent-based modeling, it is essential to consider the reconstruction in the optimization scenario when utilizing the AS to identify important search directions. The main idea of the swarm intelligence optimization algorithm based on the AS is to make each individual in the population evolve along the directions of global optima, individual optimum, and AS (the function fluctuates most strongly) simultaneously, thereby avoiding the dimension disaster problem and improving the convergence speed. After randomly initializing a certain number of individuals, before updating the position and velocity vectors, we project the individuals in the original variable space into the established AS,

$$x_{\text{AS}} = \mathbf{W}_1^{\text{T}} x. \tag{8.14}$$

The subscript AS indicates that the variable is projected into the AS. The velocity v_{AS} and position x_{AS} updates should be rewritten as

$$v_{i,\text{AS}}^{k+1} = \omega_k v_{i,\text{AS}}^k + c_1 r_1 \left(p_{i,\text{AS}}^k - x_{i,\text{AS}}^k\right) + c_2 r_2 \left(p_{g,\text{AS}}^k - x_{i,\text{AS}}^k\right), \tag{8.15}$$

$$x_{i,\text{AS}}^{k+1} = x_{i,\text{AS}}^k + v_i^{k+1}, \tag{8.16}$$

$$x_{i,\text{AS},m}^{k+1} = x_{i,\text{AS}}^{k+1} \left(1 + \text{Gauss}(\theta)\right). \tag{8.17}$$

The symbol is consistent with the standard PSO algorithm but is updated in the AS. Before re-evaluating the fitness function, the position vector reflection should also be projected into the original variable space,

$$x = \mathbf{W}_1 x_{\text{AS}} + \mathbf{W}_2 \eta, \tag{8.18}$$

where η can be considered as the information loss in the dimension reduction process, and we can reconstruct the original space through the local sampling method of the objective function orientation under a certain probability distribution.

Algorithm 8.1. ASPSO with mutation operator

Input: acceleration factor c_1 and c_2, initial coefficient ω, max iteration number Maxiter, sampling amount M, differential step h, dimension of AS n_{AS}, population size N, bounds of input variables VarBound, bounds of velocity VelBound, mutation factor η and mutation scale σ.

1. M samples are taken from the input space through the Latin hypercube sampling method.
2. Construct the AS and select the reduced dimension through equation (8.3) and (8.4).
3. Randomly initialize the position and the velocity of the population. Evaluate the initial fitness value.
4. Update the global best of the whole population and personal best of each particle. AS projection of each particle.
5. Update the velocity of each particle in the AS through equation (8.15).
6. Update the position of each particle through equation (8.16).
7. Mutation operation of each particle in the AS through equation (8.17)
8. Reversely map to the original space. Evaluate the fitness of each particle.
9. Update the damping coefficient ω through equation (8.13)
10. Repeat 2–7 until the stopping criterion is satisfied.
11. Return the global best in the last iteration.

8.3 Applications and analysis

8.3.1 Benchmark problem test for ASPSO

In order to verify the effectiveness of the proposed ASPSO, PSO, GA, ant colony optimization (ACO), active subspace genetic algorithm (ASGA), and ASPSO are compared based on the typical benchmark problems of the Rastrigin, sphere, Rosenbrock, Ackley, and Bohachevsky functions. Figure 8.2 shows the contour plot of four benchmark problems, where the Rastrigin function and Ackley function have several local minima, and the sphere function and Rosenbrock function represents the convex and unimodal functions.

In this experiment, the dimension of four problems is set as 40 and they have the common minimum of zero. The variable range is $[-5.12, 5.12]^d$. The optimized outcomes achieved by PSO, GA, ACO, ASGA, and ASPSO are displayed in table 8.1. The default values from the MATLAB toolbox are used for the hyperparamters in both GA and ACO. The same hyperparameters are also used for PSO and ASPSO, as well as GA and ASGA. The AS dimension is set to 3 and the population size remains fixed at 200. The average and standard deviation are computed based on 15 algorithm runs. Upon analysis, it is evident that ASPSO and ASGA outperform PSO, GA, and ACO significantly across all benchmark functions, highlighting the effectiveness of integrating the AS. Additionally, with

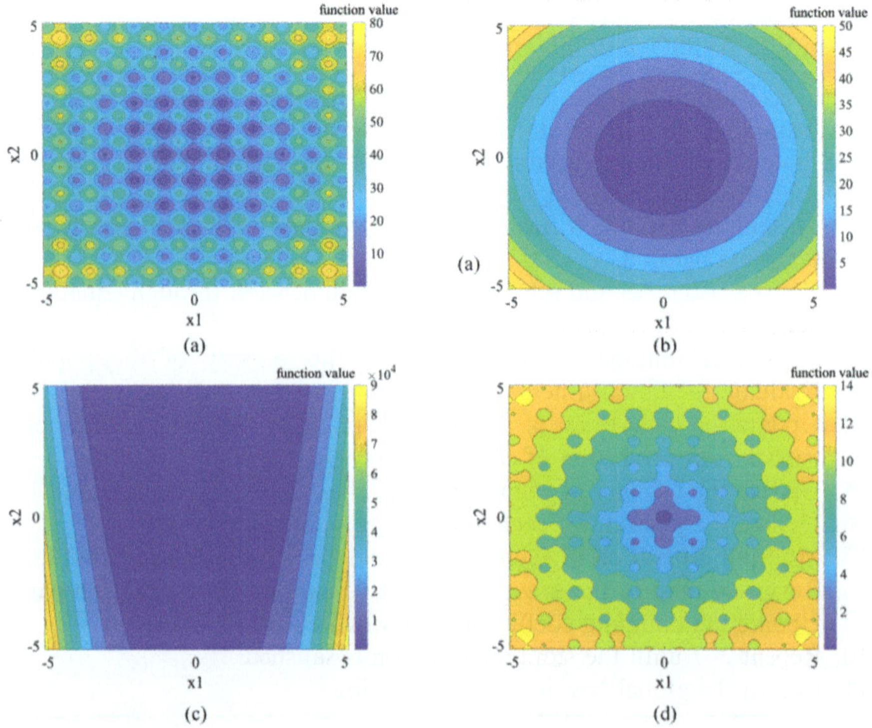

Figure 8.2. Contour plot of benchmark problems. (a) Rastrigin function, (b) sphere function, (c) Rosenbrock function, and (d) Ackley function.

subspace projection, the algorithms exhibit enhanced numerical stability, as indicated by the lower standard deviation in various experiments compared to the other population-based intelligent algorithms.

8.3.2 PDE constraints for multi-scale optimal design for RO seawater desalination

Against the global backdrop of diminishing freshwater resources, efficiently harnessing the vast seas as a source of freshwater extraction has become increasingly critical. Reverse osmosis (RO) desalination, favored for its cost-effectiveness and modular design, now captures a significant portion of the market. While advancements in membrane technology have successfully increased the water permeation flux, they have also introduced challenges, including concentration polarization and membrane fouling. Consequently, conducting multi-scale studies on RO desalination systems is crucial.

Optimal design problems, modeled through simulations, represent a classic challenge within the realm of PDE constraints. Building on prior research [26], this section details a multi-scale optimal design model specifically for RO seawater desalination, presented as follows:

Table 8.1. Benchmark test results for PSO, GA, ACO and ASGA. (Reproduced with permission from [25]. Copyright 2024 Springer Nature.)

Benchmark problem	Method	Optimal value: Best	Average	Worst	Standard deviation
Rastrigin function	PSO	58.65	93.03	147.3	21.35
	GA	96.98	140.4	188.5	24.67
	ACO	378.4	426.7	459.7	22.75
	ASGA	0	7.200×10^{-14}	6.821×10^{-13}	1.903×10^{-13}
	ASPSO	0	0	0	0
Sphere function	PSO	0.081 08	0.2603	0.5502	0.1430
	GA	9.648	13.80	21.04	2.955
	ACO	28.26	33.13	44.56	4.213
	ASGA	6.123×10^{-18}	3.930×10^{-16}	3.851×10^{-15}	9.632×10^{-16}
	ASPSO	8.179×10^{-25}	5.201×10^{-21}	3.732×10^{-20}	1.235×10^{-20}
Rosenbrock function	PSO	165.0	230.3	510.1	83.96
	GA	1339	3141	5751	1331
	ACO	6565	14 774	25 256	5010
	ASGA	38.95	38.95	38.95	2.459×10^{-5}
	ASPSO	38.86	38.95	38.99	$0.035\,46 \times 10^{-2}$
Ackley function	PSO	0.3399	1.008	1.916	0.4694
	GA	2.972	3.576	4.133	0.2954
	ACO	4.971	5.575	6.131	0.3173
	ASGA	2.616×10^{-10}	1.028×10^{-8}	7.135×10^{-8}	1.824×10^{-8}
	ASPSO	8.566×10^{-13}	6.546×10^{-11}	6.731×10^{-10}	1.702×10^{-10}

$$\begin{aligned} &\min_{\overline{\gamma}} (\text{SEC} + B \cdot A_{\text{tot}}) \\ &\text{s.t.} \quad \mathbf{F}(\mathbf{u}, c; \mathbf{x}) = \mathbf{0}, \\ &\qquad \mathbf{H}(\Delta P, J_{\text{W}}, Q; \mathbf{x}) = \mathbf{0}, \\ &\qquad \mathbf{C}(\mathbf{u}, c; \Delta P, J_{\text{W}}, Q; \mathbf{x}) \leqslant \mathbf{0}, \end{aligned} \tag{8.19}$$

where the specific energy consumption of the RO seawater desalination system is denoted as SEC, while A_{tot} represents the system's total membrane area. The coefficient B is used for multi-objective optimization. The system incorporates an equality constraint, labeled $\mathbf{F}$, which is a PDE relevant to the submillimeter scale model of the RO system; here, $\mathbf{u}$ and c serve as the state variables representing velocity and salt concentration, respectively. Another constraint, $\mathbf{H}$, is a one-dimensional differential algebraic equation set that operates at the meter scale. This constraint involves ΔP, J_{W}, and Q as state variables, which signify changes in transmembrane pressure, flow rate, and water permeation flux across the spatial dimension. Design variables are defined as $\mathbf{x} = [\mathbf{x}_1, \mathbf{x}_2]$. The geometric parameters $\mathbf{x}_1 = [D_{1,\text{A}}, D_{2,\text{A}}, D_{1,\text{B}}, D_{2,\text{B}}, D_{\text{tot}}, \alpha]$ (see figure 8.3) and the system model parameters $\mathbf{x}_2 = [L_{\text{mem}}, n_{\text{sp}}, n_{\text{mem}}, N_{\text{pv}}]$, with the latter three being integers subject to specific constraints.

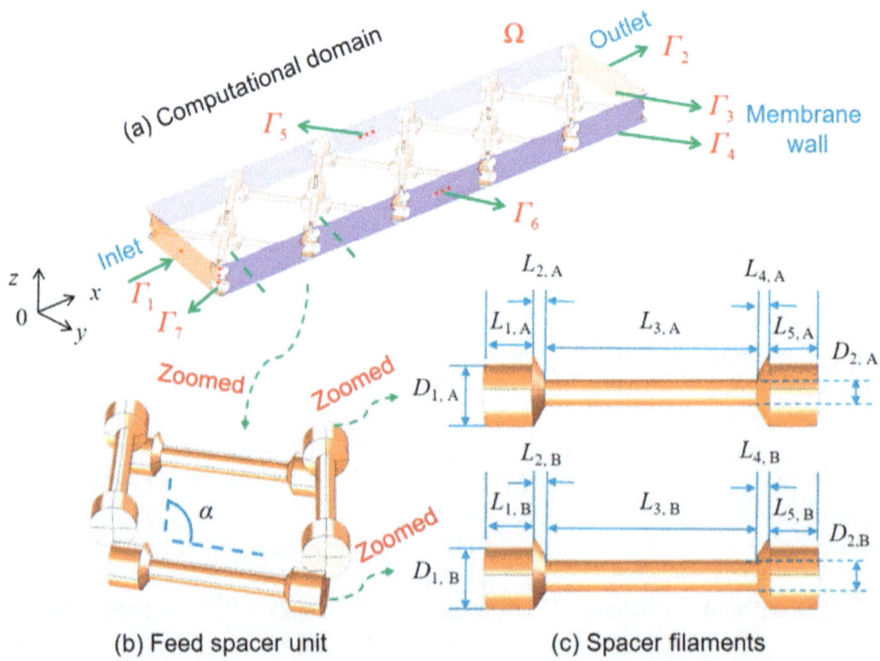

Figure 8.3. RO desalination sub-millimeter scale 3D CFD model: (a) the calculation domain of five gasket units, (b) a single gasket unit, and (c) a gasket unit filament. (Reproduced with permission from [25]. Copyright 2024 Springer Nature.)

The expression for the optimization target ratio energy consumption SEC is shown below:

$$\mathrm{SEC} = r\left[\frac{Y(P_0 - P_{\mathrm{in}})}{\eta_{\mathrm{HPP}}} + \frac{(1-Y)(P_0 - P_{\mathrm{in}})}{\eta_{\mathrm{BP}}} - \frac{(1-Y)\eta_{\mathrm{PX}}(P_{X=1} - P_{\mathrm{in}})}{\eta_{\mathrm{BP}}}\right], \tag{8.20}$$

In the given equation, the coefficient $r = Q_0/(36\eta_{\mathrm{EM}}\overline{J}_{\mathrm{W}}A_{\mathrm{tot}})$, where $\overline{J}_{\mathrm{W}}$ denotes the average water permeation flux, and Y is the freshwater recovery rate, defined as $Y = \overline{J}_{\mathrm{W}} \cdot A_{\mathrm{tot}}/Q_0$. Here, Q_0 signifies the inlet flow rate. The terms P_0 and $P_{X=1}$ correspond to the hydraulic pressures in the main body and the concentrate, respectively, while P_{in} and P_{out} indicate the inlet and outlet pressures. The coefficients η_{EM}, η_{HPP}, η_{BP}, and η_{PX} represent efficiencies of the electric motor, high-pressure pump, booster pump, and isobaric exchanger, respectively. The expression for the average water permeation flux can be defined as

$$\overline{J}_{\mathrm{w}} = \frac{\int_0^1 [J_{\mathrm{w}} \cdot A_{\mathrm{tot}}]\mathrm{d}X}{\int_0^1 A_{\mathrm{tot}}\mathrm{d}X}, \tag{8.21}$$

where J_{w} is obtained by solving the one-dimensional system of differential algebraic equations (equation (8.24)).

The submillimeter scale computational fluid dynamics (CFD) model of the membrane component of the RO desalination system is shown in figure 8.3(a). $\Gamma_1 - \Gamma_7$ represent the inlet, outlet, and penetration wall of the computing domain, and the geometric form of the calculation domain is determined by the geometric design parameter x_1. In the computational domain we establish the Navier–Stokes equation, the continuity equation, and the convection–diffusion equations describing the fluid flow and matter transfer, and the corresponding boundary conditions,

$$\mathbf{F} = \begin{cases} \rho(c)(\mathbf{u} \cdot \nabla)\mathbf{u} = \nabla \cdot [-p\mathbf{I} + \mu(c)(\nabla\mathbf{u} + (\nabla \mathbf{u})^{\mathrm{T}})], & \text{in } \Omega, \\ \nabla \cdot \rho(c)\mathbf{u} = 0, & \text{in } \Omega, \\ \nabla \cdot (D_{\mathrm{s}} \nabla c) - \mathbf{u} \cdot \nabla c = 0, & \text{in } \Omega, \end{cases} \tag{8.22}$$

where $\mathbf{u} = (u, v, w)$, is the velocity vector along the spatial coordinate $\boldsymbol{x} = (x, y, z)$, p is the pressure, $\rho(c)$ and $\mu(c)$, respectively, represent the fluid density and viscosity related to the salt concentration, and D_{s} is the diffusion coefficient at room temperature. The corresponding boundary conditions are given in equation (8.23), the laminar boundary condition $f(U_0)$ [26] is applied to the inlet Γ_1, and U_0 is the inlet velocity parameter. The L_{p}, ΔP, f_{os}, c_0 in the boundary conditions represent membrane permeability, inlet transmembrane pressure, van Hove factor, and inlet concentration, respectively. The information of velocity field, pressure field and concentration field obtained after solving the fluid flow and mass transfer equations will be used to calculate the pressure drop F_1 and mass transfer coefficient F_2 to assist in the subsequent meter-scale modeling and computing of RO seawater desalination system:

$$\begin{cases} u = f(U_0),\ v = w = 0,\ c = c_0, & \text{on } \Gamma_1, \\ P = 0,\ \mathbf{n} \cdot (-D_{\mathrm{s}} \nabla c + c\mathbf{u}) = 0, & \text{on } \Gamma_2, \\ w = L_{\mathrm{p}}\left[\Delta P_0 - f_{\mathrm{os}} \cdot c\right],\ \mathbf{n} \cdot (-D_{\mathrm{s}} \nabla c + c\mathbf{u}) = 0, & \text{on } \Gamma_3, \\ w = -L_{\mathrm{p}}\left[\Delta P_0 - f_{\mathrm{os}} \cdot c\right],\ \mathbf{n} \cdot (-D_{\mathrm{s}} \nabla c + c\mathbf{u}) = 0, & \text{on } \Gamma_4, \\ \mathbf{u}_5 = \mathbf{u}_6,\ P_5 = P_6,\ c_5 = c_6, & \text{on } \Gamma_5 \cup \Gamma_6, \\ \mathbf{u} = \mathbf{0},\ \mathbf{n} \cdot (-D_{\mathrm{s}} \nabla c + c\mathbf{u}) = 0, & \text{on } \Gamma_7. \end{cases} \tag{8.23}$$

The meter-scale model of RO seawater desalination system membrane components can be described by the following differential algebraic equation:

$$\mathbf{H} = \begin{cases} \dfrac{\mathrm{d}Q}{\mathrm{d}X} = -J_{\mathrm{w}} \cdot A_{\mathrm{tot}}, & X = 0,\ Q = Q_0, \\ \dfrac{\mathrm{d}(\Delta P)}{\mathrm{d}X} = -F_1 \cdot L_{\mathrm{mem}} \cdot n_{\mathrm{mem}}, & X = 0,\ \Delta P = \Delta P_0, \\ J_{\mathrm{W}} = L_{\mathrm{P}}\left[\Delta P - Q_0\Delta\pi_0/Q \cdot \exp(J_{\mathrm{W}}/F_2)\right], & \\ A_{\mathrm{tot}} = N_{\mathrm{pv}} n_{\mathrm{mem}}\left(n_{\mathrm{sp}}/n_{\mathrm{sp},\,0}\right)\left(L_{\mathrm{mem}}/L_{\mathrm{mem},\,0}\right)A_0, & \\ Q = 3600 \cdot N_{\mathrm{pv}} n_{\mathrm{sp}} U_{\mathrm{avg}} \varepsilon, & \end{cases} \tag{8.24}$$

where $X(X \in [0, 1])$ represents the dimensionless length of the membrane assembly, $\Delta\pi_0$ represents the inlet transmembrane osmotic pressure, and L_P represents the water permeability of the RO membrane. $n_{sp,0}$, $L_{mem,0}$, A_0 are the number of gaskets, membrane assembly length, and membrane area in the classical commercial RO membrane assembly, all are constant. ε is the membrane porosity with changing geometric parameters, and U_{avg} is the system mean flow rate. Pressure drop and mean mass transfer coefficients F_1 and F_2 can be obtained by solving the CFD model (equation (8.22)).

In order to make the optimization results have some practical significance, the inequality constraints related to geometric parameters, concentration polarization factor (CPF), and mean flow rate need to be satisfied:

$$\mathbf{C} \leqslant \mathbf{0}: = \begin{cases} \max\{D_{1,A}, D_{1,B}\} - D_{tot} \leqslant 0, \\ D_{tot} - (D_{1,A} + D_{1,B}) - D_{max} \leqslant 0, \\ Y_{min} - Y \leqslant 0, \\ \mathrm{CPF}(X) - \mathrm{CPF}_{max} \leqslant 0, \quad 0 \leqslant X \leqslant 1, \\ \overline{U}_{arg,in} - \overline{U}_{avg,max} \leqslant 0. \end{cases} \tag{8.25}$$

Among them, the geometric design parameter $\mathbf{x}_l$ needs to meet certain conditions, and the freshwater recovery rate Y needs to meet the constraint of the minimum recovery rate. In order to reduce membrane pollution, the CPF and the inlet average velocity cannot exceed a certain range.

Solving the multi-scale optimal design model for RO seawater desalination involves not only the CFD mathematical model at sub-millimeter scale but also differential algebra equations for meter-scale modeling, which increases the optimization difficulty of this problem.

8.3.3 Simulation experiments for multi-scale optimal design for RO seawater desalination

This section uses the ASPSO framework to solve the optimization design problem of RO seawater desalination, and the optimization variables are the continuity variables and the integer variables with a specific range,

$$\mathbf{x} = \left[D_{1,A}, D_{2,A}, D_{1,B}, D_{2,B}, D_{tot}, \alpha; L_{mem}, n_{sp}, n_{mem}, N_{pv}\right]. \tag{8.26}$$

Since this is an MIPDECO problem, we apply the penalty function method to treat the integer variable [27].

The algorithm parameters of the ASPSO used are shown in table 8.2, and the AS technique aims to rotate the high-dimensional input space to reveal the direction of maximum fluctuations. To better construct and understand the AS method, we scaled the input vector to between [0,1]:

$$\breve{x}_i = \frac{x_i - x_{i,\,min}}{x_{i,\,max} - x_{i,\,min}}, \quad i = 1, 2, \cdots, 10. \tag{8.27}$$

Table 8.2. ASPSO algorithm parameters for the optimal design problem of RO seawater desalination.

Parameters	Value
Sampling factor, γ	5
Active subspace dimension, n_{AS}	10
The original space dimension, n_{Var}	10
Differential step length, h	10^{-3}
Maximum iteration times, Maxiter	100
Initial damping coefficient, ω_0	0.5
Individual acceleration factor 1, c_1	0.3
Individual acceleration factor 2, c_2	0.3
Variant scale, σ	0.5
Probability of variation, η	0.3
Damage strategy parameters, β	0.5
Stop iterative tolerance, tol	10^{-4}

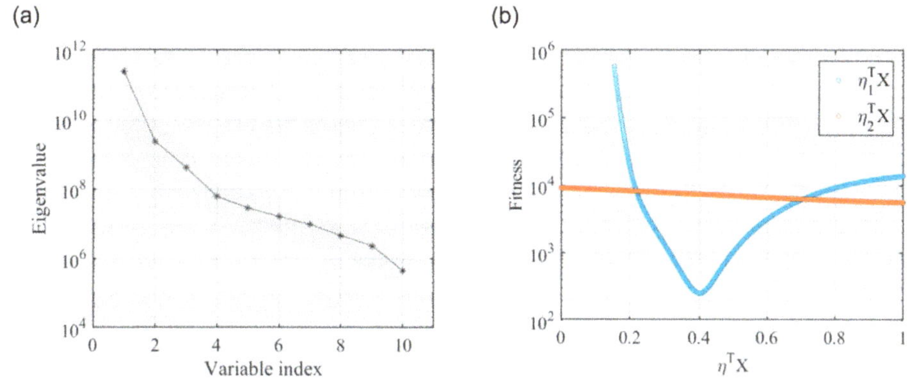

Figure 8.4. (a) Eigenvalue decay of the AS with the shaded region representing the bootstrap resampling interval. (b) Sufficient summary plot of the first eigenvector η_1 and the last eigenvector η_2 in **W**, where **x** represents the input variables varying along the specific eigenvector direction. (Reproduced with permission from [25]. Copyright 2024 Springer Nature.)

The Latin hypercube sampling strategy [18] was also used to approximate the expectation of the gradient covariance matrix. The decay of the AS is based on bootstrap resampling [28], and the decay is shown in figure 8.4(a). As can be seen from the figure, there is no obvious gap between the eigenvalues of matrix **C**, and the eigenvalues are above 10^4, which means that the function value fluctuates greatly along all the eigenvectors. Therefore, the weight-based spatial projection method is not significant, and the subspace dimension is selected as the original spatial variable dimension. As shown in figure 8.4(b), where the eigenvectors η_1 and η_2 corresponding to the maximum eigenvalues and minimum eigenvalues change, the obvious eigenvector direction function fluctuates greatly (see the blue line in figure 8.4(b)), while the curve is relatively constant (see the red line in figure 8.4(b)). The AS method explores the univariate features of the original high-dimensional data.

Using the ASPSO framework proposed in this chapter, we optimize the optimal design problem of multi-scale RO desalination systems. In this example, the tolerance value of the stop iteration is set to 10E-5, or when there are 50 iterations, the iteration stops. Figures 8.5(a)–(c) show the comparison of the ASPSO optimization results with the energy consumption, total membrane area, and water penetration flux of the modified feed spacer structure [29], respectively. According to the figure, the energy consumption, total membrane area and penetration flux achieved performance improvements of 9%, 23%, and 30%, respectively. From figure 8.5(a), it is known that the ASPSO optimization results are very close to the theoretical optimal value of an RO seawater desalination system, but there is still a certain gap, which may be because the theoretical optimal value ignores the effects of friction loss and concentration polarization.

Figure 8.6(a) shows that when the multi-target optimization parameter B is 0, ASPSO shows an improvement in convergence rate relative to PSO, where the solid line represents the average global optima for each iteration, and the shaded area represents the standard deviation interval of 15 runs of different initialized populations. As we can see from the figure, as the number of iterations increases, the shaded overlap of the PSO and ASPSO gradually decreases. The standard deviation of ASPSO gradually tends to be zero, while the standard deviation of the PSO is still very large. This suggests that ASPSO has better stability than PSO. At the same time, the global optimal performance in this problem is also notably improved. When considering the membrane area in the objective function, the performance improvement of ASPSO is stronger (see figure 8.6(b)), and over 25 iterations, the standard deviation interval between ASPSO and PSO can be fully separated.

Table 8.3 compares the performance of the improved algorithm from several dimensions: the optimal result, the number of iterations needed to achieve the optimal result, and the time at different population sizes. At a population size of 120,

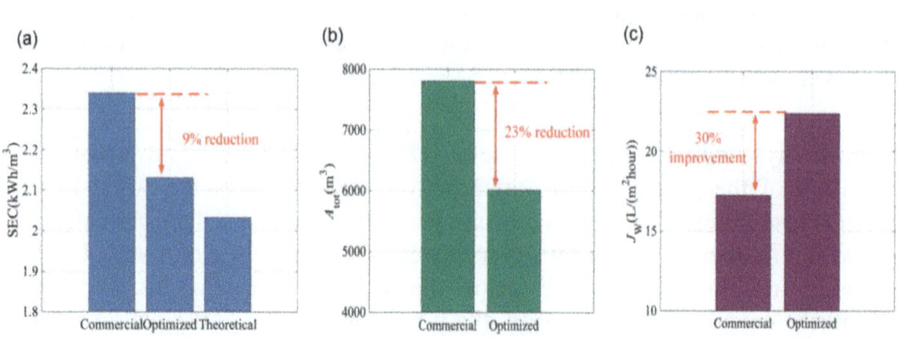

Figure 8.5. (a) Comparison of SEC between the commercial, optimized, and theoretical results. (b) Comparison of A_{tot} between the commercial and optimized results. (c) Comparison of J_w between the commercial and optimized results. (Reproduced with permission from [25]. Copyright 2024 Springer Nature.)

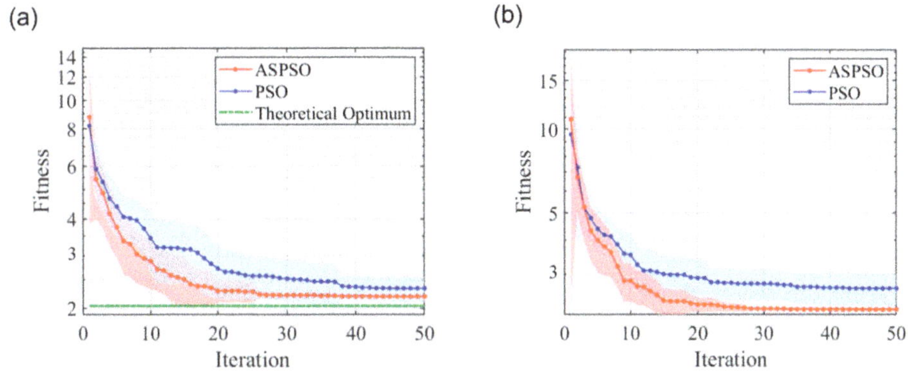

Figure 8.6. (a) Convergence plot of PSO and ASPSO with multi-objective optimization parameter B equal to zero. (b) Convergence plot of PSO and ASPSO with multi-objective optimization parameter B equal to 10^{-5}. (c) Convergence plot of ASPSO with different dimensions of AS and B equal to 10^{-5}. (Reproduced with permission from [25]. Copyright 2024 Springer Nature.)

Table 8.3. Optimization results of PSO and ASPSO when B equals 10^{-5}. (Reproduced with permission from [25]. Copyright 2024 Springer Nature.)

		Optimal value					
Population size	Method	Best	Average	Worst	Standard deviation	Generation	Time (s)
40	PSO	2.2361	2.4395	2.9344	0.2161	120.0 ± 37.5	276.9 ± 83.8
	ASPSO	2.1931	2.2236	2.2695	0.0170	149.9 ± 42.2	286.0 ± 89.5
120	PSO	2.2469	2.5694	3.1972	0.3156	96.9 ± 33.6	595.4 ± 196.8
	ASPSO	2.2160	2.2160	2.2328	0.0103	123.8 ± 33.8	697.0 ± 200.4
200	PSO	2.1957	2.2643	2.4405	0.0653	138.5 ± 49.0	1115.9 ± 391.6
	ASPSO	2.2022	2.2211	2.2702	0.0153	146.5 ± 39.0	1210.2 ± 328.0

the average optimum of ASPSO is 14% smaller than that of the PSO, indicating the strong global optimization of the ASPSO framework for high-dimensional computational problems. Meanwhile, the standard deviation of the 15 runs of the ASPSO framework is nearly ten times lower than that of PSO, which indicates the strong numerical stability of the proposed ASPSO algorithm. The mean and worst optimum of ASPSO stabilized between 2.19 and 2.27, while for PSO they fluctuated between 2.19 and 2.9. The calculation time of ASPSO is slightly longer than that of PSO, which may be caused by projection of particles from the original space into the active subspace in ASPSO and the premature convergence of PSO.

Figure 8.7 shows the convergence rate of ASPSO using different dimensions of AS. From the graph, it can be seen that as the dimension of the AS decreases, the convergence rate of ASPSO gradually declines. Notably, the convergence rate of

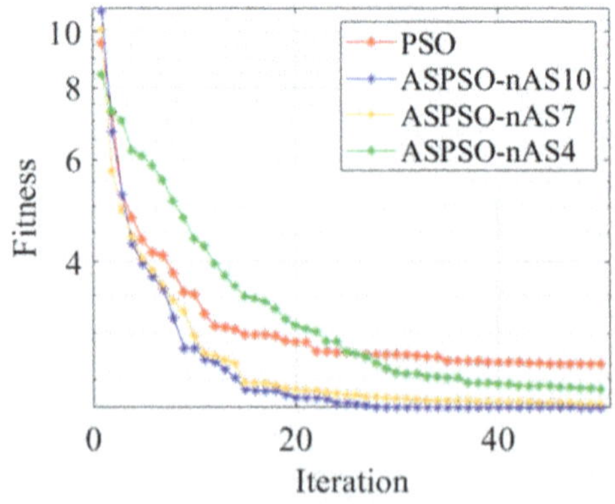

Figure 8.7. Convergence plot of ASPSO with different dimensions of AS and B equal to 10^{-5}. (Reproduced with permission from [25]. Copyright 2024 Springer Nature.)

ASPSO with four-dimensional AS is significantly slower than that of PSO. However, in general, the ASPSO algorithm converges faster, has better global optimization performance, and is less likely to fall into a local optimum.

8.4 Conclusions

Parts of this section have been reproduced with permission from [25]. Copyright 2024 Springer Nature.

Solving high-dimensional PDE-constrained optimization problems is usually computationally very expensive owing to the intractable and nonlinear nature of the complex PDEs and complex computational domains to be tackled. To enhance the efficiency of optimization design processes, the deployment of accurate and stable surrogate models or faster solvers is essential. This chapter introduces a universal optimization method that integrates AS techniques with the PSO algorithm, enabling effective, accurate, and stable resolution of optimization challenges.

In general, the ASPSO algorithm achieves better performance than the classic PSO and has abundant potential for application in many other high-dimensional optimal design problems. The future improvement of this framework could be focused on the information loss when high-dimensional input space is reduced to a lower structure. A better strategy for constructing the original space and the selection of the dimension of the AS should be considered for greater accuracy.

References

[1] Egger H *et al* 2009 Efficient solution of a three-dimensional inverse heat conduction problem in pool boiling *Inverse Prob.* **25** 095006

[2] Hernandez M 2021 Efficient momentum conservation constrained PDE-LDDMM with Gauss–Newton–Krylov optimization, semi-Lagrangian Runge–Kutta solvers, and the band-limited parameterization *J. Comput. Sci.* **55** 101470

[3] Nhu V H 2022 Levenberg–Marquardt method for ill-posed inverse problems with possibly non-smooth forward mappings between banach spaces *Inverse Prob.* **38** 015007

[4] Morrison D R *et al* 2016 Branch-and-bound algorithms: a survey of recent advances in searching, branching, and pruning *Discrete Optim.* **19** 79–102
[5] Marchand H *et al* 2002 Cutting planes in integer and mixed integer programming *Discrete Appl. Math.* **123** 397–446
[6] Manns P and Kirches C 2020 Improved regularity assumptions for partial outer convexification of mixed-integer PDE-constrained optimization problems *ESAIM: COCV* **26** 32
[7] Liu J and Wang Z 2019 Non-commutative discretize-then-optimize algorithms for elliptic PDE-constrained optimal control problems *J. Comput. Appl. Math.* **362** 596–613
[8] Wilcox L, Stadler G, Bui-Thanh T and Ghattas O 2015 Discretely exact derivatives for hyperbolic PDE-constrained optimization problems discretized by the discontinuous Galerkin method *J. Sci. Comput.* **63** 138–62
[9] Dontchev A L, Hager W W and Veliov V M 2000 Second-order Runge–Kutta approximations in control constrained optimal control *SIAM J. Numer. Anal.* **38** 202–26
[10] Becker R and Vexler B 2007 Optimal control of the convection–diffusion equation using stabilized finite element methods *Numer. Math.* **106** 349–67
[11] Habeck O 2020 Mixed-integer optimization with ordinary differential equations for gas networks *Dissertation* Technische Universität Darmstadt
[12] Wang X *et al* 2019 Well control optimization using derivative-free algorithms and a multiscale approach *Comput. Chem. Eng.* **123** 12–33
[13] Young C, Zheng Y, Yeh C and Jang S 2007 Information-guided genetic algorithm approach to the solution of MINLP problems *Ind. Eng. Chem. Res.* **46** 1527–37
[14] Tometzki T and Engell S 2009 Hybrid evolutionary optimization of two-stage stochastic integer programming problems: an empirical investigation *Evol. Comput.* **17** 511–26
[15] Michaels T C T, Weber C A and Mahadevan L 2019 Optimal control strategies for inhibition of protein aggregation *Proc. Natl Acad. Sci. USA* **116** 14593–8
[16] Abo-Hammour Z *et al* 2014 An optimization algorithm for solving systems of singular boundary value problems *Appl. Math. Inf. Sci.* **8** 2809
[17] Sahoo L *et al* 2013 An alternative approach for PDE-constrained optimization via genetic algorithm *J. Inform. Comput. Sci.* **8** 41–54
[18] Helton J C and Davis F J 2003 Latin hypercube sampling and the propagation of uncertainty in analyses of complex systems *Reliab. Eng. Syst. Saf.* **81** 23–69
[19] Romor F *et al* 2022 Kernel-based active subspaces with application to computational fluid dynamics parametric problems using the discontinuous Galerkin method *Int. J. Numer. Methods Eng.* **123** 6000–27
[20] Bridges R A *et al* 2019 Active manifolds: a non-linear analogue to active subspaces arXiv: 1904.13386
[21] Constantine P G, Dow E and Wang Q 2014 Active subspace methods in theory and practice: applications to kriging surfaces *SIAM J. Sci. Comput.* **36** A1500–24
[22] Constantine P G, Eftekhari A and Wakin M B 2015 Computing active subspaces efficiently with gradient sketching *IEEE 6th Int. Workshop on Computational Advances in Multi-Sensor Adaptive Processing (CAMSAP)* **2015** pp 353–6
[23] Constantine P G 2015 Discover the active subspace *Active Subspaces* SIAM Spotlights (Philadelphia, PA: Society for Industrial and Applied Mathematics) ch 3 21–44
[24] Kennedy J and Eberhart R 1995 Particle swarm optimization *Proc. Int. Conf. on Neural Networks* **4** 1942–8

[25] Chen K *et al* 2024 A rapid-convergent particle swarm optimization approach for multiscale design of high-permeane seawater reverse osmosis systems *Comms. Eng.* **3** 149

[26] Luo J, Li M and Heng Y 2020 A hybrid modeling approach for optimal design of non-woven membrane channels in brackish water reverse osmosis process with high-throughput computation *Desalination* **489** 114463

[27] Lucidi S and Rinaldi F 2010 Exact penalty functions for nonlinear integer programming problems *J. Optim. Theory Appl.* **145** 479–88

[28] Efron B 1979 Bootstrap methods: another look at the jackknife *Ann. Stat.* **7** 1–26

[29] Bucs S S *et al* 2014 Effect of different commercial feed spacers on biofouling of reverse osmosis membrane systems: a numerical study *Desalination* **343** 26–37

IOP Publishing

High-Performance Computing and Artificial Intelligence in Process Engineering

Mingheng Li and Yi Heng

Chapter 9

Supercomputing and machine-learning-aided optimal design of high permeability seawater reverse osmosis membrane systems

Jiu Luo, Mingheng Li and Yi Heng

Concentration polarization limits the energy and cost-reducing benefits of high-permeability seawater reverse osmosis (SWRO) membrane systems. This chapter introduces a multiscale optimization framework that integrates membrane permeability, feed spacer design, and system design using computational fluid dynamics and system-level modeling. Leveraging advanced supercomputing and machine learning, the framework suggests a potential 27.5% reduction in energy consumption, highlighting the impact of high-permeability membranes and optimized system configurations on SWRO performance.

9.1 Introduction

Clean water scarcity is one of the most pressing global problems. The reuse of wastewater and the desalination of seawater, wastewater, and brackish groundwater are effective ways to increase local fresh water supplies [1]. Reverse osmosis (RO) membranes are the most commonly employed desalination technology [2]. The cost of electricity to power high-pressure pumps in RO operation accounts for 25%–40% of the total cost of seawater desalination [3]. Hence, reducing energy consumption is crucial for bringing down the total cost of water as well as its carbon footprint.

In recent years, high permeability RO membranes made variously from carbon nanotubes (CNTs) [4, 5], graphene [6], graphene oxide (GO) [7], aquaporin [8],

doi:10.1088/978-0-7503-6174-3ch9

graphene/CNTs [9], improved polyamide [10, 11], boron nitride nanotubes [12], zeolite/thin-film nanocomposites [13], and fluorous oligoamide nanorings [14] have attracted widespread attention due to their superior water permeability over state-of-the-art commercial RO membranes. However, operating high permeability RO membranes at higher than normal fluxes exacerbates concentration polarization (CP), which leads to a reduction of driving force and an acceleration of membrane fouling, thereby adversely affecting overall performance. This is partly attributed to the limitation of fluid flow, salt mass transfer, and spacer geometry within commercial spiral wound membrane modules [15]. Elimelech and Phillip [16] pointed out that alternative RO membrane module designs could help enable high-flux operation. Patel *et al* [17] suggested that system design might yield even greater gains than the development of novel membrane materials for reducing the energy requirements and cost of desalination. Zhu *et al* [18] used RO process models to demonstrate that it is practicable to run an SWRO system with highly permeable membranes at a low operating pressure, even near to the brine osmotic pressure at the outlet. The simulated results by Cohen-Tanugi *et al* [19] and Lim *et al* [20] using a high permeability membrane of water permeability (3 l m^{-2} h^{-1} bar^{-1}), which is triple that of most commercial SWRO membranes, indicate that the specific energy consumption (SEC) could drop by 15% and 16% and the number of pressure vessels could reduce by 44% and 57% for one-stage SWRO. The conditions simulated were feed salinity 42 000 ppm, recovery rate 42% in the former [19] and feed salinity 32 000 ppm, recovery rate 50% in the latter [20]. Here, we further evaluate the potential applications of high permeability SWRO membranes by optimizing membrane, module, and system designs for high-flux operation.

Previous studies focused on feed spacer design and optimization to reduce CP and enhancing permeation flux [21–23]. Gu *et al* [21] predicted the performance of four different types of feed spacers with 20 different geometries using three-dimensional (3D) computational fluid dynamics (CFD) to find an optimized configuration. Johannink *et al* [22] proposed an optimization-based workflow for model identification. The obtained surrogate models for woven and non-woven spacers show strongly nonlinear relations between pressure drop and spacer geometric parameters. In terms of RO system design, Li [24] developed a model-based optimization method for an industrial brackish water RO (BWRO) desalination process, and the simulation results indicated that SEC can be reduced by about 10% by manipulating operating conditions, which was validated in the plant. Guillen and Hoek [25] developed a multiscale modeling approach that coupled microscopic and macroscopic transport models for RO and nanofiltration (NF) processes. It was applied to evaluate the effects of feed spacer shape and filament spacing on overall RO/NF system performance; their conclusions were that the spacer porosity dominated the pressure drop (not shape) and that in high-pressure applications (BWRO and SWRO) the spacer played very little role in determining permeate quality or SEC. Luo *et al* [26] established a hybrid model coupling a 3D CFD model and a one-dimensional (1D) system-level model for BWRO process, and proposed a heuristic optimization framework for feed spacer design. The simulated results suggest that the average permeation flux can be enhanced about 9% by using an optimized spacer.

Precise 3D local CFD modeling in RO desalination has to consider nonlinear channel flow and mass transport. Traditional optimization methods suffer from solving many nonlinear CFD models iteratively and hence are both mathematically complex and computationally expensive. In recent years, machine learning has been widely applied in membrane-based processes [27–32]. Rall *et al* [32] proposed an optimization approach for the NF process that couples a 1D ion transport model at the nanoscale and a hybrid mechanistic model at the macro-scale with the use of a deterministic global optimization algorithm. The approach may shed light on the inverse design of membrane materials and system-level optimization for the NF process. To the best of our knowledge, there is no general method applicable to address the multiscale design optimization problem that consists of module design (sub-millimeter scale) and system design (meter scale), which is crucial to suppress CP and to render feasible engineering solutions for high permeability RO membrane desalination systems. In this work we aim to develop a facile, multiscale optimization design framework (figure 9.1) supporting computer modeling that integrates CFD, machine learning, global optimization, and high throughput computing. The proposed multiscale design scheme addresses optimization of the RO process design configuration, operating conditions, and module arrangement of each stage, the feed spacer, and membrane properties.

9.2 Potential evaluation of module and system design

If the effect of CP is not taken into consideration, the SEC in SWRO normalized by the feed osmotic pressure decreased as $\gamma = A_{\text{tot}} L_{\text{p}} \pi_0 / Q_0$ (A_{tot} is total membrane area, L_{p} is water permeability, π_0 is feed osmotic pressure, and Q_0 is volumetric feed flow) or the number of stages increases [33]. A large γ allows the system to be operated near the thermodynamic limit. Using the optimization model proposed in our previous work [33], we calculate the theoretical minimum SEC of typical SWRO (feed salinity 35 000 ppm, recovery rate 50%) with respect to various γ for one-, two-, and three-stage SWRO systems considering energy recovery efficiency (0.95) and pump efficiency (0.85), but ignoring the effect of CP and axial pressure drop. The trend follows the law of diminishing returns (figure 9.1(a)). A state-of-the-art SWRO process is chosen as the baseline (one-stage SWRO, $\gamma = 0.73$, marked as an asterisk in figure 9.1(a)). A slight increment of SEC (from 2.13 to 2.28 kWh m^{-3}) is observed after CP and axial pressure drop are taken into consideration.

In this work, employing the two-stage design and the high permeability membranes ($\gamma = 4.78$, marked as a circle in figure 9.1(a)) reduces the SEC by about 27.5%, of which 14.5% is due to the two-stage design and 12.2% is due to an increased γ (i.e. RO membrane permeability). The simulated results agree with the reported data under the same conditions (feed salinity 35 000 ppm, recovery rate 50%) that the two-stage SWRO has an advantage over the traditional one-stage SWRO for SEC [34, 35] and the energy saving can reach 15% [34].

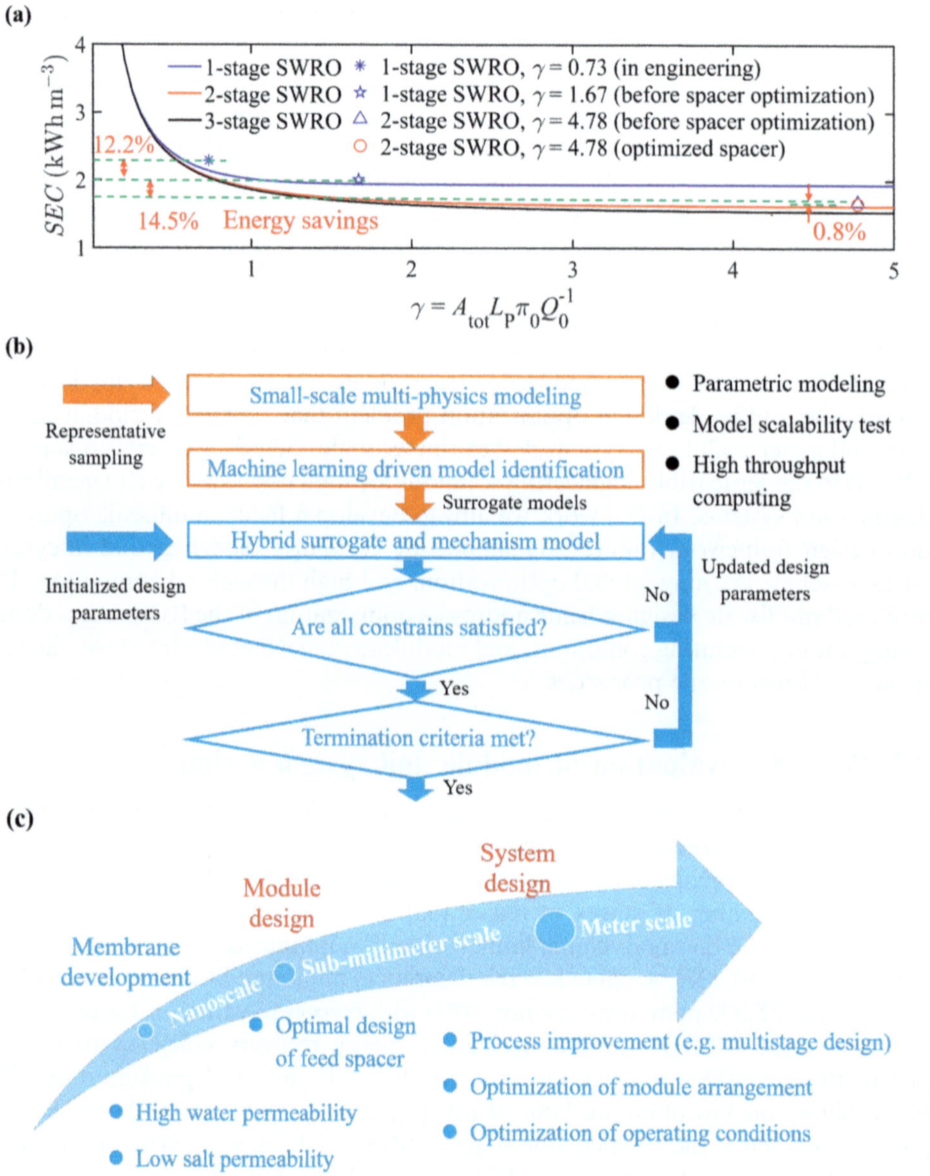

Figure 9.1. Optimization of high permeability SWRO systems. (a) Evaluation of the energy savings potential (feed salinity 35 000 ppm, recovery rate 50%, efficiencies of pump 0.85 and ERD 0.95). The three lines denote the calculated SEC with respect to various γ for one-, two-, and three-stage SWROs using an optimization model [33], ignoring the effect of CP and pressure drop. The asterisk, star, triangle, and circle denote results in this work that take into consideration the effect of CP and pressure drop. (b) Multiscale design workflow. (c) Optimized scheme consisting of system design, feed spacer geometries, and membrane properties. (Reproduced with permission from [48]. Copyright 2023 Elsevier.)

Table 9.1. Estimation of CPF in the lead element when high permeability membrane is used. (Reproduced with permission from [48]. Copyright 2023 Elsevier.)

	Current membrane	High permeability membrane	
π_0 (bar)	28	28	28
P_0 (bar)	60	45	45
L_p (l m^{-2} h^{-1} bar^{-1})	1	3	10
J_W in the lead element (l m^{-2} h^{-1})	29.2	41.2	114.8
k in the lead element (m s^{-1})	8.1×10^{-5}	9.8×10^{-5}	1.6×10^{-4}
CPF in the lead element (–)	1.10	1.12	1.20

Any further reduction in SEC is minimal using a three-stage design or an even larger γ. The CP factor (CPF) is defined as, $\text{CPF} = \exp(J_W/k)$ (J_W is permeation flux, k is mass transfer coefficient). The maximum CPF is controlled to no more than 1.20 [15] by multi-stage design and spacer optimization. The permeation flux in RO depends on both the membrane permeability and the driving force, or $J_W = L_p[\Delta P - \text{CPF} \cdot \Delta\pi]$ where ΔP and $\Delta\pi$ are transmembrane hydraulic pressure and transmembrane osmotic pressure. Apparently, the use of high permeability membranes will not increase the flux in proportion to the membrane permeability if the applied pressure is lowered at the same time. Therefore, it is not necessary to enhance the mass transfer coefficient to the same degree as the membrane permeability in order to control the CPF within the acceptable range. A rough calculation of the CPF in the lead element when increasing the membrane permeability threefold and tenfold and reducing the applied pressure by 25% is shown in table 9.1.

If the CPF in the lead element is 1.10 using state-of-the-art membranes, it becomes only 1.12 for the membrane with a permeability of 3 l m^{-2} h^{-1} bar^{-1}, provided that the mass transfer coefficient is enhanced by 20%. If a more permeable membrane (e.g. 10 l m^{-2} h^{-1} bar^{-1}) is adopted, doubling the mass transfer coefficient may suppress the CPF no more than 1.20. In this work, we aim to explore the upper bounds of membrane performance based on spacer optimization and system design.

9.3 Multiscale optimization design framework

The multiscale model developed in this work consists of three sections: small-scale multi-physics modeling, model identification with multilayer artificial neural networks (MLN), and system-level modeling at the meter scale (figure 9.2(a)). Detailed schematic diagrams of the one-stage SWRO system with an energy recovery device (ERD), the computational domain of the CFD model, the feed spacer unit, and spacer filaments are shown in figures 9.2(b)–(e), respectively.

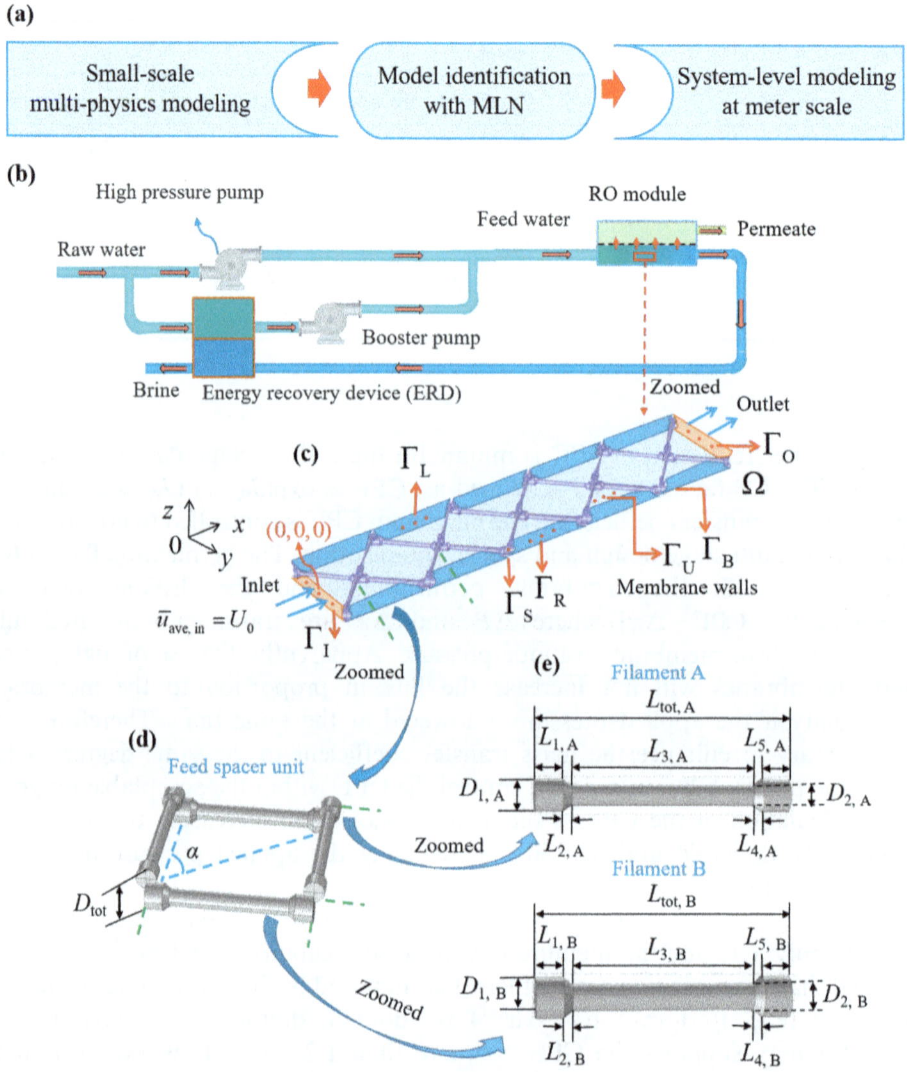

Figure 9.2. Multiscale design of high permeability SWRO systems: (a) the proposed multiscale model, (b) RO desalination systems and feed spacer, (c) computational domain of the CFD model, (d) a spacer unit, and (e) spacer filaments. (Reproduced with permission from [48]. Copyright 2023 Elsevier.)

9.3.1 Small-scale multi-physics modeling

In this work, we develop a generally applicable optimization framework using CFD, MLN, and global optimization in conjunction with advanced supercomputing for an RO desalination system. Solving the local transport phenomena in an efficient and accurate way is an important component in the computational framework. Based on the boundary conditions used on membrane walls, the CFD models for RO can be divided into the permeable wall model and impermeable wall model [36]. The permeable wall model reflects the physics occurring on the membrane surfaces better, but it is associated with a higher computational cost. In the permeable wall model, the leaking velocity of the permeate depends on the local concentration of salt on the membrane surfaces, or fluid flow and mass transfer are fully coupled. In contrast, the impermeable wall model has a higher computational efficiency by assuming a zero velocity across the membrane walls, and thus the flow is independent of the mass transport. Details of the CFD models including the boundary conditions and the applied model parameters of this work can be found in tables 9.2 and 9.3, respectively. The predicted cell-average mass transfer coefficient in the impermeable wall model is unrelated to the membrane properties

Table 9.2. Mathematical description of boundary conditions for the CFD models. (Reproduced with permission from [48]. Copyright 2023 Elsevier.)

Boundary	Permeable wall model	Impermeable wall model
Membrane walls (Γ_U and Γ_B)	$(J_W)_U = L_p\left(\Delta P_0 - f_{os} R_{salt} c_w\right)$ $(J_W)_B = -L_p\left(\Delta P_0 - f_{os} R_{salt} c_w\right)$ $\mathbf{n} \cdot (-D_s \nabla c + c\mathbf{u}) = J_W c_w (1 - R_{salt})$	$(J_W)_U = 0$ $(J_W)_B = 0$ $c = c_{w,0}$
Surface of feed spacer (Γ_S)	$\mathbf{u} = \mathbf{0};\ \mathbf{n} \cdot (-D_s \nabla c + c\mathbf{u}) = 0$	
Inlet–outlet (Γ_I and Γ_O)	$\mathbf{u}_{in} = \mathbf{u}_{out};\ (\nabla \mathbf{u})_{in} = (\nabla \mathbf{u})_{out};\ \bar{u}_{ave,\,in} = U_0$ $(\nabla P)_{in} = (\nabla P)_{out};\ P_{out} = 0$ $c_{in} = c_0;\ (-\mathbf{n} \cdot D_s \nabla c)_{out} = 0$	
Lateral boundaries (Γ_L and Γ_R)	$\mathbf{u}_R = \mathbf{u}_L;\ (\nabla \mathbf{u})_L = (\nabla \mathbf{u})_R$ $P_R = P_L;\ (\nabla P)_L = (\nabla P)_R$ $c_R = c_L;\ (\nabla c)_L = (\nabla c)_R$	

Table 9.3. The correlation parameters used in the impermeable wall model. (Reproduced with permission from [48]. Copyright 2023 Elsevier.)

	Parameters	Value
Property of feed water	Inlet concentration, c_0 (mol m^{-3})	611
	Density, ρ (kg m^{-3})	1021
	Viscosity, μ (Pa s)	9.41×10^{-4}
	Diffusion coefficient, D_s (m^2 s^{-1})	1.45×10^{-9}
	Osmotic pressure coefficient (bar)	805.1
	Reflection coefficient of solute, $\sigma(-)$	1.0
Operating conditions	Average inlet velocity magnitude, U_0 (m s^{-1})	0.05–0.2

(e.g. water permeability, L_p and salt permeability, B) and the inlet transmembrane pressure. Thus, the number of design parameters in the CFD model can be decreased greatly. In this work, we adopt the impermeable wall RO model with a constant concentration on the membrane walls and further obtain a surrogate model of cell-average mass transfer coefficient that can be converted to the cell-average mass transfer coefficient on permeable walls [37].

The Navier–Stokes equations under laminar flow conditions can be formulated as

$$\begin{cases} \rho(\mathbf{u} \cdot \nabla)\mathbf{u} - \nabla \cdot [-P\mathbf{I} + \mu(\nabla \mathbf{u} + (\nabla \mathbf{u})^{\mathrm{T}})] = 0, & \text{in } \Omega, \\ \nabla \cdot (\rho \mathbf{u}) = 0, & \text{in } \Omega. \end{cases} \tag{9.1}$$

The diffusion–convection equation can be expressed as

$$\nabla \cdot (D_s \nabla c) - \mathbf{u} \cdot \nabla c = 0, \quad \text{in } \Omega, \tag{9.2}$$

where ρ and μ are the density and viscosity at coordinates $\mathbf{x} \equiv (x, y, z)$ of the fluid, respectively. $\mathbf{u} \equiv (u, v, w)$ is the velocity vector, P denotes hydraulic pressure, Ω denotes the computational domain, and Γ_I, Γ_O, Γ_U, Γ_B, Γ_L, Γ_R, Γ_S denote various boundaries (figure 9.2(c)). In this work, the design parameter vectors γ in the small-scale multi-physics model consist of geometric parameters, $\beta_1 = [L_{tot}, D_{1,A}, D_{2,A}, D_{1,B}, D_{2,B}, D_{tot}, \alpha]$ (figure 9.2(d) and (e)) and average inlet velocity magnitude, $\bar{u}_{ave,in} = U_0$. $L_{i,A}$ and $L_{i,B}$ ($i = 1, 2, 4, 5$) are assumed to be constant in this work and $L_{tot,A} = L_{tot,B} = L_{tot}$. D_s and c denote diffusivity and molar concentration. To reduce the effect of the stagnant area [39], we consider $L_{1,A} = L_{5,A} = L_{1,B} = L_{5,B} = 300\ \mu$m, and $L_{2,A} = L_{4,A} = L_{2,B} = L_{4,B} = 100\ \mu$m in this work (figure 9.2(e)).

In table 9.2, $(J_W)_U$, $(J_W)_B$ denote the local permeation flux of water on the top and bottom membrane surfaces, respectively. On the inlet plane, Γ_I (figure 9.2(c)), the boundary conditions of constant concentration (c_0), and fully developed flow with an average inlet velocity magnitude (U_0) are enforced. The periodic boundary conditions are applied in lateral boundaries (Γ_L, Γ_R) (figure 9.2(c)). Furthermore, we establish the design space that consists of the geometric parameters of the feed spacer and the average inlet velocity magnitude based on a few commercial feed spacers and typical operating conditions [39]. A Latin hypercube-like sampling method is proposed to generate representative design parameters from the design space of interest. Some important parameters used in the surrogate models, e.g. the pressure drop per unit length ($\frac{\Delta \bar{P}_c}{\Delta L}$) along the feed direction and the cell-average mass transfer coefficient ($\bar{k}_{m,\,imp}$) on membrane walls can be obtained by solving the impermeable wall model using different design parameter combinations.

9.3.2 Model identification with MLN

The computational cost involved in obtaining samples (i.e. simulation data) using the 3D CFD model is still expensive. Different to our previous work [38], we employ the Latin hypercube-like sampling method to generate random samples of parameter values from a multidimensional design space referred to in the literature [39].

The design parameters in the CFD model include the geometric parameters of the feed spacer ($\beta_1 = [L_{\text{tot}}, D_{1,\text{A}}, D_{2,\text{A}}, D_{1,\text{B}}, D_{2,\text{B}}, D_{\text{tot}}, \alpha]$) and the average inlet velocity magnitude (U_0). We use an enhanced MLN with a penalty term (or regularization term) to obtain accurate and robust surrogate models in this work. Bridging the multi-physics model at the small scale and the system-level RO models can be formulated as a model identification problem. These obtained surrogate models can provide a thorough quantitative description for pressure drop per unit length ($-\frac{\Delta \overline{P_c}}{\Delta L}$) along the feed direction, the cell-average mass transfer coefficient ($\overline{k}_{\text{m, imp}}$) on the membrane wall, and feed channel porosity (ε) with respect to a wide range of design parameters.

The surrogate models using the MLN method can be mathematically formulated as

$$\mathbf{y}_{\text{out}} = \mathbf{f}\left(\mathbf{w}^*, \ \mathbf{b}^*, \ \breve{\beta}_l, \ \breve{U}_0\right), \tag{9.3}$$

where $\mathbf{y}_{\text{out}} = \left(\frac{\Delta \overline{P_c}}{\Delta L}, \ \overline{k}_{\text{m, imp}}, \ \varepsilon\right)$, $\mathbf{f} = \left(f_1, \ f_2, \ f_3\right)$. $\breve{\beta}_1, \breve{U}_0$ denote the normalized geometric parameters and average inlet velocity magnitude, respectively. It should be noted that feed channel porosity (ε) only depends on geometric parameters. A more detailed description for function $\mathbf{f}$ is available in our previous work [38]. In network training, the weights, $\mathbf{w}^*$, $\mathbf{b}^*$, are the optimized weights and biases of the neural network by minimizing the loss function [38], which can be mathematically described as

$$\mathbf{w}^*, \mathbf{b}^* = \lambda_1 \arg\min_{\mathbf{w},\mathbf{b}} \frac{1}{N}\sum_{1}^{N}(y_k - y_{k,\text{ out}})^2 + \lambda_2 ||\mathbf{w}||^2, \tag{9.4}$$

where $\mathbf{w}$, $\mathbf{b}$ are the weights and biases of the neural network,y_k and $y_{k,\text{ out}}$ denote the actual and predicted values, and N is the number of samples obtained from the multi-physics simulations with respect to various design parameter vectors, γ. A regularization term (the second term on the right-hand side of equation (9.4)) is added to obtain a stable solution. λ_1, λ_2 are objective function parameters.

9.3.3 System-level modeling at the meter scale

To quantify the system-level performance of a kth stage ($k = 1, 2, 3$ in this work) RO, a 1D hybrid surrogate and mechanism model is established in this section:

$$\begin{cases} \dfrac{\mathrm{d}Q}{\mathrm{d}X} = -J_{\text{W}} \cdot A_k & X = k-1, \ Q = Q_{k-1}, \\ \dfrac{\mathrm{d}(\Delta P)}{\mathrm{d}X} = -f_1\,(\breve{\beta}_1, \ \breve{U}_0) \cdot (n_{\text{mem},\,k} \cdot l_x) & X = k-1, \ \Delta P = \Delta P_{k-1}, \\ \dfrac{\mathrm{d}w_{\text{b}}}{\mathrm{d}X} = J_{\text{W}} \cdot \dfrac{A_k}{Q}(w_{\text{b}} - w_{\text{p}}) & X = k-1, \ w_{\text{b}} = w_{\text{b},\ k-1}, \\ J_{\text{W}} = L_{\text{p}}(\Delta P - \sigma \cdot \varphi R_{\text{salt}} w_{\text{w}}), & \end{cases} \tag{9.5}$$

where variables, Q, ΔP, J_W, w_b, w_p, and w_w denote the flow rate, transmembrane pressure, permeation flux of water, the bulk salinities in the feed channel and permeate channel, and brine salinity on membrane surfaces that vary along the x axial direction. X denotes dimensionless length ($X \in [k - 1, k]$). The first derivative of transmembrane pressure, $\frac{d(\Delta P)}{dx}$ is approximate to pressure drop per unit length, $-\frac{\Delta \overline{P_c}}{\Delta L}$. The derivation can be found in our previous work [26]. Membrane area (A_k) for the kth stage can be calculated by $A_k = N_{pv,\,k} \cdot n_{mem,\,k} \cdot A_0 \cdot l_x/l_0$, where A_0, l_0 denote the membrane area and the length of each commercial spiral wound module, respectively. $N_{pv,\,k}$ and $n_{mem,\,k}$ denote the number of pressure vessels and the number of membrane elements, respectively, per pressure vessel in kth stage. The Q can be converted into U_0 using $U_0 = Q/(N_{pv,\,k} n_{sp} l_y H \varepsilon)$, where l_y, H, and n_{sp} denote the length of the perpendicular to the feed direction (y-direction), channel height ($H = D_{tot} - 10^{-5}$ m) [26], and number of feed spacers per element. The membrane properties include water permeability (L_p) and salt permeability (B). The relation f_1 is obtained by the MLN method. σ, φ, andR_{salt} denote the reflection coefficient, osmotic pressure coefficient, and membrane intrinsic rejection, respectively. In addition, the operating conditions consist of inlet flow rate Q_{k-1}, inlet transmembrane pressure (ΔP_{k-1}), and feed salinity ($w_{b,\,k-1}$) in kth stage RO. In this work, we propose a novel formula to solve w_p:

$$w_p = w_b / \left[\exp\left(\ln \frac{J_W}{B} - \frac{J_W}{\overline{k}_{m,\,per}} \right) + 1 \right]. \tag{9.6}$$

The brine salinity on the membrane surfaces can be calculated by

$$w_w = \frac{w_p}{1 - R_{salt}}. \tag{9.7}$$

The cell-average mass transfer coefficient on permeable walls ($\overline{k}_{m,\,per}$) can be converted by the cell-average mass transfer coefficient using impermeable wall model ($\overline{k}_{m,\,imp}$) incorporation with equations (9.8) and (9.9) [37], as below:

$$\overline{k}_{m,\,per} = \overline{k}_{m,\,imp}[\psi + (1 + 0.26\psi^{1.4})^{-1.7}], \ (\psi < 20) \tag{9.8}$$

$$\psi = \frac{J_W}{\overline{k}_{m,\,imp}}. \tag{9.9}$$

For a given normalized average inlet velocity magnitude ($\breve{U}_0$) and geometric parameters of the feed spacer ($\breve{\beta}_1$), the cell-average mass transfer coefficient $\overline{k}_{m,\,imp}$ can be calculated by

$$\overline{k}_{m,\,imp} = f_2(\breve{U}_0, \ \breve{\beta}_1), \tag{9.10}$$

based on the obtained surrogate models.

9.3.4 Optimal design of the RO system

Furthermore, the hybrid surrogate and mechanism model constrained optimization problem is mathematically described as

$$\begin{aligned}&\min_{\beta}\ \text{obj}\\&\text{s. t.}\\&\mathbf{H}\left(Q,\ \Delta P,\ J_W,\ w_b,\ w_p,\ X,\ \beta\right)=\mathbf{0},\\&\mathbf{J}\left(Q,\ \Delta P,\ J_W,\ w_b,\ w_p,\ X,\ \beta\right)\leqslant\mathbf{0}.\end{aligned}\tag{9.11}$$

The objective function (obj) represents the cost per m^3 permeate (C, \$ m^{-3}) that consists of two terms, energy cost per m^3 permeate (C_e, \$ m^{-3}) and membrane cost per m^3 permeate (C_m, \$ m^{-3}). For the system-level RO model $\mathbf{H}(Q,\ \Delta P,\ J_W,\ w_b,\ w_p,\ X,\ \beta)=\mathbf{0}$, flow rate ($Q$), transmembrane pressure (ΔP), permeation flux of water (J_W), feed bulk salinity (w_b), permeate salinity (w_b), and brine salinity on membrane surfaces (w_w) vary along the dimensionless axial coordinate (X). The inequality constraints consist of constraints of geometric parameters and constraints of estimated results by solving a 1D system described by differential algebraic equations, $\mathbf{H}(Q,\ \Delta P,\ J_W,\ w_b,\ w_p,\ X,\ \beta)=\mathbf{0}$ (equations (9.5)–(9.10)). The optimal set of parameters (β) consists of system design parameters, e.g. the number of elements per pressure vessel, the number of pressure vessels, the inlet transmembrane pressure in each stage of the RO system, the number of feed spacers per element, the geometric parameters of the feed spacer, and the membrane properties (water permeability and salt permeability).

The optimal set of parameters (β) in this work consist of the geometric parameters of the feed spacer $\beta_1=[L_{tot},\ D_{1,A},\ D_{2,A},\ D_{1,B},\ D_{2,B},\ D_{tot},\ \alpha]$ and the system design parameters $\beta_2=[N_{pv,1},\ \cdots N_{pv,k};\ n_{mem,1},\ \cdots,\ n_{mem,k};\ \Delta P_0,\ \cdots,\ \Delta P_{k-1};\ n_{sp}]$ ($k=1$ for one-stage RO; $k=2$ for two-stage RO; $k=3$ for three-stage RO) and membrane properties $\beta_3=[L_p,\ B]$.

The inequality constraints, $\mathbf{J}(Q,\ \Delta P,\ J_W,\ w_b,\ w_p,\ X,\ \beta)\leqslant\mathbf{0}$ consist of constraints of geometric parameters, $\mathbf{J}_1(\beta_1)\leqslant\mathbf{0}$ (equation (9.12)) and constraints of estimated results, $\mathbf{J}_1(Q,\ \Delta P,\ J_W,\ w_b,\ w_p,\ X,\ \beta)\leqslant\mathbf{0}$ (equation (9.17)) from a 1D system of differential algebraic equations, $\mathbf{H}(Q,\ \Delta P,\ J_W,\ w_b,\ w_p,\ X,\ \beta)=\mathbf{0}$ (equations (9.5)–(9.10)) that need to be considered to obtain the optimization results with engineering significance. The filaments A and B of the feed spacer must be partially overlapped to support the top and bottom membrane sheets [26]. Thus the geometrical parameters $D_{1,A}$, $D_{1,B}$, and D_{tot} of the feed spacer should follow the constraints of $\mathbf{J}_1(\beta_1)\leqslant\mathbf{0}$:

$$\begin{cases}\max\{D_{1,A},\ D_{1,B}\}-D_{tot}\leqslant 0,\\ D_{tot}-(D_{1,A}+D_{1,B}-10\ \mu\text{m})\leqslant 0,\\ \text{SPD}_{\min}\leqslant\text{SPD}\leqslant\text{SPD}_{\max}.\end{cases}\tag{9.12}$$

In other words, the overlapped height of the both filaments, A and B (figure 9.2 (d)), is greater than or equal to 10 μm [26]. The support point density, SPD of feed spacer can be calculated by

$$\mathrm{SPD} = \frac{2n_x n_y}{l_x \cdot l_y}, \tag{9.13}$$

$$l_x = 2 \cdot n_x \cdot L_{\mathrm{tot}} \cdot \cos(\alpha), \tag{9.14}$$

$$l_y = 2 \cdot n_y \cdot L_{\mathrm{tot}} \cdot \sin(\alpha), \tag{9.15}$$

where l_x and l_y denote the length of the spacer parallel to flow and perpendicular to flow. n_x, n_y denote the number of spacer units along the feed direction (x) and its vertical direction (y). Incorporating with equations (9.14) and (9.15), equation (9.13) can be rewritten as

$$\mathrm{SPD} = \frac{1}{L_{\mathrm{tot}}^2 \sin(2\alpha)}. \tag{9.16}$$

The constraint of SPD is mainly for the consideration of the mechanical strength of the feed spacer. Moreover, limiting the maximum CPF can effectively mitigate fouling and scaling, which is crucial for the high permeability SWRO systems. The recovery rate (R_{r}) and average permeation flux ($\overline{J}_{\mathrm{W}}$), brine salinity at outlet ($w_{\mathrm{b},\,N_{\mathrm{st}}}$), and average permeate salinity ($\overline{w}_{\mathrm{p}}$) are within typical engineering ranges. In addition, a negative driving force should be avoided in the membrane module. For a N_{st} stage RO system ($N_{\mathrm{st}} = 1, 2, 3$ in this work), the constraints are summarized as below:

$$\begin{cases} \mathrm{CPF}(X) \le \mathrm{CPF}_{\max}, & X \in [0,\ N_{\mathrm{st}}], \\ R_{\mathrm{r}} \ge R_{\mathrm{r},0}, & \\ \overline{J}_{\mathrm{W}} \ge \overline{J}_{\mathrm{W,min}}, & \\ \overline{w}_{\mathrm{p}} \le \overline{w}_{\mathrm{p,max}}, & \\ U_{k-1} = \dfrac{Q_{k-1}}{3600 N_{\mathrm{pv},\,k} n_{\mathrm{sp}} l_y H \varepsilon} \le U_{\max}, & \text{for } k = 1,\ \cdots,\ N_{\mathrm{st}}, \\ 0 \le \Delta P_{k-1} \le \Delta P_{0,\,\min}, & \text{for } N_{\mathrm{st}} = 2,\ 3,\ \cdots;\ k = 1, \\ \Delta P_{k-2} \le \Delta P_{k-1} \le \Delta P_{0,\,\min}, & \text{for } N_{\mathrm{st}} = 3,\ \cdots;\ k = 2,\ 3,\ \cdots,\ (N_{\mathrm{st}} - 1), \\ \Delta P_{0,\,\min} \le \Delta P_{k-1} \le \Delta P_{0,\,\max}, & \text{for } N_{\mathrm{st}} = 1,\ 2,\ 3,\ \cdots;\ k = N_{\mathrm{st}}, \\ N_{\mathrm{pv},\,\min} \le N_{\mathrm{pv},\,k} \le N_{\mathrm{pv},\,\max}, & \text{for } k = 1,\ N_{\mathrm{pv},\,k} \in N^*, \\ 1/3 N_{\mathrm{pv},\,k-1} \le N_{\mathrm{pv},\,k} \le N_{\mathrm{pv},\,k-1}, & \text{for } k = 2,\ 3,\ \cdots N_{\mathrm{st}},\ N_{\mathrm{pv},\,k} \in N^*, \\ J_{\mathrm{W},\,k} \ge 0, & \text{for } k = 1,\ \cdots N_{\mathrm{st}}, \end{cases} \tag{9.17}$$

which the average inlet velocity magnitude (U_{k-1}) of each feed channel at inlet of kth-stage RO system may not exceed the maximum velocity magnitude ($U_{\max}$). The CPF, $\overline{J}_{\mathrm{W}}$, and $\overline{w}_{\mathrm{p}}$ can be calculated by

$$\mathrm{CPF} = \frac{w_{\mathrm{W}} - w_{\mathrm{p}}}{w_{\mathrm{b}} - w_{\mathrm{p}}} = \exp\left(\frac{J_{\mathrm{W}}}{\overline{k}_{\mathrm{m,\,per}}}\right), \tag{9.18}$$

$$\overline{J}_{\mathrm{W}} = \frac{\sum_{k=1}^{N_{\mathrm{st}}} \int_{k-1}^{k} (J_{\mathrm{W},k} \cdot A_k) \mathrm{d}X}{\sum_{k=1}^{N_{\mathrm{st}}} \int_{k-1}^{k} A_k \mathrm{d}X}, \quad k = 1, \cdots, N_{\mathrm{st}}, \tag{9.19}$$

and

$$\overline{w}_{\mathrm{p}} = \frac{\sum_{k=1}^{N_{\mathrm{st}}} \int_{k-1}^{k} \left(J_{\mathrm{W},k} \cdot w_{\mathrm{p},k} \cdot A_k\right) \mathrm{d}X}{\sum_{k=1}^{N_{\mathrm{st}}} \int_{k-1}^{k} (J_{\mathrm{W},k} \cdot A_k) \mathrm{d}X}, \quad k = 1, \cdots, N_{\mathrm{st}}, \tag{9.20}$$

respectively. The permeation flux of water, J_{W}, $\overline{k}_{\mathrm{m,\,per}}$, and w_{p} can be iteratively solved using the hybrid surrogate and mechanism model, $\mathbf{H}(Q,\ \Delta P,\ J_{\mathrm{W}},\ w_{\mathrm{b}},\ w_{\mathrm{p}},\ X,\ \boldsymbol{\beta}) = \mathbf{0}$. For a given product water recovery, $R_{\mathrm{r},\,0}$, and feed salinity, $w_{\mathrm{b},0}$, the minimum transmembrane pressure, $\Delta P_{0,\,\mathrm{min}}$, at the inlet can be solved by

$$\Delta P_{0,\,\mathrm{min}} = \frac{\varphi w_{\mathrm{b},\,0}}{1 - R_{\mathrm{r},\,0}}. \tag{9.21}$$

9.4 Results and discussion

9.4.1 Supercomputing-based machine-learning-driven model identification

In this work, we propose a novel model identification approach using multi-physics modeling at the small scale and an MLN method combined with supercomputing (figure 9.3). A Latin hypercube-like sampling method is developed to generate combinations of representative design parameters from the considered parameter space. The CFD model couples flow and mass transport governed by the Navier–Stokes equations under laminar flow conditions (equation (9.1)) and the convection–diffusion equation (9.2), respectively. We define the computational efficiency, $E = (t_1/N)/t_N$ where t_1 and t_N denote the computational time using one and N computational nodes, respectively. As the number of computational nodes increases, the communication time between nodes increases, although the computational time may decrease, which may lead to a lower computational efficiency. The results of model scalability test (table 9.4) indicate that using four computational nodes (24 cores/node) for each CFD model can significantly reduce the computational time with a high computational efficiency (82%). We solve 970 3D CFD models with respect to various design parameter combinations using the high throughput computing strategy. The maximum computational scale can reach about 93 120 cores (3880 nodes) and the total computational time is reduced by more than 3000 times that of a workstation having a performance comparable to one node in Tianhe-2. Based on the obtained simulation results (e.g. pressure drop per unit length and cell-average mass transfer coefficient) with respect to various design parameter combinations, we adopt an MLN method to establish the surrogate

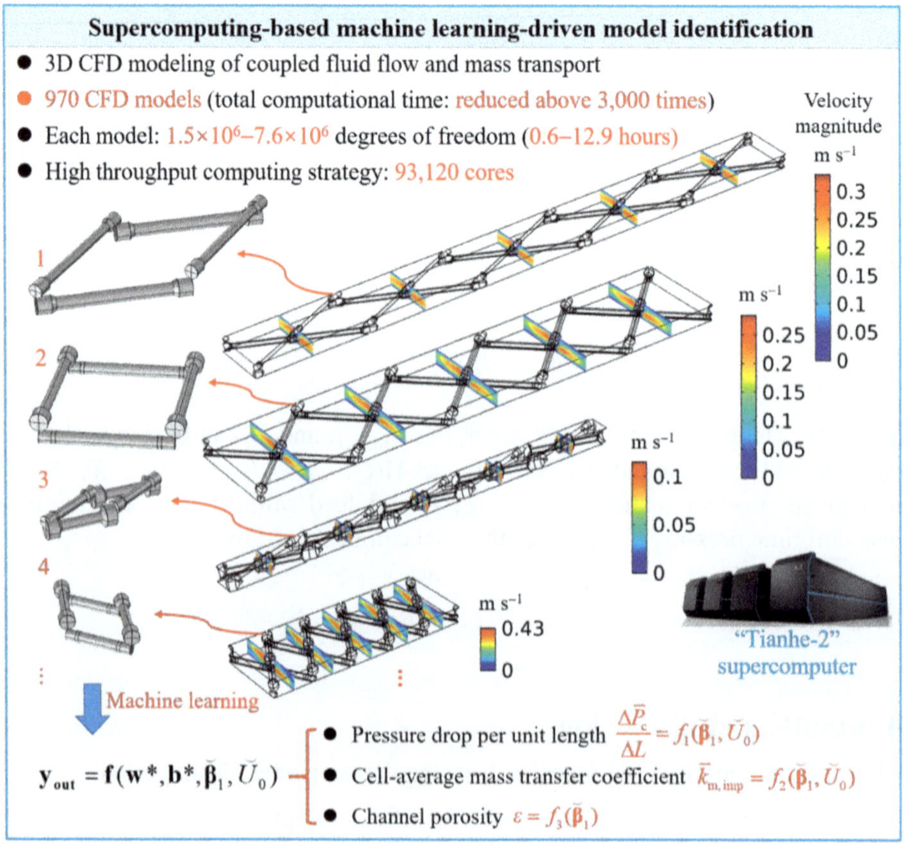

Figure 9.3. Supercomputing-based machine-learning-driven model identification. The Latin hypercube-like sampling method is used to generate design parameters that consist of spacer geometric parameters ($\beta_1 = [L_{tot}, D_{1,A}, D_{2,A}, D_{1,B}, D_{2,B}, D_{tot}, \alpha]$) and average inlet velocity magnitude (U_0). (Reproduced with permission from [48]. Copyright 2023 Elsevier.)

Table 9.4. The scalability analysis for the established impermeable wall model.

The number of cores	Computational time (s)	Computational efficiency	Speed-up ratio
24	27 529	1	1
48	14 585	0.94	1.89
96	8434	0.82	3.26
192	4961	0.69	5.55
384	3161	0.54	8.71

models. The surrogate models, as hydrodynamic and transport databases, can be used for a thorough quantitative description of the mass transfer coefficient and pressure drop within a continuous parameter space.

9.4.2 Surrogate model evaluation

Using the MLN, the obtained surrogate models fit well with the CFD simulations (figure 9.4). Correlation analyses of all pairwise combinatorial design parameters with respect to pressure drop, cell-average mass transfer coefficient, and feed channel porosity indicate strong nonlinear relationships. This is consistent with the reported results in previous work [22]. The results indicate that it may be hard to obtain an optimal solution using traditional research methods that rely on the Edisonian trial and error for spacer design. We further calculate the maximum CPF, max(CPF), (figure 9.5) in the entire RO system with respect to the geometry

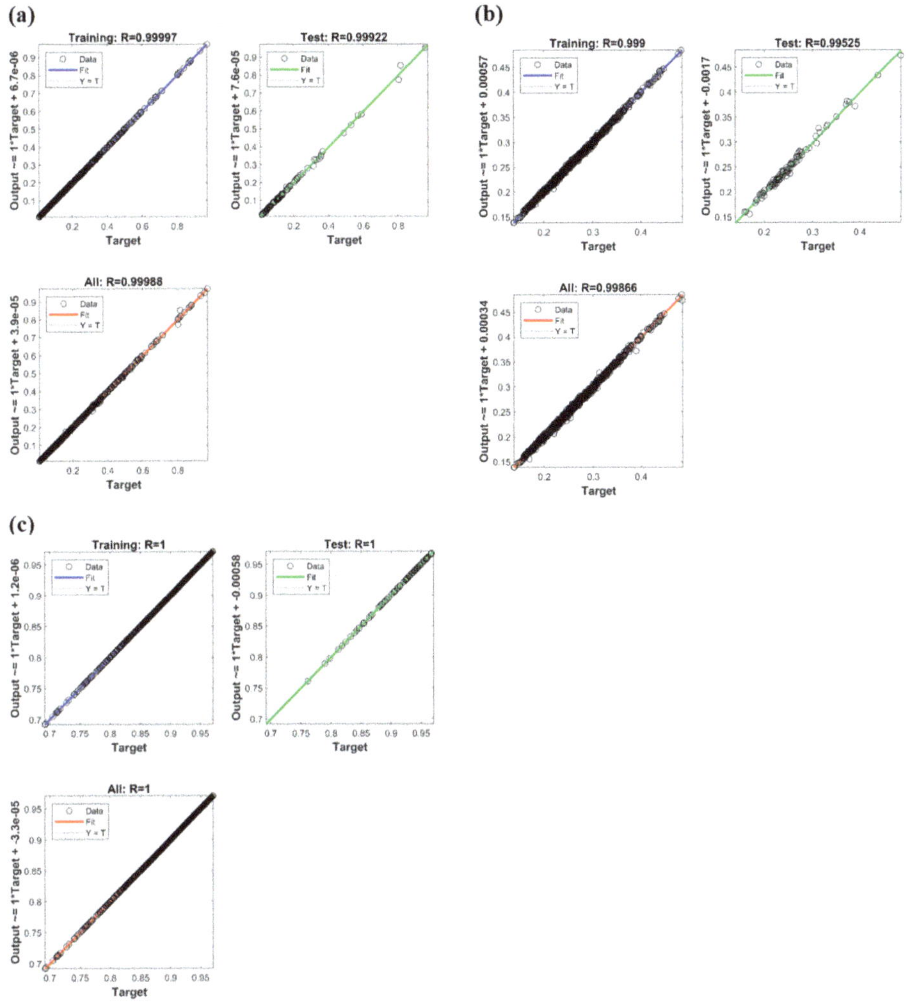

Figure 9.4. The performance of the surrogate models. (a) The pressure drop per unit length along the feed direction, $(-\frac{\Delta \overline{P}_c}{\Delta L})$. (b) Cell-average mass transfer coefficient using the impermeable wall model $(\overline{k}_{m,\,\mathrm{imp}})$. (c) Feed channel porosity (ε) for the SWRO membrane module. (Reproduced with permission from [48]. Copyright 2023 Elsevier.)

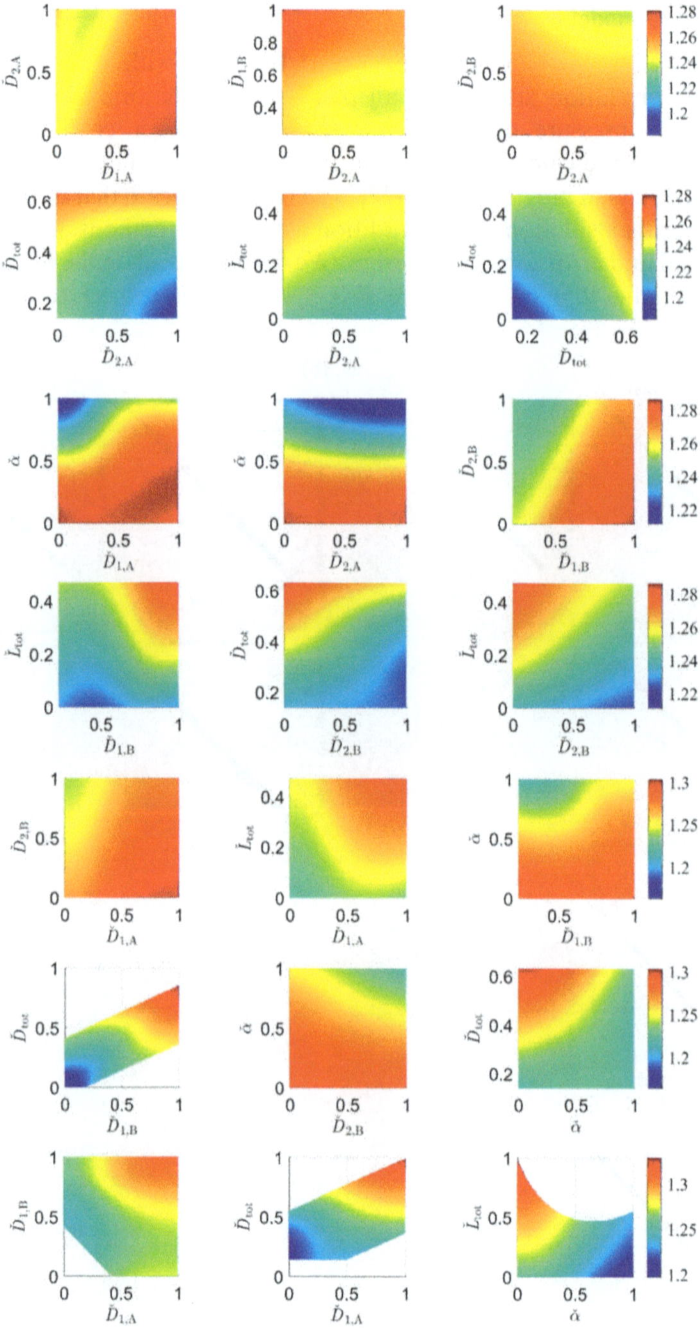

Figure 9.5. Correlation analysis of all pairwise combinational design parameters with the maximum concentration polarization factor of one-stage SWRO, max(CPF). Feed salinity: 35 000 ppm; water permeability: 3 l m^{-2} h^{-1} bar^{-1}; operating pressure: 65.8 bar. (Reproduced with permission from [48]. Copyright 2023 Elsevier.)

parameters of feed spacers (L_{tot}, $D_{1,A}$, $D_{2,A}$, $D_{1,B}$, $D_{2,B}$, D_{tot}, α). The results show that max(CPF) can be effectively controlled by decreasing the values of L_{tot} and D_{tot} and increasing the value of α.

9.4.3 Optimal design of high permeability SWRO systems

Using the proposed optimization framework of this work, we calculate the optimized results of high permeability SWRO membrane desalination systems in case 1 (feed salinity $w_{b,0}$ = 25 000 ppm, recovery rate $R_{r,0}$ = 50%), case 2 ($w_{b,0}$ = 35 000 ppm, $R_{r,0}$ = 50%), and case 3 ($w_{b,0}$ = 45 000 ppm, $R_{r,0}$ = 40%).

For a typical SWRO (case 2), the cost per m^3 permeate (objective function) for the designed one-stage, two-stage, and three-stage SWRO systems decreases about 16.0%, 27.6%, and 31.0% compared to typical commercial systems. The cost per m^3 permeate consists of the annualized capital cost of the membrane and energy cost per m^3 permeate. The parameters, e.g. the membrane cost per m^2 (40 \$ m^{-2}) and energy cost per kWh (0.15 \$ kWh^{-1}), are obtained from the literature [40, 41]. The annualized capital cost of the membrane and energy cost per m^3 permeate of the designed two-stage high permeability membrane SWRO system are 0.026 and 0.248 \$ m^{-3}, versus the state-of-the-art system costs, 0.042 and 0.337 \$ m^{-3}. In this work, the calculated γ for the commercial SWRO is 0.73, while the optimized one-, two-, and three- stage high permeability SWROs are 1.67, 4.78, and 10.26, respectively. The energy saving is relatively large by changing the design from one-stage (SEC = 2.28 kWh m^{-3} for commercial SWRO, SEC = 1.99 kWh m^{-3} for high permeability SWRO) to two-stage (SEC = 1.66 kWh m^{-3}). In the literature an SEC of 1.92 kWh m^{-3} in a single-stage SWRO has been reported using relatively permeable membranes (L_p = 2–3 l m^{-2} h^{-1} bar^{-1}) [17]. The SEC is in the range of 2.5–4 kWh m^{-3} in typical SWRO plants [2]. The contributing factors to SEC (1.66 kWh m^{-3}) for the designed two-stage SWRO consist of friction losses and CP (0.04 kWh m^{-3} 2.5%), efficiency losses of the pump (0.24 kWh m^{-3}, 14.6%) and ERD (0.08 kWh m^{-3}, 4.7%), finite flux (0.22 kWh m^{-3}, 13.1%), and thermodynamic restriction (1.08 kWh m^{-3}, 65.1%). However, the further reduction in SEC by using a three-stage design (1.58 kWh m^{-3}) is very limited. These are consistent with the results in a previous work [33]. The designed two-stage RO system is recommended in this work because the additional energy benefit of the three-stage RO system is small and is hence likely offset by the added cost of additional pumps, fittings, and valves.

For the designed two-stage high permeability membrane SWRO system (figure 9.6 (a)), the optimized spacer unit and membrane properties are shown in figure 9.6(b) and (c) respectively. More detailed information about optimized feed spacer units and a modified 28 mil feed spacer unit (as a comparison) in cases 1–3 can be found in figure 9.7. The overall and local estimated results (before and after optimization) for case 2 are shown in figures 9.8 and 9.9, respectively. The optimized cost per m^3 permeate (C = 0.204, 0.274, 0.313 \$ m^{-3}) decreases by 34.1%, 27.6%, and 24.3%, respectively, compared to similarly calculated costs for state-of-the-art RO system designs (0.309, 0.379, and 0.414 \$ m^{-3}).

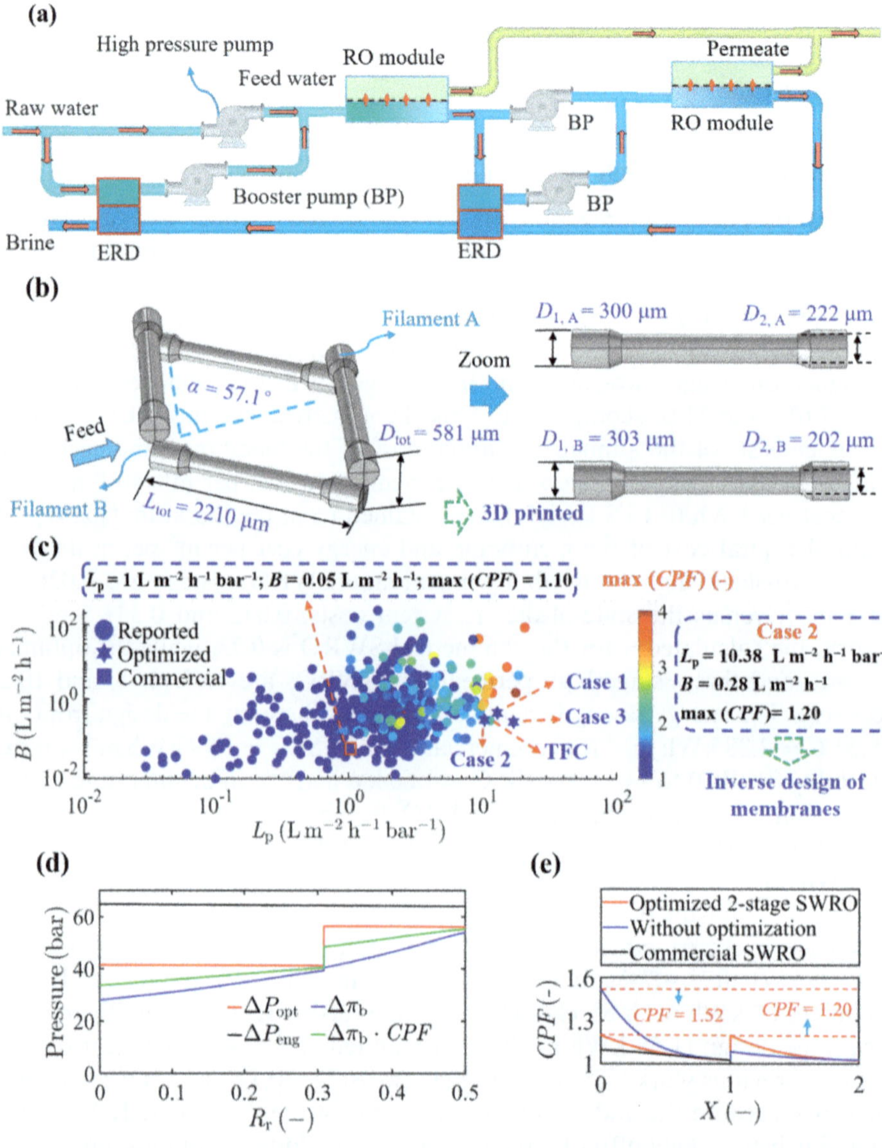

Figure 9.6. Multiscale design scheme of a two-stage high permeability membrane-based SWRO system. (a) System design consisting of the optimization of module arrangement and operating parameters in each system stage. (b) The optimized feed spacer unit. (c) The optimized, commercial, and reported [42] membrane properties, e.g. water permeability (L_P, l m^{-2} h^{-1} bar^{-1}), salt permeability (B, l m^{-2} h^{-1}), and the estimations of maximum CP factor in entire SWRO systems (max (CPF), –). (d) Transmembrane hydraulic pressures of the optimized SWRO system (ΔP_{opt}) and the baseline SWRO system (ΔP_{eng}), and transmembrane bulk osmotic pressures ($\Delta\pi_b$) and on the membrane walls ($\Delta\pi_w = \Delta\pi_b \cdot$ CPF) of the optimized SWRO system when recovery rate (R_r) varies from 0 (at the inlet) to 0.5 (at the outlet). (e) CPF (–) along the feed direction (first stage: $X \in [0, 1]$; second stage: $X \in [1, 2]$) (feed salinity 35 000 ppm, recovery rate 50%, efficiencies of pump 0.85 and ERD 0.95). (Reproduced with permission from [48]. Copyright 2023 Elsevier.)

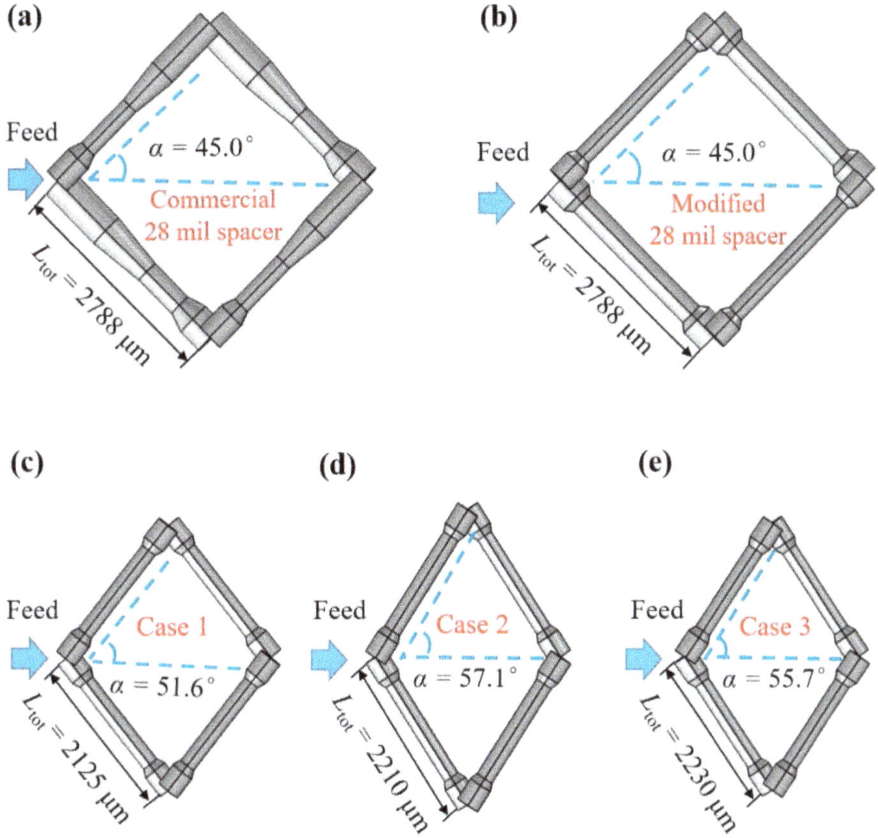

Figure 9.7. The geometric structures of feed spacer units before and after optimization. (a) Commercial 28 mil spacer unit, (b) modified 28 mil feed spacer unit, (c) optimized spacer unit in case 1 ($w_{b,0}$ = 25 000 ppm, $R_{r,0}$ = 50%), (d) optimized spacer unit in case 2 ($w_{b,0}$ = 35 000 ppm, $R_{r,0}$ = 50%), and (e) optimized spacer unit in case 3 ($w_{b,0}$ = 45 000 ppm, $R_{r,0}$ = 40%) for the designed two-stage RO desalination systems. (Reproduced with permission from [48]. Copyright 2023 Elsevier.)

Similarly, the optimized SEC is reduced by about 34.0%, 27.5%, and 21.6%. Compared with the total membrane area in a typical one-stage SWRO system (A_{tot} = 7804 m^2), the corresponding optimized results (A_{tot} = 4450 m^2, 4898 m^2, and 3790 m^2) decrease by about 43.0%, 37.2%, and 51.4%. Furthermore, the average permeation flux ($\bar{J}_W$ = 33.8, 30.7, and 31.7 l m^{-2} h^{-1}) improves by 75.6%, 59.3%, and 106.1% ($\bar{J}_W$ = 19.2, 19.3, and 15.4 l m^{-2} h^{-1} before optimization). The estimated average permeate salinity before optimization (102, 145, and 208 ppm) and after optimization (499, 498, and 500 ppm) all meet the water quality requirements in engineering (no more than 500 ppm) [41]. The permeate rate is kept constant before and after optimization. Figure 9.6(c) shows water permeability (L_p), salt permeability (B), and estimated maximum CPFs in the entire SWRO system (max(CPF)) with respect to the high permeability membranes (optimized two-stage SWROs) in cases 1–3, state-of-the-art commercial RO membrane (under typical

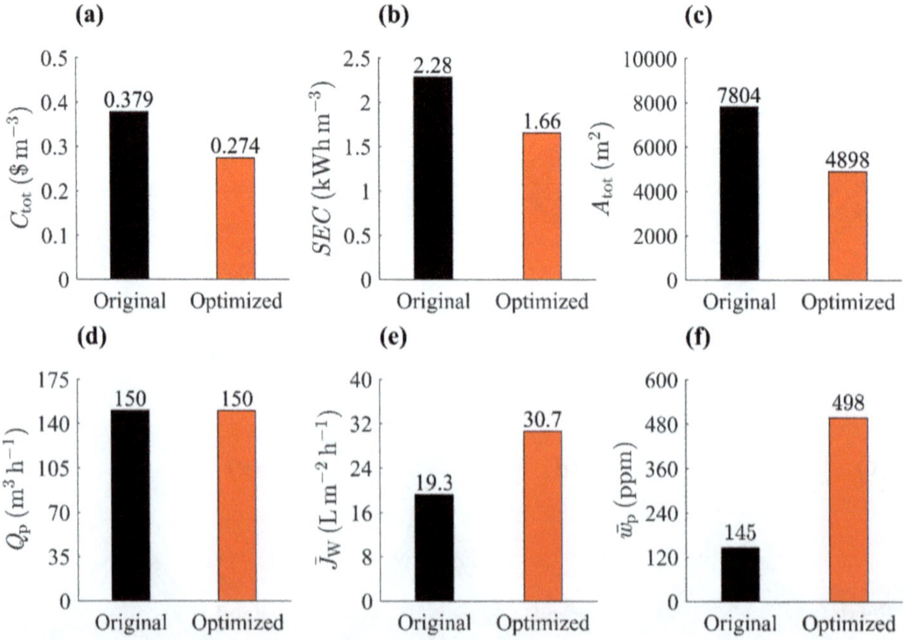

Figure 9.8. The overall performance analysis before and after optimization for the SWRO desalination systems with the feed salinity ($\overline{w}_{b,0} = 35\,000$ ppm), water recovery ($R_{r,0} = 50\%$), pump efficiency (0.85), and ERD efficiency (0.95). The estimated results consist of (a) cost per m^3 permeate, C (\$ m^{-3}), (b) specific energy consumption, SEC (kWh m^{-3}), (c) total membrane area, A_{tot} (m^2), (d) total flow rate of permeate, Q_p (m^3 h^{-1}), (e) average permeation flux of water, $\overline{J}_W$ (l m^{-2} h^{-1}), and (f) average permeate salinity, $\overline{w}_p$ (ppm). The simulated results before optimization are solved using a typical one-stage SWRO system with a modified 28 mil feed spacer (figure 9.7(b)) under typical engineering conditions and the use of a commercial RO membrane ($L_p = 1.00$ l m^{-2} h^{-1} bar^{-1}, $B = 0.05$ l m^{-2} h^{-1}). The optimized results are obtained using a two-stage SWRO system (case 2). (Reproduced with permission from [48]. Copyright 2023 Elsevier.)

operating conditions), reported above 600 RO membranes from an open membrane database [42]. The max(CPF) of over 600 RO membranes were estimated under reported experimental conditions using film theory [43]. The water and salt permeability of a reported polymeric-based thin-film composite (TFC) (figure 9.6 (c)) membrane ($L_p = 11.4$ l m^{-2} h^{-1} bar^{-1} and $B = 0.27$ l m^{-2} h^{-1}) [44] (marked in figure 9.6(c)) is comparable to that of high permeability membrane in case 2 ($L_p = 10.38$ l m^{-2} h^{-1} bar^{-1} and $B = 0.28$ l m^{-2} h^{-1}). For the optimized results in cases 1 ($L_p = 12.96$ l m^{-2} h^{-1} bar^{-1} and $B = 0.43$ l m^{-2} h^{-1}) and case 3 ($L_p = 17.12$ l m^{-2} h^{-1} bar^{-1} and $B = 0.25$ l m^{-2} h^{-1}), research and development efforts are still needed to further improve the selectivity without sacrificing water permeability [45]. However, large-scale production of high permeability membranes [9] with adequate salt rejection, mechanical strength and cost may be challenging if exotic materials are needed (e.g. GO, aquaporins, etc).

Furthermore, taking the two-stage high permeability SWRO membrane system (case 2: $w_{b,0} = 35\,000$ ppm, $R_{r,0} = 50\%$) as an example (figure 9.6(a)–(e)), the optimized feed spacer has enhanced mass transfer because the support point density

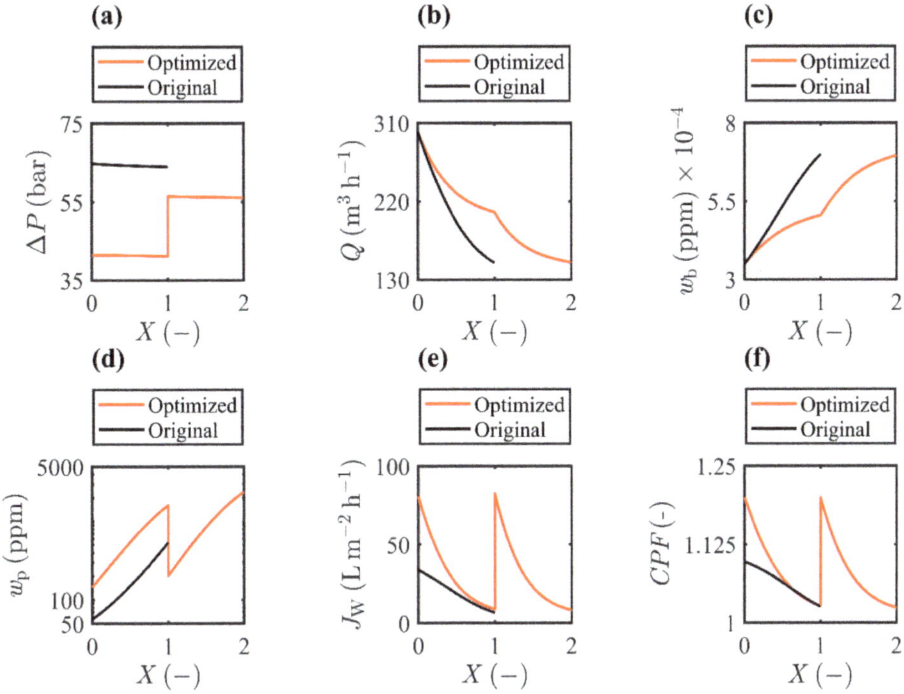

Figure 9.9. The local performance analysis before and after optimization for the SWRO desalination systems with the feed salinity ($\overline{w}_{b,0}$ = 35 000 ppm), water recovery ($R_{r,0}$ = 50%), pump efficiency (0.85), and ERD efficiency (0.95). The estimated results consist of (a) transmembrane pressure, ΔP (bar), (b) flow rate, Q (m^3 h^{-1}), (c) brine salinity, w_b (ppm), (d) permeate salinity, w_p (ppm), (e) permeation flux of water, J_W (l m^{-2} h^{-1}), and (f) concentration polarization factor, CPF(−). The simulated results before optimization are solved using a typical one-stage SWRO system with a modified 28 mil feed spacer (figure 9.7(b)) under typical engineering conditions and the use of a commercial RO membrane (L_p = 1.00 l m^{-2} h^{-1} bar^{-1}, B = 0.05 l m^{-2} h^{-1}). The optimized results are obtained using the two-stage SWRO system (case 2). (Reproduced with permission from [48]. Copyright 2023 Elsevier.)

(about 22 per cm^2) is higher than that of the commercial ones (10–12 cm^{-2}) [41]. Also, the thickness of the optimized feed spacer, D_{tot} (figure 9.6(b), 581 μm ≈ 23 mil) is less than that of the commercial 28 mil feed spacer. The cell-average mass transfer coefficient (1.23 × 10^{-4} m s^{-1}) on membrane surfaces at the inlet is enhanced by about 21.1% while the axial pressure drop per meter (0.256 bar m^{-1}) increases by about 23.4%, compared to those of a typical commercial SWRO system (1.02 × 10^{-4} m s^{-1}, 0.207 bar m^{-1}). The designed feed spacer unit (figure 9.6(b)) can be fabricated using 3D printing technology [46]. Moreover, the two-stage design allows the first stage to operate at a much lower pressure due to the use of an interstage booster pump. The inlet transmembrane pressures at the first and second stage are 41.4 and 56.4 bar, respectively, for the optimized two-stage SWRO system (figure 9.6(d)). It contributes a 27.5% energy saving compared than that of the baseline one-stage SWRO system using an inlet transmembrane pressure of 64.8 bar. The discontinuity in the transmembrane osmotic pressure between the first and second stages is due to the variation of permeate salinity (figure 9.6(d)). The estimated CPF at the inlet of the first stage without optimization (or an initialized result during optimization) is about 1.52 (figure 9.6(e)) that will intensify biological and organic fouling, which usually occur in the lead elements [47]. The CP in the system inlet can be suppressed to the maximum allowable value (1.20) [15] using the system design and spacer optimization. We further estimate the effect of the membrane cost per m^2 on the optimization function. The optimized results with the use of various membrane costs (C_m = 40, 60, 80, 100, and 120 \$ m^{-2}) indicate that the energy costs (C_e) are 0.248 \$ m^{-3} (90.5%), 0.253 (88.8%), 0.257 (86.5%), 0.263 (85.4%), and 0.267 (84.2%) respectively, and C_m is 0.026 \$ m^{-3} (9.5%), 0.031 (10.9%), 0.040 (13.5%), 0.045 (14.6%), and 0.050 (15.8%). The proportion of C_m increases with the membrane cost. The cost ($C_e + C_m$) is reduced by 27.6%, 24.9%, 21.5%, 18.8%, and 16.5% respectively, compared with that of the baseline one-stage SWRO system using a membrane cost of 40 \$ m^{-2}.

9.5 Conclusions

In recent years, high permeability SWRO membranes have attracted increasing attention in the materials science community because they offer the potential to operate at either higher water fluxes (reducing plant size and capital investment) or at lower applied pressures (reducing energy demand). However, operating high permeability SWRO membranes at higher fluxes leads to elevated CP, which defeats the energy savings and increases fouling rates. Hence, operating at significantly higher fluxes would require major changes in the designs of both SWRO modules and plant configuration. In this work, for the first time, we develop a general supercomputing-based machine-learning-driven multiscale optimization framework for high permeability SWRO membrane desalination systems that couples module improvement, system design, and optimization of membrane permeability. Using the optimization method, we lay out the technological pathway to achieving a 27.5% lower energy demand for seawater desalination with 37.2% less membrane area compared with state-of-the-art desalination systems (feed salinity 35 000 ppm, recovery rate 50%). The maximum CPF is controlled to no more than 1.20 by optimization of module and process designs, which markedly reduces the risk of fouling and scaling and takes a crucial step towards the application of high permeability membranes. This work adopts a novel interdisciplinary research method, and attempts to alter the traditional research paradigm that relies on Edisonian trial and error efforts for improvement on a single scale, utilizing advanced supercomputing techniques combined with machine learning. The proposed optimization framework is universal and rigorous from a mathematical perspective. Nevertheless, the validation, uncertainty, and quantification analysis of the multiscale model still need to be conducted with experimental data in future work.

The presented multiscale approach [48] can be potentially combined with unsteady-state shear technologies [49], which may enhance the mass transfer by 2–5 times. Furthermore, dynamic RO technologies, e.g. semi-batch RO [50] with the use of a high permeability membrane may, reduce CP and membrane fouling due to a more uniform driving force compared with a continuous RO system. Finally, the use of high permeability membranes may reduce SEC in minimal/zero liquid discharge applications [51, 52], thereby improving their economic viability. The proposed multiscale design method can be extended to these areas in future studies.

References

[1] Shannon M A *et al* 2008 Science and technology for water purification in the coming decades *Nature* **452** 301–10

[2] Qasim M *et al* 2019 Reverse osmosis desalination: a state-of-the-art review *Desalination* **459** 59–104

[3] Voutchkov N 2018 Energy use for membrane seawater desalination—current status and trends *Desalination* **431** 2–14

[4] Vahdatifar S *et al* 2021 Functionalized open-ended vertically aligned carbon nanotube composite membranes with high salt rejection and enhanced slip flow for desalination *Sep. Purif. Technol.* **279** 119773

[5] Tunuguntla R H *et al* 2017 Enhanced water permeability and tunable ion selectivity in subnanometer carbon nanotube porins *Science* **357** 792–6

[6] Wei G *et al* 2018 Direct growth of ultra-permeable molecularly thin porous graphene membranes for water treatment *Environ. Sci.: Nano* **5** 3004–10

[7] Chen L *et al* 2017 Ion sieving in graphene oxide membranes via cationic control of interlayer spacing *Nature* **550** 380–3

[8] Lee C S *et al* 2021 Aquaporin-incorporated graphene-oxide membrane for pressurized desalination with superior integrity enabled by molecular recognition *Adv. Sci.* **8** e2101882

[9] Yang Y *et al* 2019 Large-area graphene-nanomesh/carbon-nanotube hybrid membranes for ionic and molecular nanofiltration *Science* **364** 1057–62

[10] Tan Z *et al* 2018 Polyamide membranes with nanoscale turing structures for water purification *Science* **360** 518–21

[11] Culp T E *et al* 2021 Nanoscale control of internal inhomogeneity enhances water transport in desalination membranes *Science* **371** 72–5

[12] Siria A *et al* 2013 Giant osmotic energy conversion measured in a single transmembrane boron nitride nanotube *Nature* **494** 455–8

[13] Li P F *et al* 2020 Precise assembly of a zeolite imidazolate framework on polypropylene support for the fabrication of thin film nanocomposite reverse osmosis membrane *J. Membr. Sci.* **612** 118412

[14] Itoh Y *et al* 2022 Ultrafast water permeation through nanochannels with a densely fluorous interior surface *Science* **376** 738–43

[15] Fane A G, Wang R and Hu M X 2015 Synthetic membranes for water purification: status and future *Angew. Chem. Int. Ed. Engl.* **54** 3368–86

[16] Elimelech M and Phillip W A 2011 The future of seawater desalination: energy, technology, and the environment *Science* **333** 712–7

[17] Patel S K *et al* 2020 The relative insignificance of advanced materials in enhancing the energy efficiency of desalination technologies *Energy Environ. Sci.* **13** 1694–710

[18] Zhu A H, Christofides P D and Cohen Y 2009 Effect of thermodynamic restriction on energy cost optimization of ro membrane water desalination *Ind. Eng. Chem. Res.* **48** 6010–21

[19] Cohen-Tanugi D *et al* 2014 Quantifying the potential of ultra-permeable membranes for water desalination *Energy Environ. Sci.* **7** 1134–41

[20] Lim Y J *et al* 2022 Assessing the potential of highly permeable reverse osmosis membranes for desalination: specific energy and footprint analysis *Desalination* **533** 115771
[21] Gu B R, Adjiman C S and Xu X Y 2017 The effect of feed spacer geometry on membrane performance and concentration polarisation based on 3D CFD simulations *J. Membr. Sci.* **527** 78–91
[22] Johannink M *et al* 2015 Predictive pressure drop models for membrane channels with non-woven and woven spacers *Desalination* **376** 41–54
[23] Lin W C *et al* 2020 Impacts of non-uniform filament feed spacers characteristics on the hydraulic and anti-fouling performances in the spacer-filled membrane channels: experiment and numerical simulation *Water Res.* **185** 116251
[24] Li M H 2012 Optimal plant operation of brackish water reverse osmosis (BWRO) desalination *Desalination* **293** 61–8
[25] Guillen G and Hoek E M V 2009 Modeling the impacts of feed spacer geometry on reverse osmosis and nanofiltration processes *Chem. Eng. J.* **149** 221–31
[26] Luo J, Li M H and Heng Y 2020 A hybrid modeling approach for optimal design of non-woven membrane channels in brackish water reverse osmosis process with high-throughput computation *Desalination* **489** 114463
[27] Fetanat M *et al* 2021 Machine learning for design of thin-film nanocomposite membranes *Sep. Purif. Technol.* **270** 118383
[28] Hu J *et al* 2021 Artificial intelligence for performance prediction of organic solvent nanofiltration membranes *J. Membr. Sci.* **619** 118513
[29] Zhang Z *et al* 2021 Deep spatial representation learning of polyamide nanofiltration membranes *J. Membr. Sci.* **620** 118910
[30] Ignacz G and Szekely G 2022 Deep learning meets quantitative structure–activity relationship (QSAR) for leveraging structure-based prediction of solute rejection in organic solvent nanofiltration *J. Membr. Sci.* **646** 120268
[31] Barnett J W *et al* 2020 Designing exceptional gas-separation polymer membranes using machine learning *Sci. Adv.* **6** eaaz4301
[32] Rall D *et al* 2020 Multi-scale membrane process optimization with high-fidelity ion transport models through machine learning *J. Membr. Sci.* **608** 118208
[33] Li M H 2020 Effects of finite flux and flushing efficacy on specific energy consumption in semi-batch and batch reverse osmosis processes *Desalination* **496** 114646
[34] Lin S and Elimelech M 2015 Staged reverse osmosis operation: configurations, energy efficiency, and application potential *Desalination* **366** 9–14
[35] Werber J R, Deshmukh A and Elimelech M 2017 Can batch or semi-batch processes save energy in reverse-osmosis desalination? *Desalination* **402** 109–22
[36] Fimbres-Weihs G A and Wiley D E 2010 Review of 3D CFD modeling of flow and mass transfer in narrow spacer-filled channels in membrane modules *Chem. Eng. Process.* **49** 759–81
[37] Geraldes V and Afonso M D 2006 Generalized mass-transfer correction factor for nano-filtration and reverse osmosis *AIChE J.* **52** 3353–62
[38] Gu J H *et al* 2020 Modeling of pressure drop in reverse osmosis feed channels using multilayer artificial neural networks *Chem. Eng. Res. Des.* **159** 146–56
[39] Bucs S S *et al* 2014 Effect of different commercial feed spacers on biofouling of reverse osmosis membrane systems: a numerical study *Desalination* **343** 26–37
[40] Toh K Y *et al* 2020 The techno-economic case for coupling advanced spacers to high-permeance ro membranes for desalination *Desalination* **491** 114534

[41] Johnson J and Busch M 2012 Engineering aspects of reverse osmosis module design *Desalin. Water Treat.* **15** 236–48
[42] Ritt C L *et al* 2022 The open membrane database: synthesis–structure–performance relationships of reverse osmosis membranes *J. Membr. Sci.* **641** 119927
[43] Baker R W 2012 *Membrane Technology and Applications* (New York: Wiley)
[44] Cadotte J E 1981 Interfacially synthesized reverse osmosis membrane US Patent No 4277344
[45] Park H B *et al* 2017 Maximizing the right stuff: the trade-off between membrane permeability and selectivity *Science* **356** eaab0530
[46] Tijing L D *et al* 2020 3D printing for membrane separation, desalination and water treatment *Appl. Mater. Today* **18** 100486
[47] Al-Amoudi A and Lovitt R W 2007 Fouling strategies and the cleaning system of NF membranes and factors affecting cleaning efficiency *J. Membr. Sci.* **303** 4–28
[48] Luo J *et al* 2023 Supercomputing and machine learning-aided optimal design of high permeability seawater reverse osmosis membrane systems *Sci. Bullet* **68** 397–407
[49] Zamani F *et al* 2015 Unsteady-state shear strategies to enhance mass-transfer for the implementation of ultrapermeable membranes in reverse osmosis: a review *Desalination* **356** 328–48
[50] Lee T, Rahardianto A and Cohen Y 2019 Multi-cycle operation of semi-batch reverse osmosis (SBRO) desalination *J. Membr. Sci.* **588** 117090
[51] Davenport D M *et al* 2018 High-pressure reverse osmosis for energy-efficient hypersaline brine desalination: current status, design considerations, and research needs *Environ. Sci. Technol. Lett.* **5** 467–75
[52] Du Y *et al* 2021 Module-scale analysis of low-salt-rejection reverse osmosis: design guidelines and system performance *Water Res.* **209** 117936

Chapter 10

Supercomputing-based inverse identification of high-resolution atmospheric pollutant source intensity distributions

Mingming Huang and Yi Heng

Atmospheric pollution research is critical for monitoring industrial emissions, optimizing environmentally friendly processes, and understanding pollutant degradation mechanisms. This chapter presents a high-throughput parallel computing framework for solving nonlinear inverse problems in high-resolution spatiotemporal contaminant source estimation. Using the Lagrangian transport model MPTRAC for forward simulations, a million-core supercomputing strategy is employed to significantly reduce computational time while maintaining accuracy and reliability.

10.1 Introduction

Process engineering has been adopted across a diverse range of disciplines, with its application in chemical engineering representing a particularly mature and sophisticated field of study. This is driven by the inherent complexity and diversity of chemical processes, coupled with industrial demands that require process systems engineering (PSE) methodologies capable of managing multi-scale, multi-variable systems while balancing economic and environmental objectives. In chemical processes, the emission of pollutants frequently presents a complex problem, involving numerous reactions and transfers of various materials. Process optimization is an effective method for reducing pollutant emissions. Nevertheless, even when processes are optimized, certain unavoidable pollutants, such as atmospheric particulate matter, are still released into the environment. Air pollution has a

doi:10.1088/978-0-7503-6174-3ch10

significant negative impact on ecosystems, including human beings. Therefore, monitoring and understanding the distribution of air pollutants is of vital importance.

Atmospheric pollutants, including particulate matter (PM), nitrogen oxides (NOx), sulfur dioxide (SO_2), and volatile organic compounds (VOCs), etc, are associated with respiratory and cardiovascular diseases, acid rain formation, and climate variations. In addition, greenhouse gases may be released as a consequence of natural or anthropogenic disasters, including the combustion of fossil fuels, industrial processes, waste management, and forest fires, among a number of other potential sources. Greenhouse gases such as carbon dioxide (CO_2) and methane (CH_4) have been identified as significant contributors to global warming, with their heat-trapping capacity leading to increased global temperatures and disrupted weather patterns. CO_2 is an exemplary greenhouse gas for scientific investigation, offering a number of advantages including a prolonged atmospheric residence time, extensive observational data accessibility, and a high degree of public interest. A comprehensive understanding of the dispersion of these pollutants in the atmosphere, which is closely linked to the Earth's atmospheric dynamics, is a natural extension of process engineering in industry and other fields and essential to further investigate and reduce impacts on the environment and public health.

A thorough knowledge of Earth's atmospheric dynamics is contingent upon an appreciation of the interrelationship between chemistry and atmospheric science. The chemical properties and reactions of atmospheric constituents are of significant importance with regard to the processes occurring within the atmospheric environment. The transport and transformation of atmospheric particles involve not only physical processes such as advection, dispersion, and sedimentation, but also complex chemical reactions such as oxidation and the formation of secondary aerosols, all of which together influence air quality and contribute to climate change. By simulating the particle dispersion, researchers can gain deeper insights into their environmental impact, thereby underscoring the intrinsic link between atmospheric science and atmospheric chemistry.

Atmospheric particle dispersion models are a common tool in this context. These models simulate the transport, deposition, and transformation of particulate matter in the atmosphere, thereby facilitating the prediction of the spatial distribution and concentration of pollutants. A sensitivity analysis can be employed to elucidate the pivotal influencing factors. Moreover, inverse source identification techniques can be employed to enhance the identification of primary emission sources. Nevertheless, the completion of these tasks may prove challenging due to factors such as the mathematical ill-posedness of the problem and the high computational cost associated with solving it. Furthermore, the distribution of pollutants in the atmosphere is a function of both space and time, which makes it a high-dimensional problem. The dispersion of pollutants is influenced by a number of factors, including wind patterns, temperature, humidity, and chemical reactions. The reliability of estimates is contingent upon the availability of high-quality observational data, which are often sparse or noisy. Notwithstanding these challenges, considerable efforts have been made, resulting in the development of numerous effective solutions.

Both the Lagrangian particle dispersion model and the Eulerian model are utilized for atmospheric particle dispersion simulation. Lagrangian models are more preferable for examining small-scale structures, filamentary transport, and mixing processes in the atmosphere because they are not constrained by the fixed spatial resolution of Eulerian grid cells, and exhibit minimal numerical dispersion. However, Lagrangian transport simulations are often expensive to execute due to the necessity of introducing stochastic perturbations to statistically represent subgrid-scale processes, including dispersion, and mesoscale wind fluctuations. The necessity for high-performance computing arises from the need to process the vast quantities of data and to perform the intensive calculations required for the accurate modeling of a large number of particle behaviors over extensive spatial and temporal scales.

This chapter is organized as follows. Section 10.2 provides a concise overview of the diverse range of Lagrangian models and studies conducted on the MPTRAC platform, which serves as the foundation for the high-performance computing systems utilized in the case studies presented in this chapter. Section 10.3 delineates the methodologies and theoretical frameworks employed in the high-throughput parallel forward computing framework and the supercomputing-based inverse computing strategy. In section 10.4 we elucidate the dispersion simulation of CO_2 and SO_2 released by forest fires and volcanic eruptions, as previously investigated by our research team [1–3]. Discussions and conclusions can be found in section 10.5.

10.2 Lagrangian models

Various transport models play a crucial role in the study of pollutant dispersion. The Lagrangian particle dispersion model tracks individual air parcels to describe airflow, while Eulerian models use a grid-based approach to present fluid flows between regular grid boxes. The Lagrangian model, which provides a microscopic perspective by assuming unique velocities and pressures for each air parcel, excels at capturing small-scale features and minimizing numerical dispersion. The independence of particle motions within this model allows for the parallel computation of their trajectories, enhancing computational efficiency and scalability.

Lagrangian models are widely applied in atmospheric chemistry and dynamics research. Notable examples include the flexible particle dispersion model (FLEXPART) [4], the hybrid single-particle Lagrangian integrated trajectory model (HYSPLIT) [5], the California puff model (CALPUFF) [6], and MPTRAC [7]. These models differ in their numerical methods for solving the advection and dispersion problem, the complexity of the flow field environment during particle transport, and the accuracy of their simulation results, etc. MPTRAC is selected for this study primarily due to its high computational efficiency in large-scale simulations, which yield high-quality output. MPTRAC is an open-source software package for Lagrangian particle dispersion simulation developed by the Jülich Supercomputing Center in Germany [7]. It is capable of simulation and analysis of trace gas and aerosol transport in the free troposphere and stratosphere. The model is suitable for studying fine-scale structures, filamentary transport, and mixing processes in the atmosphere. It allows for the accurate estimation of high-resolution spatiotemporal distributions of atmospheric pollutants, which is crucial for understanding and mitigating their impacts.

Since its initial description in 2006, the MPTRAC central processing unit (CPU) and graphics processing unit (GPU) code has been validated in various studies. As illustrated in figure 10.1, the principal functions of MPTRAC are classified into three categories: input, processing, and output. While the input and output operations make use of the CPU memory, the processing functions have been ported to GPUs, significantly reducing the computational time. The dispersion of anthropogenic and natural emissions and the atmospheric dynamics of the free troposphere and stratosphere have been studied on the MPTRAC platform. Heng *et al* developed an inverse transport model and applied the 'product rule' to simulate volcanic SO_2 emissions utilizing sequential importance resampling and parallel computing [1]. Rößler *et al* investigated the model accuracy influenced by numerical integration schemes, and recommend a midpoint scheme and third-order Runge–Kutta method for efficient simulations [8]. Hoffmann *et al* ported the model to GPUs and achieved a speed-up factor of 16 for large-scale Lagrangian transport simulations, demonstrating its effectiveness for near-real-time applications on the supercomputer [7]. Wu *et al* found that the Asian summer monsoon anticyclone (ASMA) plays an important role in aerosol transport between the tropics and extra-tropics [9]. Hoffmann *et al* optimized the baseline codes through data restructuring and memory alignment, significantly reducing simulation runtime and making them suitable for future exascale high-performance computing systems [10]. Clemens *et al* found that the Asian tropopause aerosol layer (ATAL) is primarily influenced by continental air masses from the Tibetan Plateau, the Indo-Gangetic Plain, and maritime air from the western Pacific, with variations depending on the simulation scenarios [11].

To address the challenges outlined in section 10.1, advanced approaches are employed within the MPTRAC platform. Advanced algorithms coupled with substantial computational resources are deployed to tackle the high-dimensional nature of these problems. Additionally, in contrast to regularization techniques that consider mathematical ill-posedness, analytical inversion algorithms are utilized for the reconstruction of volcanic ash or SO_2 emission rates [12]. Furthermore, more detailed data, integrating information from diverse sources such as ground-based stations, satellites, and aircraft, are incorporated. These combined strategies effectively alleviate the challenges and significantly improve the accuracy of the estimated spatiotemporal distributions of atmospheric pollutants.

The MPTRAC simulation employs a hybrid MPI–OpenMP–OpenACC parallelization strategy, allowing for execution on not only single workstations but also

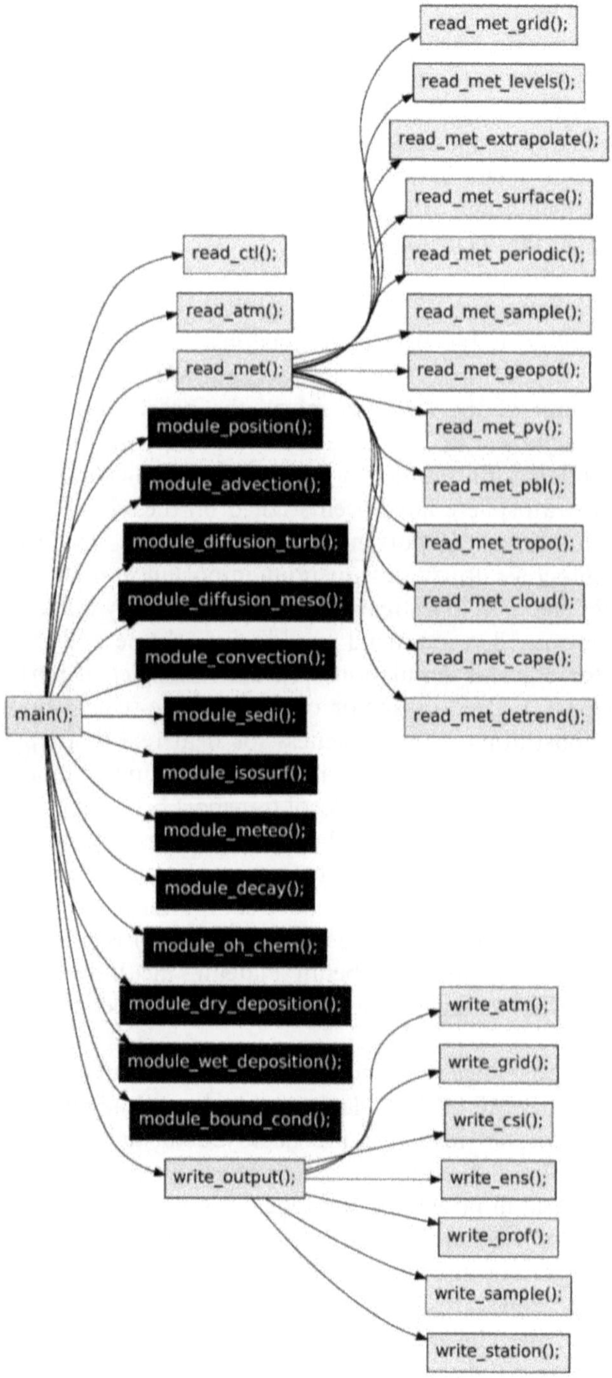

Figure 10.1. Overview of MPTRAC modules. (Reproduced from [7]. CC BY 4.0.)

high-performance computing (HPC) and GPU clusters. This chapter presents the results of simulations conducted on two national supercomputer platforms: the Jülich Wizard for European Leadership Science (JUWELS) HPC system at the Jülich Supercomputing Centre in Germany and the Tianhe-2 supercomputer platform at the National Supercomputer Center in Guangzhou (NSCC-GZ) in China [13]. In the 2010 and 2011 editions of the TOP500 supercomputing rankings (TOP500), the JUWELS Booster Module was placed in the seventh to eighth position. Since 2010, the Tianhe-2 supercomputer has been ranked first in the TOP500 list for a total of six times. Table 10.1 shows the most recent TOP500 list in 2024, with the only two exascale machines ranking in the top two positions and the two supercomputers employed in this chapter ranking sixteenth and twenty-first, respectively.

10.3 Methods and theories

10.3.1 Forward simulation framework

The Lagrangian transport model MPTRAC is employed for forward simulations to meet the requirements of large-scale atmospheric transport calculations. It contains many modules for simulating the effects of convection, sedimentation, dispersion, subgrid-scale wind fluctuations, dry and wet deposition, chemical decomposition, etc. Among them, when simulating the chemical reactions with hydroxyl, the chemical module specifies the Arrhenius factor and the ratio of activation energy to the universal gas constant as control parameters, as well as the exponents in the reaction rate formula, which correspond to bimolecular and termolecular reactions, respectively.

Given the horizontal wind and vertical velocity fields $v(x, t)$ at time t, the position $x(t)$ of an air parcel at time t is given by the following trajectory equation:

$$\frac{\mathrm{d}x}{\mathrm{d}t} = v(x, t), \tag{10.1}$$

where $v = (u, v, \omega)$ is a velocity vector comprising the zonal wind component u, the meridional wind component v, and the vertical velocity ω. In the meteorological coordinate system, $x = (\lambda, \phi, p)$ is represented by the values of longitude λ, latitude ϕ, and pressure p, respectively. Previous analysis of different numerical integration schemes has revealed that the midpoint method can achieve a balance between accuracy, stability, and computational efficiency in the trajectory calculation tasks [8]. Consequently, MPTRAC uses the explicit midpoint method to compute the particle trajectories, as shown in

$$x(t + \Delta t) = x(t) + \Delta t \cdot v\left\{x(t) + \frac{\Delta t}{2}v[x(t), t], t + \frac{\Delta t}{2}\right\}. \tag{10.2}$$

Here, the time step Δt should be selected in a manner that ensures the error control of non-stiff problems. The inequality $\Delta t < \Delta\lambda_{\mathrm{met}} R_E/u_{\max}$ guarantees that the change in position $\Delta x = x(t + \Delta t) - x(t)$ during Δt does not exceed the spatial

Table 10.1. Top500 supercomputing rankings in 2024 [14].

Rank	System	Cores	Rmax (PFlop/s)	Rpeak (PFlop/s)	Power (kW)
1	**Frontier**—HPE Cray EX235a, AMD Optimized 3rd Generation EPYC 64 C 2 GHz, AMD Instinct MI250X, Slingshot-11, HPEDOE/SC/Oak Ridge National Laboratory, United States	8 699 904	1206.00	1714.81	22 786
2	**Aurora**—HPE Cray EX—Intel Exascale Compute Blade, Xeon CPU Max 9470 52 C 2.4 GHz, Intel Data Center GPU Max, Slingshot-11, IntelDOE/SC/Argonne National Laboratory, United States	9 264 128	1012.00	1980.01	38 698
16	**Tianhe-2A**—TH-IVB-FEP Cluster, Intel Xeon E5-2692 v2 12 C 2.2 GHz, TH Express-2, Matrix-2000, NUDT, National Supercomputer Center in Guangzhou, China	4 981 760	61.44	100.68	18 482
21	**JUWELS Booster Module**—Bull Sequana XH2000, AMD EPYC 7402 24 C 2.8 GHz, NVIDIA A100, Mellanox HDR InfiniBand/ParTec ParaStation ClusterSuite, EVIDEN Forschungszentrum Juelich (FZJ), Germany	449 280	44.12	70.98	1764

resolution of the meteorological grid, thereby ensuring both accuracy and physical consistency. $\Delta\lambda_{\text{met}}$, R_E, andu_{max} stand for the longitudinal grid spacing of the meteorological data, the mean radius of the Earth, and maximum zonal wind speed, respectively.

The impact of random perturbations on particle dispersion, including turbulent dispersion and subgrid-scale wind fluctuations, are considered. The dispersion rates in the free troposphere and stratosphere remain poorly understood. In order to simulate the effects of turbulent dispersion, the MPTRAC model incorporates random disturbances in the following manner:

$$\Delta x_i(t + \Delta t) = x_i(t) + \sqrt{2D_i\Delta t}\,\xi_i. \tag{10.3}$$

Here, elements of the diffusivities vector $D_i = \left(D_x, D_y, D_z\right)$ are the horizontal and vertical diffusivities. The vector $\xi = (\xi_x, \xi_y, \xi_z)$ denotes a vector comprising random variables, which are drawn from the standard normal distribution for each air parcel at each time step. The smooth transition of diffusivities between the troposphere and stratosphere is achieved through linear interpolation within a ±1 km log-pressure altitude range around the tropopause.

Subgrid-scale wind fluctuations are simulated using a random walk methodology. Random walk describes stochastic Brownian motion or dispersion processes that adhere to the Markov property. In this property, the subsequent state of the process is dependent solely on the current state, reflecting the physical reality that underlies the process. The subgrid-scale wind fluctuations are calculated based on the Langevin equation,

$$v = \overline{v} + v' \tag{10.4}$$

$$v'_i(t + \Delta t) = rv'_i(t) + \sqrt{1 - r^2}(f\sigma_i)^2\xi_i \tag{10.5}$$

$$r = 1 - 2\frac{\Delta t}{\Delta t_{\text{met}}}, \tag{10.6}$$

where v, $\overline{v}$, and v' represent the horizontal wind and vertical velocity vector, the grid-scale mean, and the subgrid-scale perturbations, respectively. $r, f, \sigma,$ and ξ stand for the correlation coefficient, the scaling factor, the square root of the grid-scale variance, and the vector of random variates, respectively.

For the forward simulation results, the critical success index (CSI) can be calculated and compared with satellite observation results to measure the goodness-of-fit of the forward simulation [15]:

$$\text{CSI} = C_x/(C_x + C_y + C_z), \tag{10.7}$$

where C_x and C_y denote the number of forecasts that are positive and negative, respectively, while both have positive observations. In comparison, C_z represents the number of positive forecasts that have negative observations. A comparative analysis of satellite observation data and MPTRAC simulation data mapped on discrete grids allows the calculation of temporal CSI values, which can be employed in the inverse modeling and simulation system [2].

Based on this forward simulation framework, the nonlinear inverse problem of pollutant source estimation is solved in an appropriate manner, as illustrated in the next subsection.

10.3.2 High-throughput parallel inverse computing strategy

A million-core supercomputing-based inverse computation strategy is implemented, which drastically reduces the simulation run times while ensuring the accuracy and reliability required for real-time engineering-level prediction tools. This strategy leverages high-performance computing resources to meet the needs of future atmospheric pollutant modeling and real-time prediction.

The simulation of atmospheric dispersion incorporates data pertaining to time and spatial dimensions. The computational domain is defined as

$$E: =\left(t_0,\, t_f\right) \times \Omega. \tag{10.8}$$

The computational domain is uniformly discretized into $N = n_t \times n_h$ subdomains, where t and h represent the temporal and spatial dimension, respectively, and the latter is expressed in the form of log-pressure altitude. Consequently, during the iteration process, there are N parallel simulation units. The initialization conditions assign the preset total number of air parcels to each subdomain with equal probability, referred to as the 'mean rule', thus ensuring that each simulation unit contains a certain number of air parcels. Then the mass unit simulations are performed, and the importance weights are estimated based on the weight-updating strategy. The estimated weight parameters are employed to redistribute air parcels to each subdomain until the termination condition is satisfied. Ultimately, the source intensity distribution of the entire space–time region is obtained, according to the final importance weight.

The aforementioned importance weight estimation employs a heuristic strategy, which is in fact equivalent to particle filtering. This encompasses the following procedures: the total particle mass is provided, the posterior particle release probability is estimated as the importance weight w_i, which satisfies the following relationship with CSI_k^i in time series:

$$\sum\nolimits_{i=1}^{n_t}\sum\nolimits_{j=1}^{n_h} w_{ij} = 1 \tag{10.9}$$

$$w_{i,j}^{z-1} = \begin{cases} \dfrac{1}{N}, & z = 1 \\ \dfrac{m_{ij}}{\sum\limits_{a=1}^{n_t}\sum\limits_{b=1}^{n_h} m_{ab}}, & z = 2, 3, \ldots \end{cases} \tag{10.10}$$

$$m_{ij} = \frac{\sum_{k=1}^{n_k} \mathrm{CSI}_k^{ij}}{n_k}, \tag{10.11}$$

where $k(k = 1, \ldots, n_k)$ corresponds to the minimum time interval t_k, which is necessary to ensure comprehensive coverage of the specified area in the satellite observation data. The variables i $(i = 1, \ldots, n_t)$ and j $(j = 1, \ldots, n_t)$ represent the uniform discrete temporal and spatial grid boxes sequences of the simulation units, respectively. m_{ij} is the probability of the release source air parcels falling into the (i, j) subdomain. In order to more accurately solve the complex problem of particle dispersion simulation, the 'mean rule' of equation (10.11) in the resampling process is modified, and the 'product rule' of equation (10.12) is used to calculate m_{ij}:

$$m_{ij} = \left(\sum_{k=1}^{n_k'} \mathrm{CSI}_k^{ij}/n_k' \right) \cdot \left(\sum_{k=n_k'+1}^{n_k} \mathrm{CSI}_k^{ij}/(n_k - n_k') \right), 1 \leqslant n_k' \leqslant n_k, \tag{10.12}$$

where n_k' represents a specific split point within the time series. The enhanced resampling algorithm is more effective at excluding local emissions with fewer possibilities. This algorithm is illustrated in table 10.2.

MPTRAC employs the MPI–OpenMP–OpenACC hybrid parallelization scheme, thereby ensuring high computational efficiency for large-scale simulations. The two parallelization techniques, Open Multi-processing (OpenMP) and Message Passing

Table 10.2. Resampling strategy in the inverse transport modeling.

Input n_t, n_k, $d_{\min}$

1. Calculate the number of subdomains, i.e. the number of iterations $N = n_t \times n_h$, and set $z = 1$.
2. **Do**
3. Perform forward simulation of N units in parallel and calculate the corresponding CSI in time series.
4. Update the importance weights w_{ij} according to equations (10.9), (10.10), and (10.12).
5. Update the importance weight matrix W and relative difference $d(W^{z+1}, W^z)$ by the following equation,

$$\|W^z\| = \sqrt{\sum_{i=1}^{n_t}\sum_{j=1}^{n_h} \left| w_{ij}^z \right|^2} \quad W^z = \left(w_{ij}^z \right)_{i=1, \ldots, n_t; } j = 1, \ldots, n_h \tag{10.13}$$

$$d(W^{z+1}, W^z) = \frac{\|W^{z+1} - W^z\|}{\max(\|W^{z+1}\|, \|W^z\|)} \quad z \geqslant 1 \tag{10.14}$$

6. **While** $d \leqslant d_{\min}$.
7. If the stopping rule is fulfilled, returned W, and redistribute the air parcels throughout the initialization area accordingly.

The default value of $d_{\min}$ is 0.01.

Interface (MPI), each have distinct characteristics. OpenMP, as a shared-memory programming model, is directly supported by compilers such as GNU and Intel. It facilitates parallel processing within a single computing node, maximizing the utilization of multi-core resources, with the maximum number of threads constrained by the physical cores available on the node. MPI, on the other hand, is a distributed-memory parallel programming model that relies on a specific MPI environment and software stack. It enables computational tasks to run across multiple nodes, facilitating data exchange and collaboration, and allows integrated simulations for sensitivity analysis, with the maximum number of processes limited by the total compute nodes available under the user's account [16]. Open Accelerators (OpenACC) is a programming model, designed to be a high-level, platform independent language for programming accelerators. It employs high-level compiler directives to expose parallelism in the code and parallelizing compilers to construct code for a range of parallel accelerators [17].

In this chapter, the effectiveness of MPTRAC with these parallelization techniques is further demonstrated on two of the most advanced supercomputer systems in the world. The two cases presented, namely the dispersion simulation of CO_2 and SO_2, were conducted on the Tianhe-2 supercomputer platform at the NSCC-GZ, China, and the JUWELS HPC system at the Jülich Supercomputing Centre, Germany [14, 18, 19]. The Tianhe-2 supercomputer has a total of 4 981 760 cores. It has a peak performance of 100.68 PFlops. Each node is equipped with two Intel Xeon E5-2692 v2 CPUs and three Xeon Phi coprocessors. The nodes are interconnected via a custom high-speed TH Express-2 network. The JUWELS supercomputer consists two main modules: the JUWELS Cluster and the JUWELS Booster. The JUWELS Cluster is equipped with Intel Xeon Skylake-SP processors and Mellanox EDR InfiniBand, while the JUWELS Booster uses second generation AMD EPYC processors, NVIDIA Ampere GPUs and NVIDIA/Mellanox HDR InfiniBand. The JUWELS Booster consists of 449 280 cores and delivers a peak performance of over 70.98 PFlops.

To enhance computational efficiency and to circumvent the utilization of excessive computing resources during operation, MPTRAC incorporates additional settings that facilitate the promotion of efficient parallel computing. To illustrate, the input data utilized in the model are employed to calculate the supplementary meteorological data necessitated, thus obviating the need to compute and store these additional data points in storage media prior to their actual utilization. The input data undergo preprocessing through a linear interpolation methodology that fulfills the requisite accuracy standards, circumventing the deployment of a high-order interpolation scheme. The interpolation weight and the change of mesh ratio of air parcels are stored in the cache for expedient retrieval.

10.4 Applications and analysis

10.4.1 Data product

Input data for atmospheric transport models may come from ground-based, airborne, or space-based observation platforms. Although susceptible to atmospheric conditions such as cloud cover, space-based observations offer clear relative advantages, including global coverage, continuity, and regularity. Gases or solid particles released into the atmosphere from natural or anthropogenic sources can be detected by space-based observing technologies, such as satellites in low Earth orbit (LEO) and geostationary Earth orbit (GEO) [20]. Currently, regular monitoring of harmful pollutants and greenhouse gas emissions globally, including SO_2 and CO_2, relies primarily on satellite observations, supplemented by ground-based and airborne data [21–25]. Below is an overview of the main data sources used in this study.

The ERA-Interim and the ERA5 reanalysis data are provided by the European Centre for Medium-Range Weather Forecasts (ECMWF) [26, 27]. Atmospheric reanalysis has been carried out successively at ECMWF since the First Global Atmospheric Research Program (GARP) Global Experiment (FGGE) in 1979. The ERA-Interim, ECMWF's fourth generation reanalysis, provides 4D-Var data from January 1979 to August 2019. The ERA-Interim is ECMWF's fifth-generation reanalysis available from 2020, and allows the extension of the reanalysis 4D-Var data from 1950 to the present. The assimilated observations and measurements have bene generated by satellites, radiosondes, dropsondes, aircraft, scatterometers, and altimeters, etc. For the ERA-Interim and the ERA5 data, the temporal resolutions are 6 h and 1 h, and the grid resolutions are 80 km and 31 km, respectively. The ERA5 data provide wind and velocity fields for CO_2 trajectory calculations, and both of them are utilized for the SO_2 transport simulations in this chapter.

The Meteosat Visible and InfraRed Imager (MVIRI) instrument aboard the Meteosat-7 satellite of the European Organization for the Exploitation of Meteorological Satellites (EUMETSAT) provided data in the three spectral regions, including the visible range (VIS) with spectral ranges of $0.5-0.9\ \mu m$, the thermal infrared region (TIR) of $10.5-12.5\ \mu m$, and the water vapor absorption bands (WV) of $5.7-7.1\ \mu m$ [1, 28]. Meteosat-7 is one of the Meteosat series of satellites operated by the EUMETSAT as part of the Meteosat Transition Programme (MTP), also with a mission to support Indian Ocean Data Coverage (IODC). It is one of the two longest-serving operational satellites serving for twenty years in EUMETSAT's history, from its launch in September 1997 to its retirement in April 2017 [29]. MVIRI's near-real-time imagery data, with half-hour temporal resolution and 5 km grid resolution, serves as validation data in this SO_2 transport simulation research.

The Cloud-Aerosol Lidar with Orthogonal Polarization (CALIOP), a near-nadir viewing two-wavelength polarization-sensitive lidar, is a primary instrument aboard the Cloud-Aerosol Lidar and Infrared Pathfinder Satellite Observations (CALIPSO) satellite [30–32]. CALIPSO is part of the Earth System Science Pathfinder (ESSP) program operated by the National Aeronautics and Space Administration (NASA) and the Centre national d'études spatiales (CNES). It was launched in April 2006 to fill existing gaps in the ability to observe the global distribution and properties of aerosols and clouds. Different to passive techniques, CALIPSO provides its own illumination and yields a more complete dataset over the full globe in both day and night time conditions. In this chapter, the CALIOP data, with a spatial resolution of 1.67 km (horizontal) $\times$ 60 m (vertical) at an altitude of $8-20$ km, is utilized to verify the altitude distribution of the volcanic emissions. The CALIPSO Level 2

Lidar Vertical Feature Mask (VFM) offers tropospheric aerosol detections near the site of interest in this CO_2 transport simulation research.

The Michelson Interferometer for Passive Atmospheric Sounding (MIPAS) instrument on board the Environmental satellite (Envisat) provides global vertical profiles of the atmospheric composition via mid-wave infrared (MWIR)-TIR with spectral ranges of $4.45-14.6\ \mu m$ [33–35]. Operating from March 2002 to April 2012, Envisat aimed to endow Europe with an enhanced capability for remote sensing observation of Earth from space, further enhancing the study and monitoring of the Earth and its environment. As a high-resolution Fourier transform infrared spectrometer, MIPAS retrieves trace gas vertical profiles of over 20 species day and night. The spatial resolution of MIPAS is 410 km (horizontal) $\times$ 1.5 m (vertical) at an altitude of $6-21$ km in the optimized resolution phase from 2005 to 2012. Compared to CALIOP, MIPAS has a lower spatial resolution but a higher sensitivity to low aerosol concentrations due to its limb geometry. It is also used to verify the altitude distribution of volcanic emissions.

The OCO-2 GEOS Level 3 data are produced base on the Orbiting Carbon Observatory-2 (OCO-2) satellite observations [36, 37]. OCO-2 was launched in July 2014, and joined the A-Train constellation of satellites including Aqua, Aura, CloudSat, CALIPSO, and PARASOL in August 2014 [38]. The OCO-2 mission provides the highest quality space-based XCO_2 retrievals to date, incorporating three high-resolution spectrometers that simultaneously measure reflected sunlight in the near-infrared CO_2 at $1.61-2.06\ \mu m$ and the molecular oxygen (O_2) A-Band at $0.76\ \mu m$. The Goddard Earth Observing System (GEOS) Constituent Data Assimilation System (CoDAS), a modeling and data assimilation system maintained by NASA's Global Modeling and Assimilation Office (GMAO), produces OCO-2 GEOS Level 3 data by ingesting OCO-2 L2 retrievals every 6 h. The OCO-2 GEOS Level 3 daily data are used to analyse the simulation results in this chapter, with a resolution of $0.5° \times 0.625°$ and a temporal resolution of 24 h.

The Atmospheric Infrared Sounder (AIRS) on NASA's Aqua satellite is an infrared sounder operating in synchrony with the microwave instruments AMSU-A1, AMSU-A2, and HSB achieving radiosonde accuracy retrievals [39, 40]. It has spectral ranges of $3.7-4.61\ \mu m$, $6.20-8.22\ \mu m$, and $8.8-15.4\ \mu m$ and spatial resolution of 13.5 km. AIRS aims to measure temperature and water vapor profiles with the same degree of accuracy as radiosondes, while offering a lower vertical resolution and global daily coverage. Since its launch in May 2002, it has consistently delivered datasets with unparalleled radiometric and spectral sensitivity and stability. These datasets comprise land-surface temperature, sea-surface temperature, atmospheric temperature, fractional cloud cover, column abundances of minor atmospheric gases such as CO_2, CH_4, CO, and N_2O, precipitation rate, and other variables. AIRS provides cloud or aerosol information in the atmosphere for this study.

10.4.2 Case study: application to SO_2 transport from volcanic eruptions

10.4.2.1 Data preparation

The Nabro volcano (13°220 N, 41°420 E) in Eritrea's Afar Depression first erupted on record on 12 June 2011, resulting in a significant humanitarian crisis, aviation

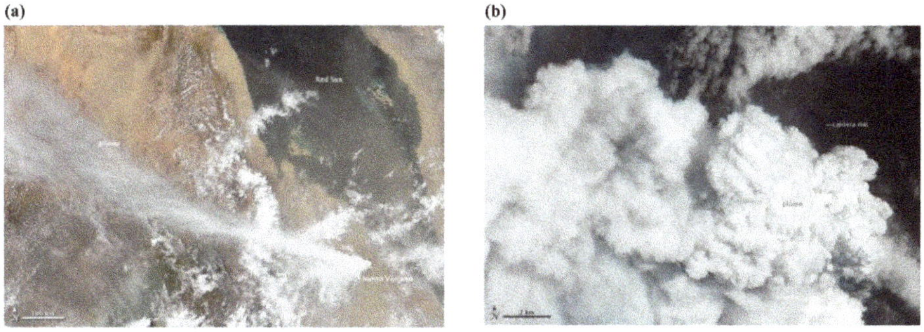

Figure 10.2. Natural-color satellite images of the eruption at Nabro Volcano: (a) 13 June 2011 and (b) 14 June 2011. (Credit: NASA image courtesy Jeff Schmaltz, MODIS Rapid Response Team. CC BY 4.0 [44].)

disruptions, and a substantial sulfur release, ranking as one of the largest atmospheric injections since the 1991 eruption of Mount Pinatubo [41]. Two of the available natural-color satellite images of this eruption are shown in figure 10.2. This eruption has been linked to discernible impacts on global tropospheric temperature, sea-surface temperature, and precipitation, contributing to the observed deceleration in tropospheric warming since 1998 [42]. Based on measurements by the infrared atmospheric sounding interferometer (IASI), the reported total SO_2 mass in the upper troposphere/lower stratosphere (UT/LS) region was approximately 1.5×10^9 kg [43]. Generally, the Nabro volcano has complex emission patterns, which presents a significant challenge for the simulation of atmospheric particle trajectories.

It is necessary to determine a number of parameters for simulation in advance. The minimum time period for the CSI analysis is set at 12 h, which allows for sufficient satellite coverage of the volcanic plume. This is because satellites typically pass over mid- and low-latitude regions twice daily at 01:30 and 13:30 LT, providing optimal satellite coverage. In the computational domain $E: =[t_0, t_f] \times \Omega$, $[t_0, t_f]$ represents the time range of possible emissions, and the spatial range $\Omega: = [\lambda_c - 0.5\Delta_\lambda, \lambda_c + 0.5\Delta_\lambda] \times [\phi_c - 0.5\Delta_\phi, \phi_c + 0.5\Delta_\phi] \times [h_l, h_u]$ defines a rectangular column oriented vertically and centered over the volcano. λ_c, ϕ_c, h_l, and h_u are geographic longitude, geographic latitude, and the lower and upper boundaries of the altitude range. The horizontal cross-sectional area of the column may be modified by varying the values of Δ_λ and Δ, thereby enabling regulation of this parameter in a given simulation. The following serves to illustrate the sensitivity tests for certain of the key parameters.

Two distinct types of simulations are the focus of this investigation. Unit simulations are employed with the goal of reconstructing altitude-dependent time series data for Nabro SO_2 emissions. These simulations are followed by final forward simulations conducted based on the estimates obtained from the unit simulations. The construction of models incorporating parameter estimation will be conducted subsequent to the two types of simulations.

10.4.2.2 Sensitivity tests on model parameters in inverse modeling

In this case, we use the inverse modeling approach to determine the total mass of the source term for final forward simulation. The unit simulations assume SO_2 emitted by the Nabro volcano from 12 June 2011, 12:00 UTC to 18 June 2011, 00:00 UTC. An initial guess for the emission reconstruction is 1.5×10^9 kg. AIRS detected volcanic SO_2 in nearly 75 000 satellite footprints, which provides constraints for the inversion of emissions during this period. The emission domain is discretized with a one-hour time step and a 250 m altitude step, resulting in 15 840 subdomains. The AIRS data from 13 June to 23 June 2011 are employed for the reconstruction of SO_2 emission rates, with 15 840 unit simulations conducted on the JuRoPA supercomputer.

1. Initialized altitude for SO_2 emission

In order to determine the optimal initialized altitude for SO_2 emission, three representative examples have been established for sensitivity tests on emission altitude, as illustrated in figures 10.3–10.5. In the initial example, the assigned initialization in a specific subdomain results in SO_2 air parcel trajectories that align well with satellite observations.

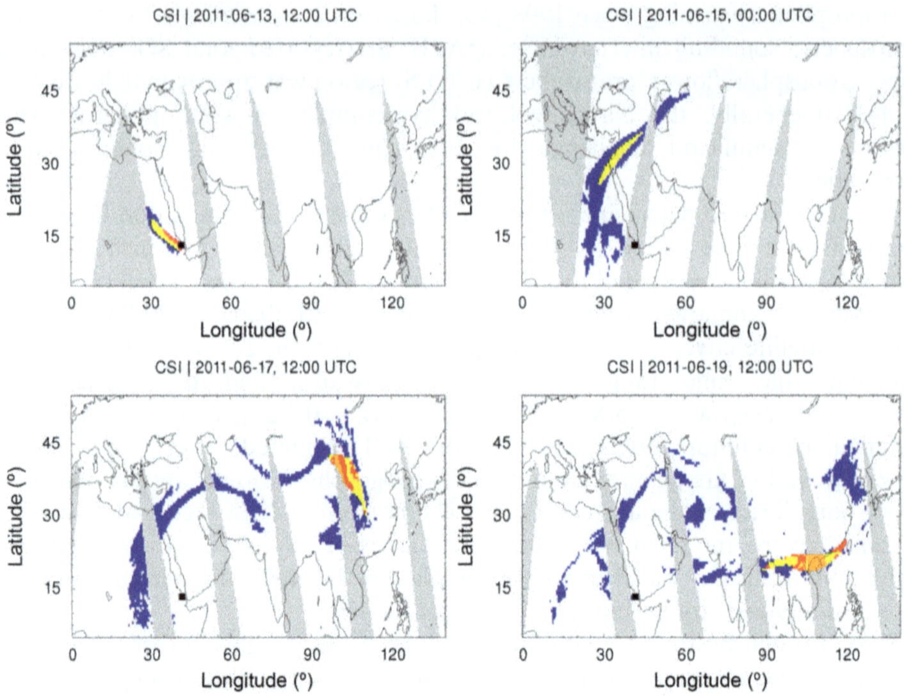

Figure 10.3. The first example of unit simulation at an initialized altitude of (16.5 ± 0.125) km. The gray indicates the absence of satellite data. The orange corresponds to positive model forecasts, but a lack of satellite data. The yellow indicates positive forecasts and positive satellite observations. The blue corresponds to negative forecasts with positive observations. The red corresponds to positive forecasts with negative observations. The black square shows the location of the Nabro volcano. (Reproduced from [1]. CC BY 3.0.)

In figure 10.3, a unit simulation with emissions initiated on 13 June 2011 at 00:00 UTC ± 30 min at an altitude of (16.5 ± 0.125) km demonstrates remarkable concordance with satellite data throughout the entire simulation period. This suggests that emissions likely occurred within that specific temporal and spatial subdomain.

The second example is illustrated in figure 10.4. Model forecasts rapidly diverge from satellite observations, with forecasts for emissions at an altitude of (29 ± 0.125) km. The results align with observations shortly after the eruption but diverge after 12 h, indicating that emissions were unlikely in that subdomain.

The third example includes successful forecasts that persist for a longer duration than those observed in the second example, as evidenced in figure 10.5. However, air parcels released at an altitude of (20 ± 0.125) km exhibit a correlation with observations for approximately two days before deviating, suggesting that emissions were also unlikely to have occurred in this instance.

Notably, the inversion modeling approach distinguishes the three categories of examples to assign appropriate importance weights. While the mean rule effectively captures transport dynamics for the first category, it assigns small weights to the second and third categories. In contrast, the product rule excludes unrealistic cases and assigns proper weights by selecting a suitable split point in the CSI time series, making it a superior strategy both qualitatively and quantitatively.

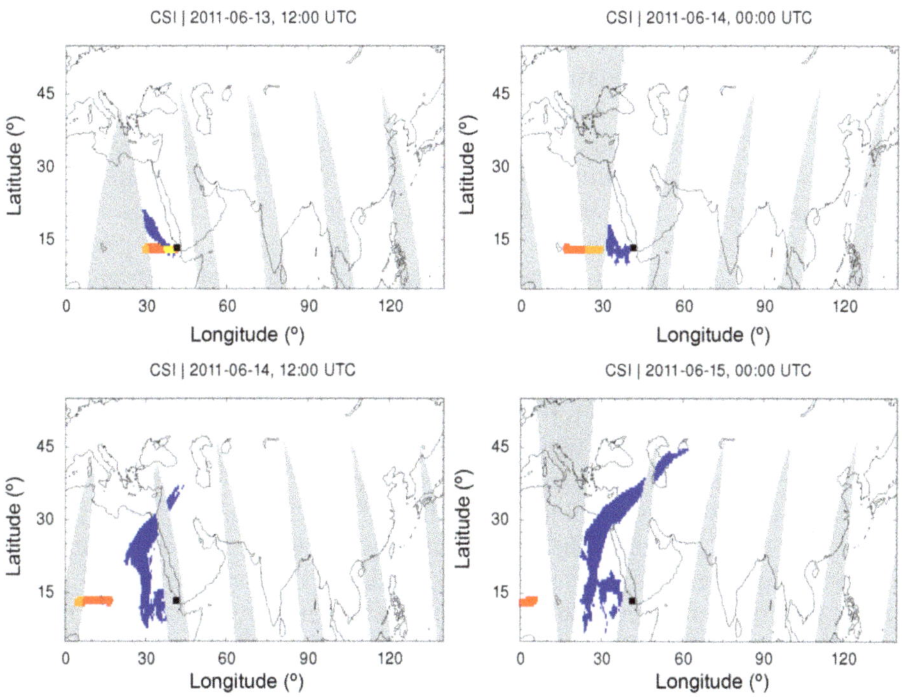

Figure 10.4. The same as figure 10.3, with the exception of the initialized altitude set at (29 ± 0.125) km. (Reproduced from [1]. CC BY 3.0.)

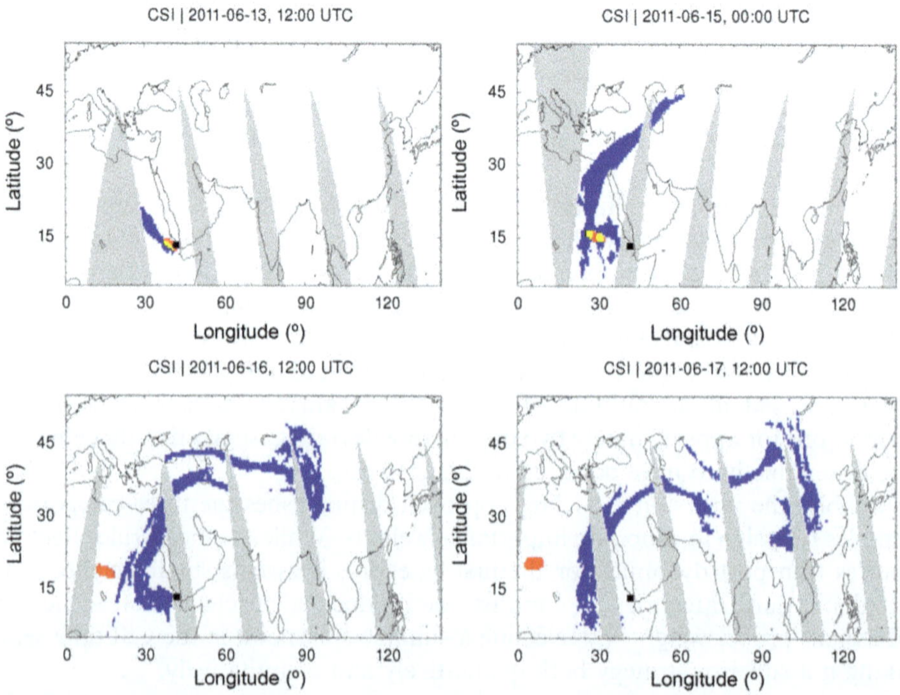

Figure 10.5. The same as figure 10.3, with the exception of the initialized altitude set at (20 ± 0.125) km. (Reproduced from [1]. CC BY 3.0.)

2. Importance weight-updating schemes for emission rate estimation

The total mass of the source term across the entire initialization domain represents a fundamental input for forward simulation. This may be determined by empirical methods based on observations or by the inverse modeling approach described in section 10.3 and the subsequent subsection. It is essential to consider the update strategy for the importance weights in CSI during the resampling process. Simulations are conducted with varying weight-updating schemes, beginning with equal-probability assumptions for emissions across subdomains. This serves as a baseline for estimating more realistic emission rates through iterative inversion.

The initial scheme assumes that the probability of SO_2 emissions is equal across the entire initialization domain. Furthermore, it assumes that all subdomains are of equal importance, with importance weights $w_{ij} = 1/15840$. This results in constant vertical emission rates of approximately 0.1052 kg m^{-1} s^{-1}. While this assumption is not realistic, it provides a useful initial condition for estimating the final importance weights with other updating schemes. The iterative inversion process, which utilizes the mean and product rules, reconstructs volcanic SO_2 emission rates that are more realistic in terms of both time and altitude compared to the equal-probability scheme. Figure 10.6 presents the estimated SO_2 emission rates corresponding to the mean rule and the product rule.

Figure 10.6 shows that the mean rule results in the generation of broader emission areas with lower rates, up to 1.5 kg m^{-1} s^{-1}, and the inclusion of unlikely cases. The

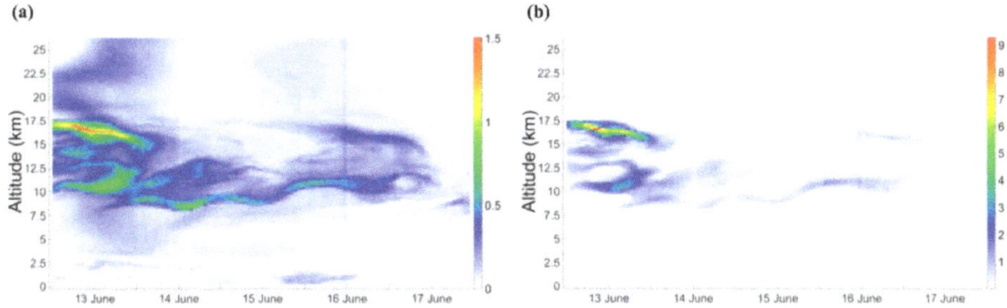

Figure 10.6. SO_2 emission rate estimation ($kg\,m^{-1}\,s^{-1}$): (a) the mean rule and (b) the product rule. (Reproduced from [1]. CC BY 3.0.)

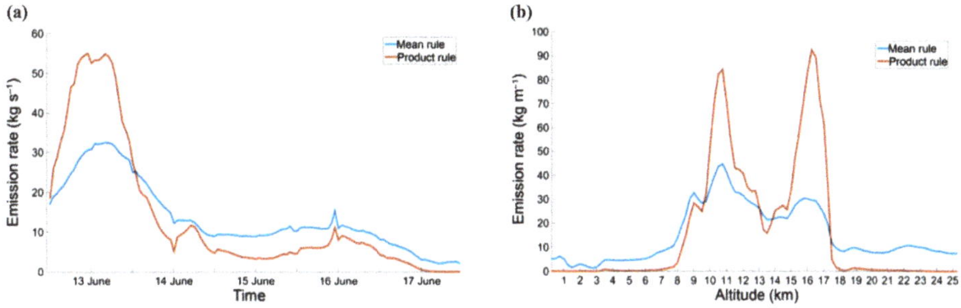

Figure 10.7. SO_2 emission rates estimation: (a) the mean rule and (b) the product rule. (Reproduced from [1]. CC BY 3.0.)

product rule is designed to focus on probable emissions, with peak rates of 9.28, 0.57, and 0.70 $kg\,m^{-1}\,s^{-1}$ on 13 June 00:00 UTC, 14 June 15:00 UTC, and 16 June 10:00 UTC in 2011, which are approximately six times higher than the values produced by the mean rule. Both methods utilize the same total emission, however, the mean rule underestimates likely emissions and overestimates unlikely ones. The results are in qualitative alignment with the previous backward-trajectory approach [45]. Figure 10.7 presents estimated SO_2 emission rates integrated over time and altitude corresponding to the mean rule and the product rule.

The product rule is demonstrated to be effective in capturing higher peak rates for the main eruption on 13 June 2011, as well as a more constrained altitude distribution, particularly between 10–12 km and 15–17 km. The following analysis seeks to refine the crucial parameter, namely the splitting point of the product n_k', with the objective of achieving more optimal results, while the investigation of the total number of discrete-time intervals of the mean rule will be discussed in the forward modeling section.

3. The splitting point for CSI analysis of the product rule

Five different split points $n_k' = 24, 36, 48, 60,$ and 72 h are after the start of the simulation, 13 June 2011 00:00 UTC, among which 48 h is set as the reference. The other four n_k' give importance weights close to the reference, leading to relative differences in importance weights of 23.1%, 11.3%, 8.7%, and 13.7%. The forward

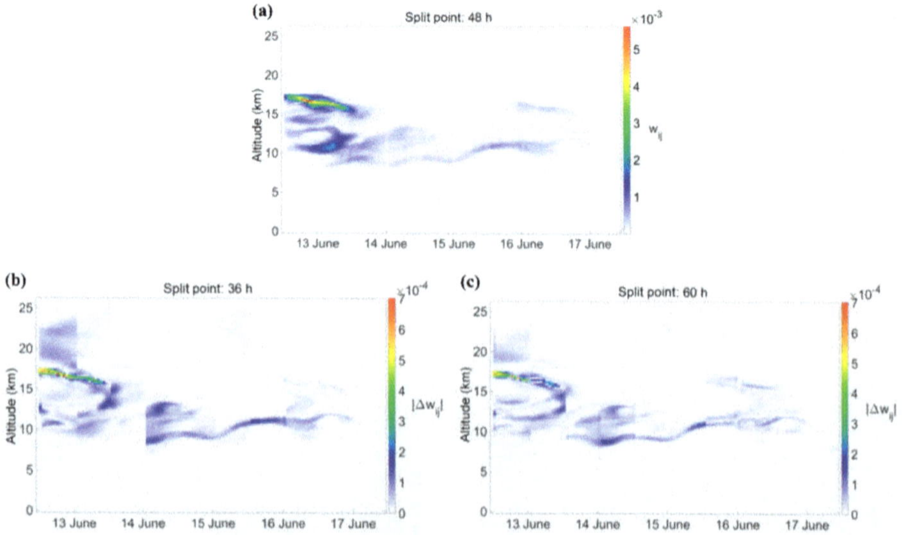

Figure 10.8. Analysis of importance weights with varying n_k^l: (a) results with $n_k^l = 48$ h and (b)–(c) the absolute differences between the estimated importance weights of the reference and other split points $n_k^l = 36$, 60 h, respectively. (Reproduced from [1]. CC BY 3.0.)

simulation results do not vary much with small perturbations of ± 12 h with respect to the chosen reference for the other split points, except for 24 h, which is too short to constrain the time and altitude distribution. The results of the two split points within ± 12 h with respect to the reference are shown in the figure above (figure 10.8).

Note that the optimal split point may vary for different volcanic eruptions, with 48 h being suitable for the Nabro case study (figure 10.8).

10.4.2.3 Validation with satellite emission time series and eruption altitudes

Validation data can be found in figure 10.9 and table 10.3.

The WV channel shows high-altitude eruptions, while the IR channel shows low-altitude plumes. The strongest eruptions occurred between 00:00 and 12:00 UTC on 13 June 2011, with smaller events continuing until 16 June 2011 15:00 UTC. Simulated emission time series is in good agreement with MVIRI observations. MIPAS and CALIOP detected sulfate aerosols as indicators of the SO_2 position. The reconstructed emissions reflect inhomogeneous eruption altitudes, as illustrated in table 10.3. This consistency indicates that the estimated emission time series and eruption altitudes are reliable.

On this basis, the resolution of the volcanic SO_2 emission rates is extended for the first time to 30 min in time and 100 m in height, using the inverse algorithm and parallel strategy. The largest computation used 60 250 nodes on Tianhe-2, with each node calculating the trajectories of 1 million air parcels using 24 cores. At this scale, the inverse reconstruction and final forward simulation took about 22 min and required approximately 530 000 core hours. The results are shown in figure 10.10.

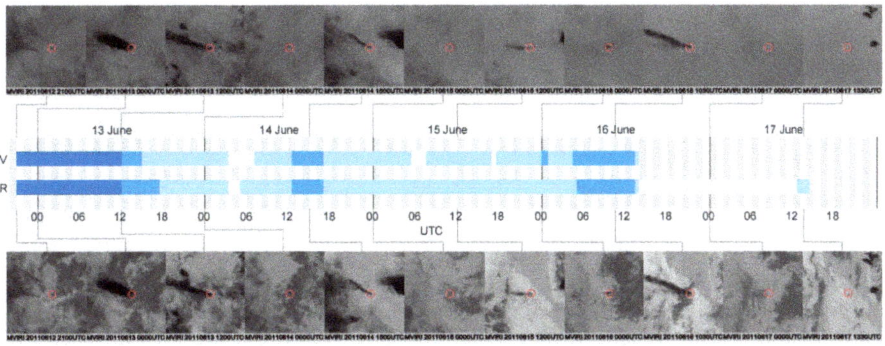

Figure 10.9. Timeline based on Meteosat-7 MVIRI(IODC) IR and WV measurements of the 2011 Nabro eruption. Satellite images were used to roughly estimate the level of volcanic activity: white, none; light blue, low; blue, moderate; dark blue, high. (Reproduced from [1]. CC BY 3.0.)

Table 10.3. Major eruption altitudes of the Nabro volcano on different days. (Reproduced with permission from [2]. Copyright 2020 Springer Nature.)

	13 June	14 June	16 June
CALIOP and MIPAS data	19 km	9–13 km	
Fromm *et al* (2013)	15–19 km		
Fromm *et al* (2014)			17.4 km
Result from this work	15–17 km	9–13 km	17 km

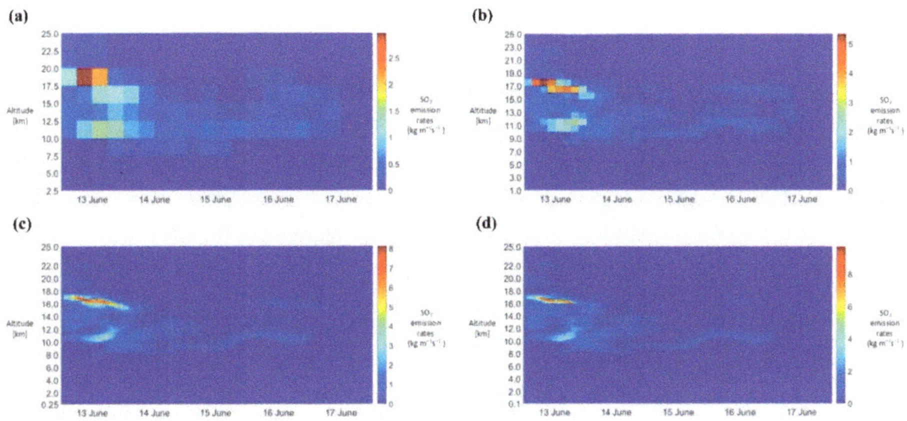

Figure 10.10. Reconstructed volcanic SO_2 emission rates from the June 2011 Nabro eruption (kg $m^{-1}s^{-1}$). (a)–(d) Simulation on 210 compute nodes (5040 processes), 1025 compute nodes (24 600 processes), 12 100 compute nodes (290 400 processes), and 60 250 compute nodes (1 446 000 processes), respectively. The x-axis refers to time, the y-axis refers to altitude (km), and the color bar refers to the emission rate. (Reproduced with permission from [2]. Copyright 2020 Springer Nature.)

10.4.2.4 Sensitivity tests on model parameters in forward modeling: the total number of discrete-time intervals for CSI analysis of the mean rule

In order to provide a reference, the final time was set to 00:00 UTC on 23 June 2011. This resulted in the selection of the parameter value $n_k = 21$. The final times of 22 June 00:00 UTC, 22 June 12:00 UTC, 23 June 12:00 UTC, and 24 June 00:00 UTC in 2011, corresponding to $n_k = 19, 20, 22,$ and 23, yield relative differences of approximately 9.5%, 7.2%, 6.2%, and 10%, respectively. The results of the two temporal endpoints exhibiting the greatest discrepancies with respect to the reference are illustrated in figure 10.11.

A visual inspection of the results indicates that the various importance weights yield relatively similar results in the final forward simulations.

10.4.2.5 High-throughput parallel computing of the final forward simulations

The final forward simulations are executed on the JUWELS HPC system and the Tianhe-2 supercomputer platform. This case study presents an inversion approach that employs sequential importance resampling to reconstruct volcanic emission rates from satellite observations. This approach is independent of the forward transport model and is suitable for massive parallel computing, while capable of providing reliable emission rates as key data for final forward simulations. The final simulation was also performed on the Tianhe-2 supercomputer using the product rule, based on a reconstruction of the emission data with 1 h temporal and 250 m vertical resolution. The results are shown in figure 10.12.

The results of the research conducted on different supercomputing systems belonging to two distinct periods are only aligned by date, due to the nature of

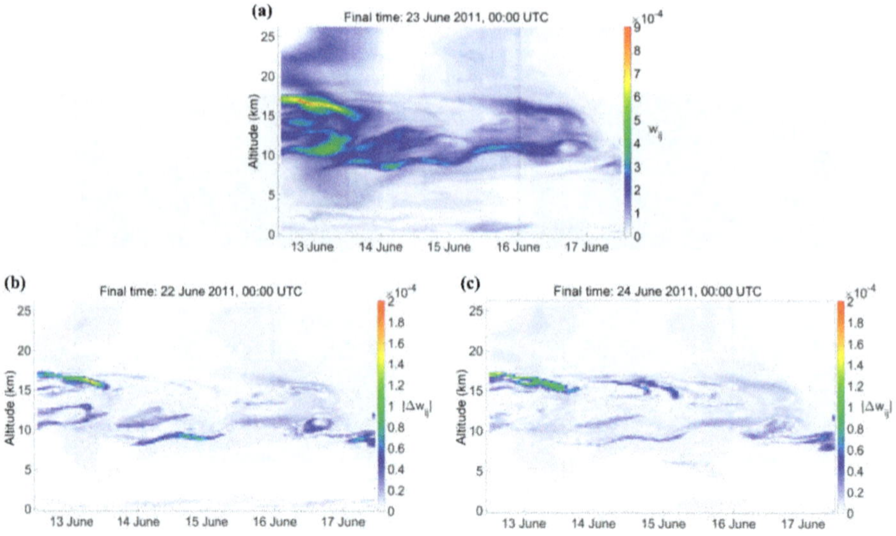

Figure 10.11. Analysis of importance weights with varying n_k: (a) results with $n_k = 21$ and (b)–(c) the absolute differences between the estimated importance weights of the reference and other final times with $n_k = 19$, 23, respectively. (Adapted from [1]. CC BY 3.0.)

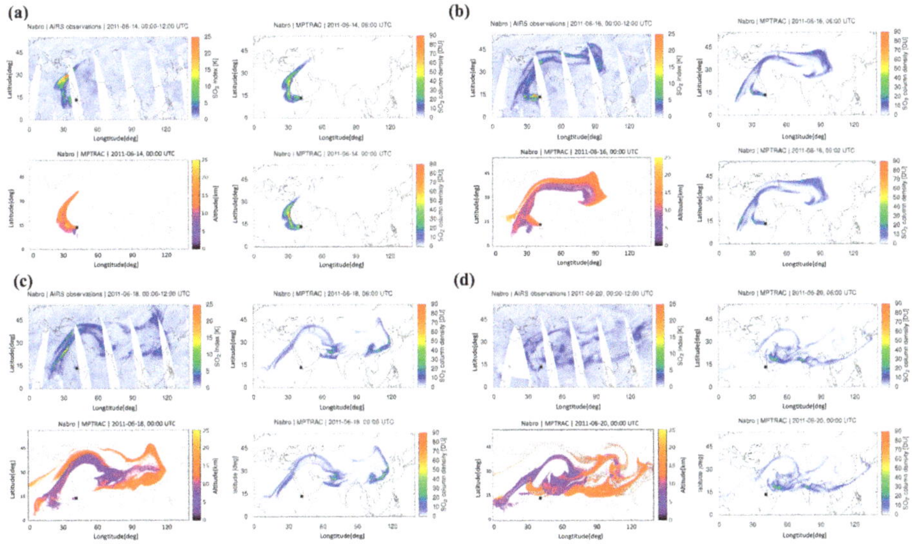

Figure 10.12. Comparison of satellite observations and simulation results. Top left: AIRS satellite observations; top right: MPTRAC simulation results using the product rule on the JUWELS HPC system; bottom left: MPTRAC simulation results using the product rule on the Tianhe-2 supercomputer platform, showing the altitude distribution of air parcels; bottom right: same as bottom left, but showing SO_2 vertical column density. (a) 14 June 2011; (b) 16 June 2011; (c) 18 June 2011; (d) 20 June 2011. (Adapted from [1]. CC BY 3.0. Adapted with permission from [2]. Copyright 2020 Springer Nature.)

the research. Notwithstanding the aforementioned discrepancies, the output results for the same day remain consistent. Simulation results on both of the supercomputer systems are comparable to the AIRS observations, indicating that the simulation is stable and accurate.

10.4.3 Case study: application to greenhouse gas CO_2 transport from forest fires

10.4.3.1 Data preparation

To gain further insights into the influence of intense forest fires on the long-distance transport of CO_2 emissions, a representative fire incident was selected for numerical simulation in the present study. Despite Canada's mostly subarctic climate, severe droughts in recent years have led to more forest fires in Canada, with a peak in 2021 concentrated between May and August [46, 47]. Our study focuses on simulating CO_2 emissions from a major forest fire in Saskatchewan in mid-May 2021, which affected 5470 hectares and was human-induced. The fire shown in figure 10.13, named 21PA-CLOVERDALE, was extinguished in June, and the data were last updated in April 2022 [47].

This case study employs two distinct data types. The meteorological input data are the ERA5 reanalysis provided by the ECMWF, while the CO_2 concentration data are derived from NASA's OCO-2 GEOS model. The ERA5 data are horizontally resolved at 0.3° × 0.3°, and vertically at 137 model levels extending from the surface up to 0.01 hPa [27]. Both datasets have been subjected to extensive

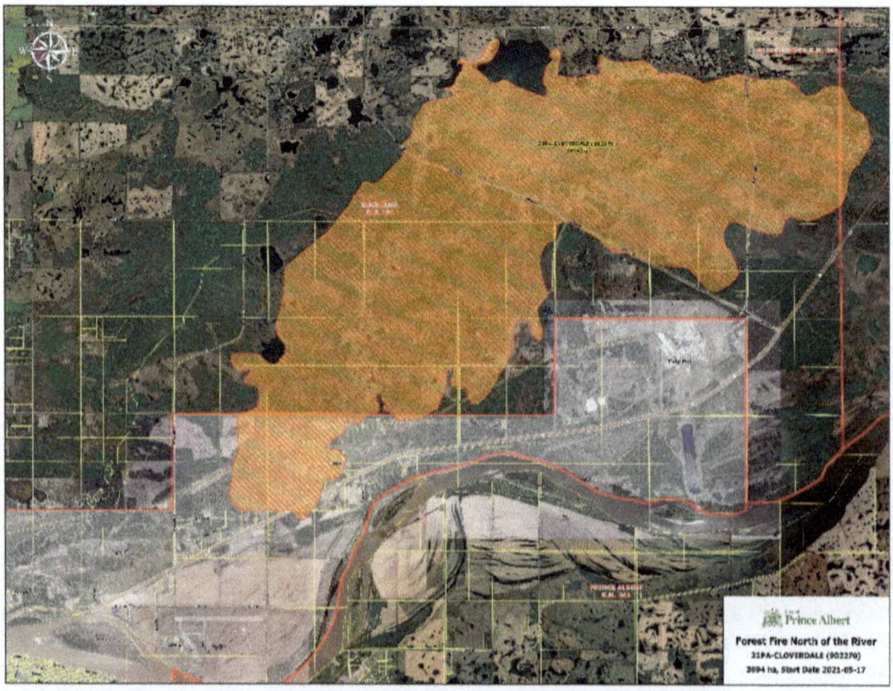

Figure 10.13. Map of forest fire 21PA-CLOVERDALE. (Reproduced with permission from [48].)

validation procedures and are widely employed in environmental research to confirm the reliability of studies.

10.4.3.2 Baseline experiment

Similar to the forward simulation of SO_2, MPTRAC simulated the trajectory of CO_2 released from the aforementioned forest fire. This process entails the implementation of several pivotal parameter configurations and data preprocessing procedures during the initialization phase, which are indispensable for guaranteeing the integrity and reliability of the resulting output. The initialization dataset includes source information for CO_2, such as the total mass, the number of air parcels, initial positions and times, computational domain details including spatial and temporal ranges, and meteorological data. The source term is represented by a Gaussian distribution with a mean at the coordinates of the fire occurrence and a full width at half maximum (FWHM) of 200 km. In regard to the model output settings, the grid output was selected, with a grid size of 240×120, wherein each grid box encompasses an area of approximately 55 km $\times$ 55 km. Moreover, data from the CALIPSO satellite indicate that aerosol emissions following the forest fire are primarily concentrated within the altitude range of 8–12 km, as illustrated in figure 10.14 [30]. Accordingly, the initial height of CO_2 emissions was established at 10 km [30].

CO_2 trajectories are simulated for the period from 00:00 UTC on 17 May to 00:00 UTC on 25 May. In the model, 'air parcels' refer to specific amounts of CO_2

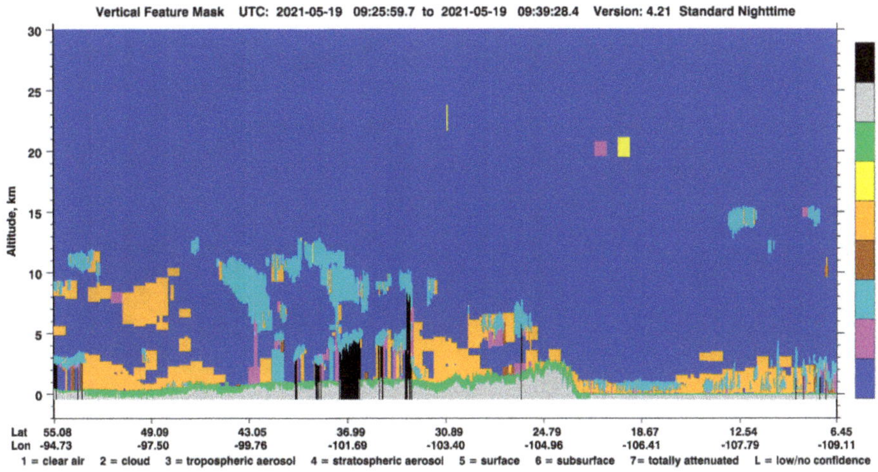

Figure 10.14. CALIPSO Level 2 Lidar Vertical Feature Mask (VFM) (2021–5–19 09:26–09:39 UTC, over the case study area, Saskatchewan, Canada). (Reproduced from [3]. CC BY 4.0.)

particles, each corresponding to a transport trajectory, and a series of these trajectories together represent the entire CO_2 dispersion process, with each trajectory carrying a specific amount of CO_2 particles [8].

To validate the accuracy of the simulation results, the platform compared the horizontal transport of CO_2 simulated by MPTRAC with NASA's OCO-2 Level 3 assimilated data. To eliminate background variations in the CO_2 concentration, the preprocessing of the NASA OCO-2 CO_2 data subtracted the daily average concentration within the observation area, resulting in CO_2 anomalies due to forest fires and other causes, thereby improving the accuracy of the comparison.

As shown in figure 10.15, MPTRAC captured the dispersion dynamics of CO_2 released by wildfires and demonstrated consistency between the simulation results and the assimilated observational data.

The visualization of the MPTRAC simulation results shows that on 18 May, the large amount of CO_2 emitted by forest fires was mainly concentrated in the region of 45°N to 55°N and 80°W to 110°W, forming a conveyor belt in a general south-eastern direction. On 21–22 May, after the development of gradual diffusion, the transport distribution of CO_2 changed, with a narrow band extending eastward, and at the same time, a circular conveyor belt formed in the range of 25°N to 50°N and 40°W to 75°W; with the passage of time, CO_2 continued to be transported eastward and gradually covered a larger area. On 24 May, the transport distribution of CO_2 was clearly divided into two main parts: one part was located in the range of 40°N to 55°N and 60°W to 90°W, forming a trajectory toward the southeast and the other part was widely dispersed over a wide area ranging from 30°N to 60°N and from 0°W to 60°W.

At the same time, the visualization of the OCO-2 observation data for reference and comparison shows that from 21 May to 22 May, an obvious narrow conveyor belt was formed in the source term of the red triangle, and a circular distribution of

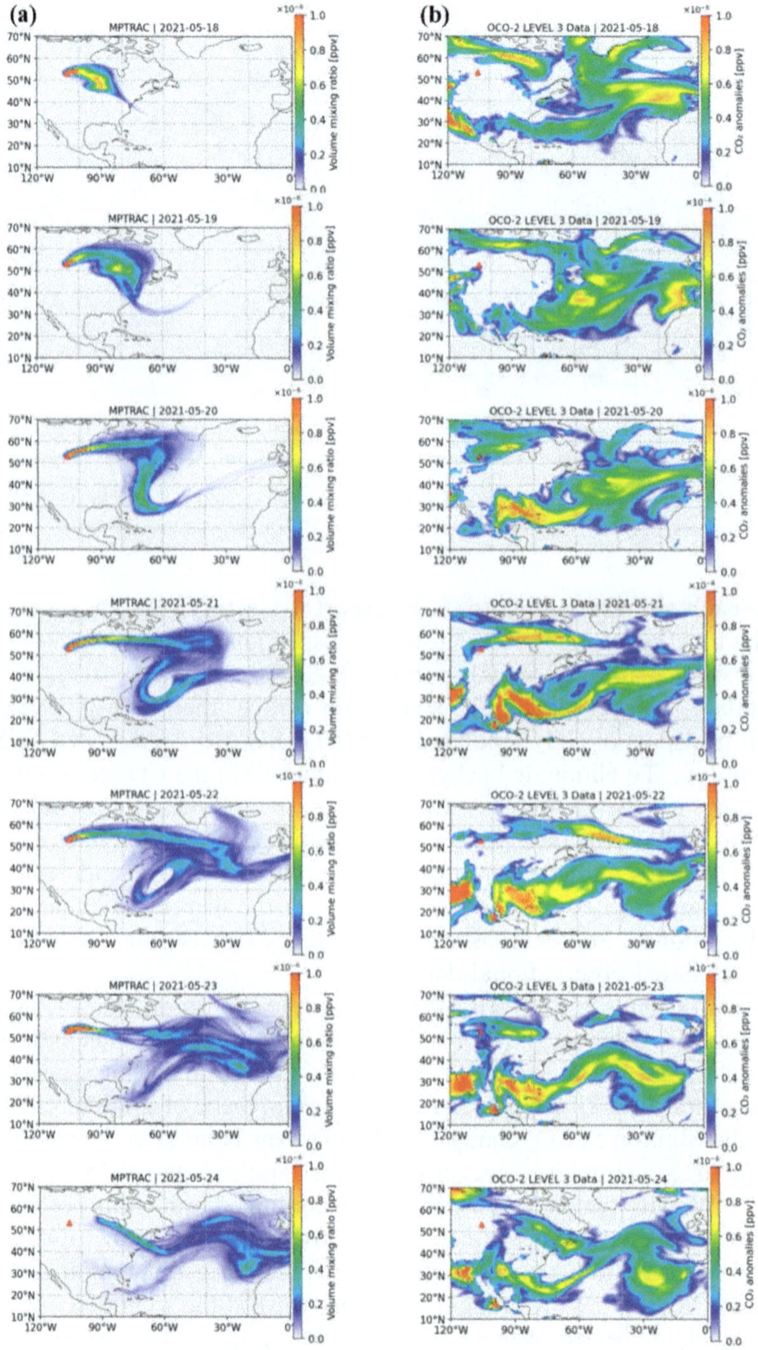

Figure 10.15. Distribution of CO_2 from forest fires in Canada. (a) Results of the MPTRAC simulation. (b) CO_2 anomalies based on the NASA OCO-2 Level 3 assimilated data product. The red triangle indicates the position of the particle source term. (Reproduced from [3]. CC BY 4.0.)

low concentration center and high concentration around the middle area of the image was found, which was consistent with the MPTRAC simulation results. From 23 May to 24 May, the distribution of CO_2 near the source term was consistent with the simulation results. In the whole process of eastward transport, the observed results also show a wide range of diffusion phenomena, which further verifies the accuracy of the simulation results.

The comparative analysis showed the inconsistency between the simulated results and the observed results. In the specific regions of the observed data, namely 20°N to 40°N and 60°W to 120°W, there was an abnormal CO_2 peak distribution, which was not reflected in the simulation results. It is speculated that the reasons for this inconsistency may be as follows. First, the accuracy of the OCO-2 satellite assimilation data is limited in some regions due to local observational obstructions, and the CO_2 anomalies in the corresponding regions can normally be reflected in the simulation data. Since the assimilated OCO-2 data from 18 May to 19 May did not show a significant CO_2 signal around the fire site, by combining cloud index data from the AIRS at the corresponding time, the presence of a large number of clouds or aerosols in the atmosphere can be inferred, which hinder the observation of the area near the fire site [49]. Figure 10.16 shows a plot of AIRS cloud index data in the simulated region at about 13:30 local time on 18–19 May as a reference. Known spectral window region measurements can be used to detect cloud and convective conditions. Higher brightness temperature (BT) cloud index values correspond to more clouds and aerosols in the atmosphere. Second, the low-level CO_2 anomalies that are not reflected in the simulation results can be observed in the satellite assimilation data. It is speculated that emission events from CO_2 sources not included in the simulation, such as fires in other locations, have been captured by satellite observations. Similar situations can be further analysed and processed, and comparative analysis can be performed when relevant data are complete for the same time period.

10.4.3.3 High sensitivity test for CO_2 emission

In order to investigate the sensitivity of the simulated output to the particle release height parameter, this case utilizes parameters that encompass the aerosol

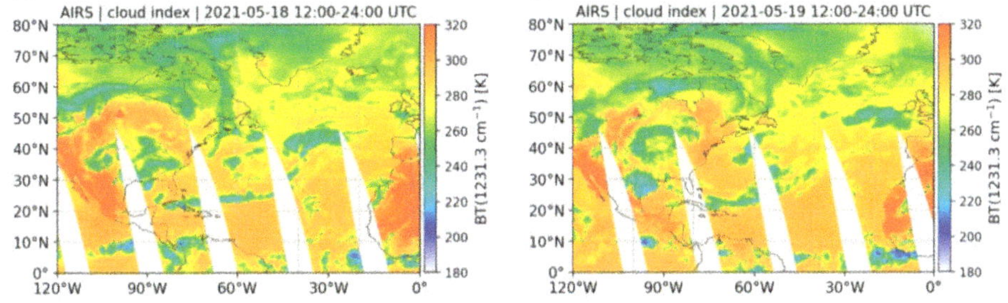

Figure 10.16. AIRS 8.1 μm BTs cloud index data: (a) data on 18 May and (b) data on 19 May. (Reproduced from [3]. CC BY 4.0.)

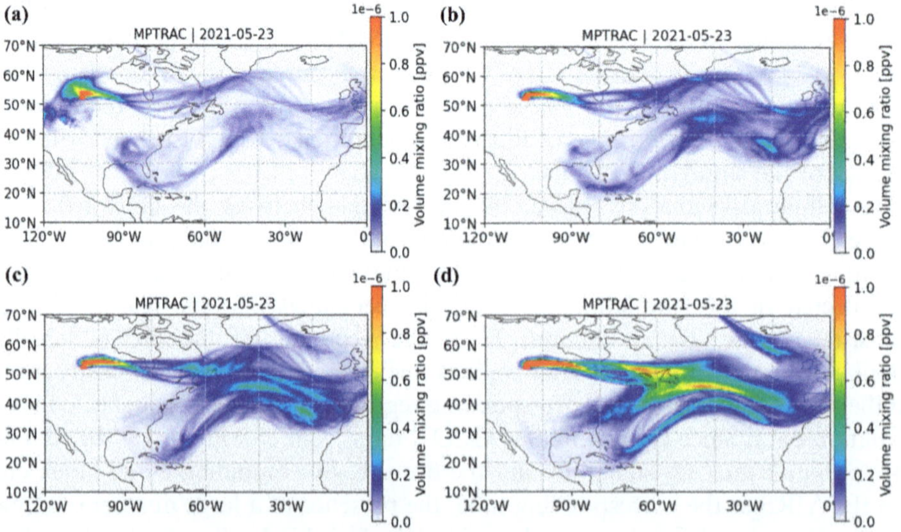

Figure 10.17. Results of CO_2 transport simulation on 23 May with varying release heights. (a)–(d) The simulated outputs corresponding to the release of air parcels at altitudes of 3, 7, 10, and 13 km, respectively. (Reproduced from [3]. CC BY 4.0.)

observation height range provided by the atmospheric particle observation station. The air parcel trajectory is simulated with varying release heights, centered at 3, 7, 10, and 13 km, respectively [30]. To ensure the consistency of the experiment, the remaining parameter settings from the baseline experiment were utilized in the simulation, and the convection parameterization settings were disabled to prevent any potential influence of vertical mixing on CO_2 transportation. Figure 10.17 depicts the simulated transport results for CO_2 released at varying heights. The selected display date is 23 May, which allows for a more comprehensive observation of diffusion. A comparative analysis reveals a clear correlation between the horizontal transmission path of CO_2 and the height of particle release. For instance, when released at a higher altitude, the plume of CO_2 flows faster towards the east coast of North America, where Canada is located, and the Atlantic Ocean. This phenomenon is hypothesized to be related to the strong westerly winds in the upper troposphere.

10.4.3.4 Comparison of the simulation results for two HPC systems

In order to evaluate the potential discrepancies in the simulation output of MPTRAC across different HPC systems, this case study compares the simulation results of the Tianhe-2 supercomputer platform at the NSCC-GZ, China, and the JUWELS HPC system at the Jülich Supercomputing Centre, Germany. To calculate the mean CO_2 concentration disparity between each grid box in the two HPC system simulations, the following equation was employed:

$$\text{Difference} = \frac{1}{n}\sum \frac{x_{\text{JUWELS}} - x_{\text{Tianhe-2}}}{\max(x_{\text{JUWELS}}, x_{\text{Tianhe-2}})}, \tag{10.15}$$

where $x_{\text{Tianhe-2}}$ and x_{JUWELS} refer to the simulated output of the Tianhe-2 supercomputer platform and JUWELS HPC system, respectively. The value of n represents the total number of non-empty grid boxes. It is necessary to exclude grid boxes devoid of particles from the calculations in order to circumvent the potential for zero-division errors. A comparative analysis reveals discrepancies in the output simulation results of the two supercomputing systems. The spatial distribution of these differences is illustrated in figure 10.18.

The primary plumes of the two systems exhibit high consistency, with the exception of some minor discrepancies in diffusion characteristics. The overall simulation results on the Tianhe-2 supercomputer platform demonstrate a greater prevalence of diffusion characteristics than those observed on JUWELS HPC system. This discrepancy is postulated to be attributable to the divergence in the

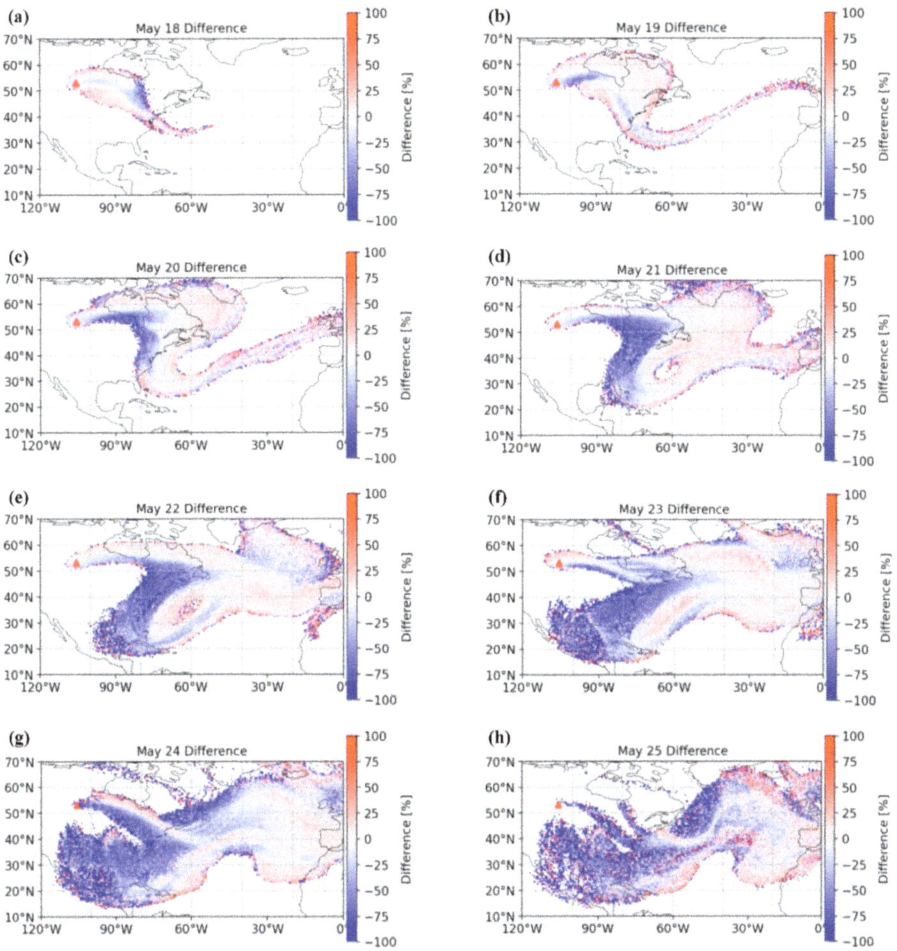

Figure 10.18. The spatial distribution of differences in CO_2 distributions between the Tianhe-2 and JUWELS HPC supercomputer systems. (Unit: ppmv). (a)–(h) Differences on 18–25 May. (Reproduced from [3]. CC BY 4.0.)

random number generators employed by the two systems. Further in-depth analysis and discussion of this topic will be conducted in subsequent studies.

10.5 Conclusions

The computational tools and methodologies developed in this study are crucial for designing environmentally friendly industrial processes, exploring pollutant degradation mechanisms, and understanding the causes of global climate change. By integrating high-performance computing and advanced modeling techniques, this research contributes to more accurate and reliable predictions of atmospheric pollution and its impacts.

It is theoretically expected that the computation time will decrease in a nearly linear fashion with an increasing number of computing processes. Consequently, adequate computational performance can significantly reduce the costs associated with computation time, thereby facilitating simulations involving hundreds of millions of air parcels on supercomputing systems and achieving real-time or near-real-time atmospheric particles transport process prediction.

The proposed high-throughput parallel computing framework provides a robust solution for estimating high-resolution spatiotemporal distributions of atmospheric pollutants. By leveraging the capabilities of the Lagrangian transport model MPTRAC and advanced data assimilation techniques, the framework addresses the computational and mathematical challenges inherent in this problem. A novel modeling system has been developed, employing the Lagrangian transport model MPTRAC for efficient simulations of SO_2 transport. This system is independent of the forward transport model and is suitable for massive parallel computing.

This approach will enhance our ability to monitor, understand, and mitigate atmospheric pollution, contributing to improved public health and environmental protection. It paves the way for potential future work which may including near-real-time forecasting, an adaptive initialization strategy, the consideration of pollutant kernel functions, and an analysis of data uncertainties.

References

[1] Heng Y, Hoffmann L, Griessbach S, Rößler T and Stein O 2016 Inverse transport modeling of volcanic sulfur dioxide emissions using large-scale simulations *Geosci. Model Dev.* **9** 1627–45

[2] Liu M, Huang Y, Hoffmann L, Huang C, Chen P and Heng Y 2020 High-resolution source estimation of volcanic sulfur dioxide emissions using large-scale transport simulations *Computational Science—ICCS* **2020** pp 60–73

[3] Liao Y, Deng X, Huang M, Liu M, Yi J and Hoffmann L 2024 Tracking carbon dioxide with Lagrangian transport simulations: case study of Canadian Forest Fires in May 2021 *Atmosphere* **15** 429

[4] Pisso I *et al* 2019 The Lagrangian particle dispersion model FLEXPART version 10.4 *Geosci. Model Dev.* **12** 4955–97

[5] Cohen M D, Stunder B J B, Rolph G D, Draxler R R, Stein A F and Ngan F 2015 NOAA's HYSPLIT atmospheric transport and dispersion modeling system *Bull. Am. Meteorol. Soc.* **96** 2059–77

[6] Abdul-Wahab S, Sappurd A and Al-Damkhi A 2010 Application of California puff (CALPUFF) model: a case study for Oman *Clean Technol. Environ. Policy* **13** 177–89
[7] Hoffmann L *et al* 2022 Massive-parallel trajectory calculations version 2.2 (MPTRAC-2.2): Lagrangian transport simulations on graphics processing units (GPUs) *Geosci. Model Dev.* **15** 2731–62
[8] Rößler T, Stein O, Heng Y, Baumeister P and Hoffmann L 2018 Trajectory errors of different numerical integration schemes diagnosed with the MPTRAC advection module driven by ECMWF operational analyses *Geosci. Model Dev.* **11** 575–92
[9] Wu X, Qiao Q, Chen B, Wang X, Hoffmann L, Griessbach S, Tian Y and Wang Y 2023 The influence of the Asian summer monsoon on volcanic aerosol transport in the UTLS region *npj Climate Atmos. Sci.* **6** 11
[10] Hoffmann L, Haghighi Mood K, Herten A, Hrywniak M, Kraus J, Clemens J and Liu M 2024 Accelerating Lagrangian transport simulations on graphics processing units: performance optimizations of Massive-Parallel Trajectory Calculations (MPTRAC) v2.6 *Geosci. Model Dev.* **17** 4077–94
[11] Clemens J, Vogel B, Hoffmann L, Griessbach S, Thomas N, Fadnavis S, Müller R, Peter T and Ploeger F 2024 A multi-scenario Lagrangian trajectory analysis to identify source regions of the Asian tropopause aerosol layer on the Indian subcontinent in August 2016 *Atmos. Chem. Phys.* **24** 763–87
[12] Seibert P 2000 Inverse modelling of sulfur emissions in Europe based on trajectories *Inverse Methods in Global Biogeochemical Cycles* **vol 114** ed P Kasibhatla *et al* (Washington, DC: American Geophysical Union) pp 147–54
[13] Krause D 2019 JUWELS: modular tier-0/1 supercomputer at the Jülich Supercomputing Centre *J. Large-scale Res. Fac. JLSRF* **5** A171
[14] Strohmaier E *et al* 2024 *Top500 List—June 2024* (Sinsheim: Prometeus) https://top500.org
[15] Schaefer J T 1990 The critical success index as an indicator of warning skill *Weather Forecast.* **5** 570–5
[16] 2017 *Scaling OpenMP for Exascale Performance and Portability* ed B R de Supinski, S L Olivier, C Terboven, B M Chapman and M S Müller (New York: Springer)
[17] OpenACC.org 2023 *OpenACC Programming and Best Practices Guide* vol 2024 https://www.openacc.org/sites/default/files/inline-files/OpenACC_Programming_Guide_0_0.pdf
[18] NSCC-GZ 2024 The National Supercomputing Center in Guangzhou (NSCC-GZ) http://www.nscc-gz.cn/index.html
[19] Jülich Supercomputing Centre *Jülich Wizard for European Leadership Science* https://www.fz-juelich.de/en/ias/jsc/systems/supercomputers/juwels
[20] Smith N, Schmit T, Loeb N, Skofronick-Jackson G, Heidinger A, L'Ecuyer T, Duncan B, Bhartia P K, Platnick S and Ackerman S A 2019 Satellites see the world's atmosphere *Meteorol. Monogr.* **59** 4.1–4.53
[21] Bluestein H B, Carr F H and Goodman S J 2022 Atmospheric observations of weather and climate *Atmos. Ocean* **60** 149–87
[22] Afe O T, Richter A, Sierk B, Wittrock F and Burrows J P 2004 BrO emission from volcanoes: a survey using GOME and SCIAMACHY measurements *Geophys. Res. Lett.* **31** L24113
[23] Khokhar M F, Frankenberg C, Van Roozendael M, Beirle S, Kühl S, Richter A, Platt U and Wagner T 2005 Satellite observations of atmospheric SO_2 from volcanic eruptions during the time-period of 1996–2002 *Adv. Space Res.* **36** 879–87

[24] He C, Ji M, Grieneisen M L and Zhan Y 2022 A review of datasets and methods for deriving spatiotemporal distributions of atmospheric CO_2 *J. Environ. Manage.* **322** 116101
[25] Wunch D, Toon G C, Blavier J-F L, Washenfelder R A, Notholt J, Connor B J, Griffith D W T, Sherlock V and Wennberg P O 2011 The total carbon column observing network *Philos. Trans. R. Soc.* A **369** 2087–112
[26] Dee D P *et al* 2011 The ERA-interim reanalysis: configuration and performance of the data assimilation system *Q. J. R. Meteorolog. Soc.* **137** 553–97
[27] Hersbach H *et al* 2020 The ERA5 global reanalysis *Q. J. R. Meteorolog. Soc.* **146** 1999–2049
[28] European Space Agency (ESA) Meteosat first generation *eoPortal* https://www.eoportal.org/satellite-missions/meteosat-first-generation#data-collection-platforms
[29] Symbios 2024 CEOs EO Handbook—Agency Summary—Eumetsat (CEOS) *The COES Database* https://database.eohandbook.com/database/agencysummary.aspx?agencyID=9
[30] N L R Center 2021 *Standard Lidar Browse Images for Production Release* [v 4.11] C.L.B.I. (ed.) https://www-calipso.larc.nasa.gov/data/BROWSE/production/V4-11/
[31] Powell K A, Hu Y, Omar A, Vaughan M A, Winker D M, Liu Z, Hunt W H and Young S A 2009 Overview of the CALIPSO mission and CALIOP data processing algorithms *J. Atmos. Oceanic Technol.* **26** 2310–23
[32] Winker D M *et al* 2010 The CALIPSO mission *Bull. Am. Meteorol. Soc.* **91** 1211–30
[33] Fischer H *et al* 2008 MIPAS: an instrument for atmospheric and climate research *Atmos. Chem. Phys.* **8** 2151–88
[34] Griessbach S, Hoffmann L, von Hobe M, Müller R, Spang R and Riese M 2012 A six–year record of volcanic ash detection with Envisat MIPAS *Proc. Advances in Atmospheric Science and Applications* vol 708 *(Bruges, Belgium)* L Ouwehand
[35] Raspollini P *et al* 2013 Ten years of MIPAS measurements with ESA Level 2 processor V6—part 1: retrieval algorithm and diagnostics of the products *Atmos. Meas. Tech.* **6** 2419–39
[36] Weir B and L E Ott 2022 OCO-2 GEOS Level 3 daily, 0.5 × 0.625 assimilated CO2 V10r Goddard Earth Sciences Data and Information Services Center (GES DISC) Greenbelt, MD, USA [10.5067/Y9M4NM9MPCGH] [Dataset]
[37] Morgan K 2022 README document for OCO-2 GEOS L3 XCO_2 products *Goddard Earth Sciences Data and Information Services Center (GES DISC)* (Greenbelt, MD: National Aeronautics and Space Administration)
[38] Stephens G L *et al* 2002 The CloudSat mission and the A-Train: a new dimension of space-based observations of clouds and precipitation *Bull. Am. Meteorol. Soc.* **83** 1771–90
[39] Aumann H H *et al* 2003 AIRS/AMSU/HSB on the Aqua mission: design, science objectives, data products, and processing systems *IEEE Trans. Geosci. Remote Sens.* **41** 253–64
[40] Pagano T S, Johnson D L, McGuire J P, Schwochert M A and Ting D Z 2022 Technology maturation efforts for the next generation of grating spectrometer hyperspectral infrared sounders *IEEE J. Sel. Top. Appl. Earth Obser. Remote Sens.* **15** 2929–43
[41] Goitom B *et al* 2015 First recorded eruption of Nabro volcano, Eritrea, 2011 *Bull. Volcanol.* **77** 85
[42] Santer B D *et al* 2015 Observed multivariable signals of late 20th and early 21st century volcanic activity *Geophys. Res. Lett.* **42** 500–9
[43] Clarisse L, Hurtmans D, Clerbaux C, Hadji-Lazaro J, Ngadi Y and Coheur P F 2012 Retrieval of sulphur dioxide from the infrared atmospheric sounding interferometer (IASI) *Atmos. Meas. Tech.* **5** 581–94

[44] Earth Observatory 2011 Eruption at Nabro Volcano *NASA* https://earthobservatory.nasa.gov/images/51031/eruption-at-nabro-volcano

[45] Hoffmann L, Rößler T, Griessbach S, Heng Y and Stein O 2016 Lagrangian transport simulations of volcanic sulfur dioxide emissions: impact of meteorological data products *J. Geophys. Res.: Atmos.* **121** 4651–73

[46] Peel M C, Finlayson B L and McMahon T A 2007 Updated world map of the Köppen–Geiger climate classification *Hydrol. Earth Syst. Sci.* **11** 1633–44

[47] Wang Z *et al* 2024 Severe global environmental issues caused by Canada's record-breaking wildfires in 2023 *Adv. Atmos. Sci.* **41** 565–71

[48] City Hall of Prince Albert, Saskatchewan, Canada https://www.citypa.ca/en/news/resources/2021-05-18---21PA-CLOVERDALE_FireExtent.pdf

[49] Hoffmann L, Griessbach S and Meyer C I 2014 Volcanic emissions from AIRS observations: detection methods, case study, and statistical analysis *Proc. SPIE* **9242** 924214

Chapter 11

Enhancing boiling heat transfer via model-based experimental analysis

Yi Heng, Min Hong and Dongchuan Mo

Boiling heat transfer is a highly efficient energy transfer method essential for high-heat-flux dissipation and precise temperature control. This chapter explores heuristic modeling approaches for analyzing dynamic boiling processes. A parametric 3D transient heat-conduction model, based on CT-reconstructed geometries, is used for temperature distribution estimation, while computational fluid dynamics (CFD) simulations of multi-bubble nucleate boiling provide insights into heat transfer enhancement mechanisms. These approaches contribute to optimizing porous surface fabrication and improving boiling process understanding for advanced thermal management solutions.

11.1 Introduction

11.1.1 Pool boiling applications

The boiling process occurs in numerous industrial applications as well as in our daily lives [1]. It is imperative to acquire a comprehensive understanding of the boiling phenomenon. Pool boiling refers to the phase transition between liquid and vapor that occurs on a heated solid surface within a quiescent liquid environment. This phenomenon occurs when the temperature of the heater surface surpasses the saturation temperature of the liquid. Its high heat dissipation capacity renders it suitable for addressing thermal challenges in high-flux energy applications, such as spacecraft, electronic cooling, nuclear power plants, distillation, lithography machines, and so on [2–4]. The experimental set-up for pool boiling comprises a chamber housing a heated surface at the bottom, a power source for the heater, a

doi:10.1088/978-0-7503-6174-3ch11

data acquisition system, and thermocouples deployed to monitor the temperature of both the heating surface and the bulk liquid [1]. During this process, a substantial heat transfer occurs owing to the phase change.

The understanding of boiling heat transfer (BHT) can be facilitated through boiling curves [5]. Two optimization indicators for measuring BHT are the heat transfer coefficient (HTC) and the critical heat flux (CHF) [6]. The heat transfer intensity is commonly assessed through the HTC. Augmenting the boiling HTC is crucial for enhancing the energy efficiency of boiling systems, thereby achieving substantial reductions in energy consumption, system size, and volume. The CHF describing the thermal limit is known as a boiling crisis. The enhancement of CHF is essential for rendering boiling systems compact and safe to withstand operation under high-heat-flux conditions. Therefore, the core objectives of enhancing BHT are to delay the appearance of CHF and to augment the maximum HTC as well [7].

11.1.2 Extensive investigations of pool boiling

Pool boiling entails a complex physical process characterized by interactions between the heating surface and working fluid. Over the past few decades, various methods including theoretical models [8–10], empirical or semi-empirical correlations [11], active and passive techniques [12–14], and computational fluid dynamics (CFD) techniques [15–17] have been considered. Significant progress has been made in understanding the pool boiling phenomenon, as evidenced primarily by passively experimental studies and numerical simulations.

Most documented experimental studies concentrate on engineering surfaces with the aim of fabricating structured surfaces with excellent heat transfer performance by offering extended surface area and increased nucleation site density. A representative example of the surface structures known to effectively enhance BHT performance is wicking surfaces. These structures enhance CHF by capillary-fed wicking through the structures [18]. In the work by Li *et al* [12], an ultrahigh CHF of approximately 400 $W \cdot cm^{-2}$ is achieved using a highly scalable, conformal, and tunable three-tier hierarchical surface deposition technique. Conceptually similar work has been carried out by Chen *et al* [19], whose sintered porous biomimetic structure with a height of 1.0 mm achieves the highest CHF of 343.1 $W \cdot cm^{-2}$ due to the biomimetic channels facilitating easier access of liquid water flow toward dry-out spots. Recent efforts by Song *et al* [20] involved innovatively designed three-tier hierarchical structures to achieve the simultaneous enhancement of HTC and CHF by up to 389% and 138% by minimizing bubble coalescence. It can be observed that significant advancements have been achieved in enhancing BHT through the modification of complex micro-nano porous structured surfaces. Distinct bubble dynamics induced by various structured surfaces serve as direct determinants in driving the enhancement of BHT.

The emergence of CFD technologies has enabled the prediction of transient behavior in local fluid flow and heat transfer. These kinds of methods facilitate the acquisition of local information regarding the boiling process and provide dynamic phase velocity, temperature distributions, local heat transfer coefficients, and phase distributions [21]. Pool boiling is accompanied by abundant bubble generation. Understanding the nucleation, growth, and detachment process of bubbles from the heated surface can facilitate the elucidation of heat transfer mechanisms. In the work by Yi *et al* [22], dynamic details of single bubble growth under varying gravity conditions is captured numerically by considering a thin superheated layer and thermocapillary effects resulting from surface tension variation. Li *et al* [15] innovatively simulated the process of growth and departure of multiple bubbles on a flat surface by adjusting the hydrophilic–hydrophobic area ratio to manipulate bubble dynamics, and an optimal area ratio was suggested. In recent years, heated surfaces featuring customized structures were employed to conduct numerical simulations. The micropillar-structured surface described in reference [23] was used to explore both corner and center nucleation sites, aiming to examine the dynamics of single bubbles and the consequent heat transfer characteristic. Conceptually similar research has been conducted by Wang *et al* [24], wherein modified heated surfaces with 36 hemispheres in various orientations were examined. The numerical results demonstrated that the surface with hemispheres facing downward exhibited the most effective dissipation performance. Although numerous numerical methods have demonstrated efficacy in investigating certain boiling phenomena, they have limits when applied to some complex situations. Also, the methods discussed above have primarily concentrated on the fluid-side processes, overlooking the coupling between heat transfer processes in the fluid and solid phases. The solid phase remains thermally decoupled from the fluid-side processes.

11.1.3 Motivation

So far, experimental investigations into bubble dynamics are predominantly constrained to macroscopic bubble statistics, lacking the capability to offer detailed local heat transfer information during the bubble growth and departure processes, such as local temperature distribution, heat-flux distribution beneath the bubbles, and the evolution of bubble morphology. On the other hand, current numerical investigations typically focus on single bubble dynamics and uniformly structured surfaces, far behind the intricate micro-nano porous structured surfaces of experimental research. Boiling performance is correlated with bubble dynamics associated with cavities [25]. Incorporating realistic geometric structures fabricated in experiments can further present more realistic numerical results. The composite micro-nano porous structured surfaces [13, 26] with increased nucleation sites and extended surface area have attracted considerable attention in the field of boiling heat transfer, wherein the honeycomb micro-nano porous structured surfaces, characterized by a naturally segregated liquid–vapor pathway that facilitates liquid replenishment and bubble escape, have been extensively explored in our previous investigations [13, 21, 27–29]. Thus, the innovation of this chapter lies in the utilization of computed tomography (CT) scanning techniques to reconstruct the realistic honeycomb micro-nano porous structured surfaces prepared in our experiments. Based on the reconstructed geometry, the numerical simulations of transient heat conduction representing the solid-side and dynamic multiple bubbles representing the fluid-side have been conducted to mimic the boiling behavior to address the problem of potential heat transfer performance improvement mechanisms.

11.2 Modeling and methods

11.2.1 Fabrication of honeycomb porous structured surfaces

Parts of this section have been reproduced with permission from [21] and [27]. Copyright 2024 and 2021 Elsevier.

In our previous studies [28, 29], experimental findings indicate that honeycomb micro-nano porous structured surfaces significantly improve boiling heat transfer performance. The size of micro and nano structures can influence both bubble nucleation and departure. For this purpose, the influence of pore size on honeycomb micro-nano porous copper surfaces is further investigated in this work. The preparation process of four samples with varying pore sizes is outlined below. The copper substrate undergoes a cleaning process involving dilute sulfuric acid, deionized (DI) water, hot alkaline solution (NaOH 40 $g \cdot l^{-1}$, Na_2CO_3 40.5 $g \cdot l^{-1}$, $Na_3PO_4 \cdot 12H_2O$ 69.52 $g \cdot l^{-1}$, and OP-10 2 $g \cdot l^{-1}$), and DI water in turn. A square copper plate with a side length of 6 cm is employed as the anode, positioned 3.5 cm above the cathode, which is a preprocessed copper block with a diameter and thickness of 19 and 8 mm, respectively. Subsequently, electrodeposition is conducted in a stable solution containing 0.4 $mol \cdot l^{-1}$ $CuSO_4$ and 1.8 $mol \cdot l^{-1}$ H_2SO_4, lasting for 1 min. A DC power supply (Maynuo 8852) is employed for the electrodeposition process, and the electrolyte solution is replaced after depositing each sample to maintain a consistent concentration of electrolyte solution. During the process, the copper ion deposition reaction (equation (11.3)) and hydrogen formation reaction (equation (11.2)) occur concurrently at the cathode, while the copper dissolution reaction (equation (11.1)) occurs at the anode, resulting in the formation of honeycomb micro-nano porous structured surfaces:

$$Cu - 2e^{-} = Cu^{2+} \tag{11.1}$$

$$2H^{+} + 2e^{-} = H_2\uparrow \tag{11.2}$$

$$Cu^{2+} + 2e^{-} = Cu. \tag{11.3}$$

Four samples with different pore sizes are obtained through electrodeposition at deposition current densities of 0.5, 1, 1.5, and 2 $A \cdot cm^{-2}$, and are called Sample#M1, Sample#M2, Sample#M3, and Sample#M4, respectively. Finally, the samples were sintered in a nitrogen–hydrogen mixture gas at 710 °C for 30 min to reinforce the prepared samples. Scanning electron micrography (SEM) images of the four samples

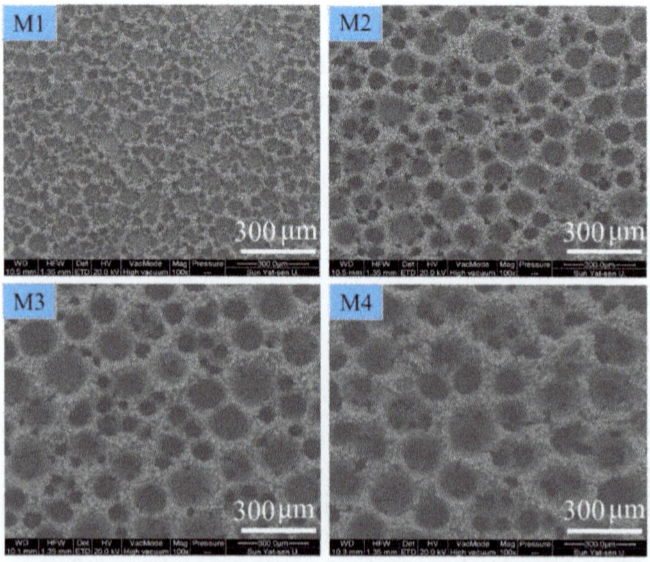

Figure 11.1. SEM images of Sample#M1, Sample#M2, Sample#M3, and Sample#M4. (Reproduced with permission from [27]. Copyright 2021 Elsevier.)

are provided in figure 11.1 to illustrate their intricate structures. The SEM images reveal that the structured surfaces exhibit honeycomb-like porous structures. The primary distinction among them lies in the distribution of micro-pore structures. Higher current densities lead to the generation of more hydrogen bubbles within a given time frame, further leading to larger honeycomb pores (ranging from 48 to 128 μm). The wettability of a micro-nano porous structure significantly influences the boiling heat transfer performance. Surface wettability measurements have been conducted on the four prepared surfaces. The results indicate that the plain copper surface exhibits a contact angle of 45°, whereas all other prepared samples exhibits super-hydrophilicity.

The detailed experimental procedure for pool boiling is delineated below. The four fabricated surfaces are individually welded to the heater block using tin–lead material to reduce contact thermal resistance. Thermal insulation cotton is utilized to envelop the copper heater block, effectively reducing heat loss into the surrounding environment. DI water, pre-boiled for 30 min, serves as the working fluid. It is subsequently poured into the confined boiling set-up and heated for an additional 30 min to minimize the presence of air within the system. Four T-type thermocouples with a measurement error of ±0.3 K are employed to monitor temperature. The first thermocouple T_1 is positioned 6.5 mm below the bottom surface of the sample. The second thermocouple T_2 is immersed in the boiling water to monitor the temperature. The third thermocouple T_3 is inserted into the copper block to gauge the temperature of the heater block. The fourth thermocouple T_4 is utilized to measure the ambient temperature. The Agilent 34970 A is utilized to collect all temperature data mentioned above. Two plain copper surfaces, following the above-mentioned procedures, are employed for boiling tests to validate the stability of the experimental set-up. A detailed description of the experimental set-up can be found in [27].

11.2.2 Reconstruction of geometric models for honeycomb surfaces

Parts of this section have been reproduced with permission from [27]. Copyright 2021 Elsevier.

Following the identical experimental preparation procedure as previously described, four samples are electrodeposited on copper sheets of the same area with a thickness of 0.05 mm, specifically for conducting CT scanning (nano Voxel-3000D). CT is a radiographic technique used for visualization [30] to furnish structural data pertaining to the 3D complex geometric morphology. As illustrated in figure 11.2, the abundant honeycomb pores are clearly captured and sections measuring 900 μm × 900 μm are intercepted to reconstruct the 3D honeycomb porous surfaces. The reconstructed surfaces retained only the macroscopic honeycomb morphology, facilitating CFD simulations. Figure 11.2 illustrates that the reconstructed 3D geometric surfaces maintain high fidelity with the original surfaces (figure 11.1).

11.2.3 Numerical simulation

11.2.3.1 The transient heat-conduction model related to the solid side

Parts of this section have been reproduced with permission from [27]. Copyright 2021 Elsevier.

Based on the micro-boundary layer proposed by Stephan and Hammer [8], as well as Van Strahlen [31], the overall heat transfer comprises the combined contributions

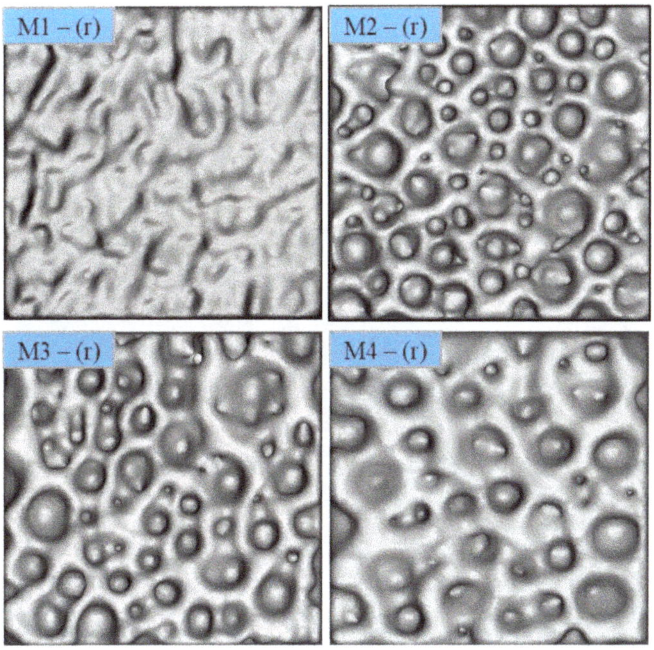

Figure 11.2. CT reconstructed 3D geometric surfaces of Sample#M1, Sample#M2, Sample#M3, and Sample#M4.

from heat transfer within the micro layer and the adjacent convective heat transfer. According to Wayner *et al* [32], the occurrence of extremely high heat fluxes in a micro region proximate to the absorbed film implies the onset of evaporation. The associated heat and mass transport phenomena within this micro region could exert a significant influence on the overall macroscopic heat transfer. Hence, in this study, we propose a parametric model that accounts for the three-phase contact lines (TPCL; micro-boundary layer) as well as neighboring convective heat transfer, aiming to address the challenge of reconstructing the Neumann boundary condition of the 3D transient heat-conduction governing equations. To the best of our knowledge, such a model-based experimental analysis method is, for the first time, employed to guide the reconstruction of the unknown boiling heat-flux distribution on physically realistic micro-nano porous surfaces, which is beyond our previous works [33, 34]. As depicted in figure 11.3(a), various mechanisms exist to enhance nucleate boiling heat transfer across both the low- and high-heat-flux regimes [35].

At low-heat-flux nucleate boiling, numerous small independent bubbles are observable. Conversely, at high-heat-flux nucleate boiling, abundant small bubbles coalesce into larger bubbles, periodically covering the boiling surface. In this scenario, instead of the typical pattern characterized by numerous small rings of heat flux, a more realistic single-ring heat-flux pattern with a larger radius is considered.

The reconstructed 3D geometry data of the four surfaces are imported into COMSOL Multiphysics 5.3a to conduct finite-element-based numerical simulations. The governing equations for the 3D transient heat conduction defined within the

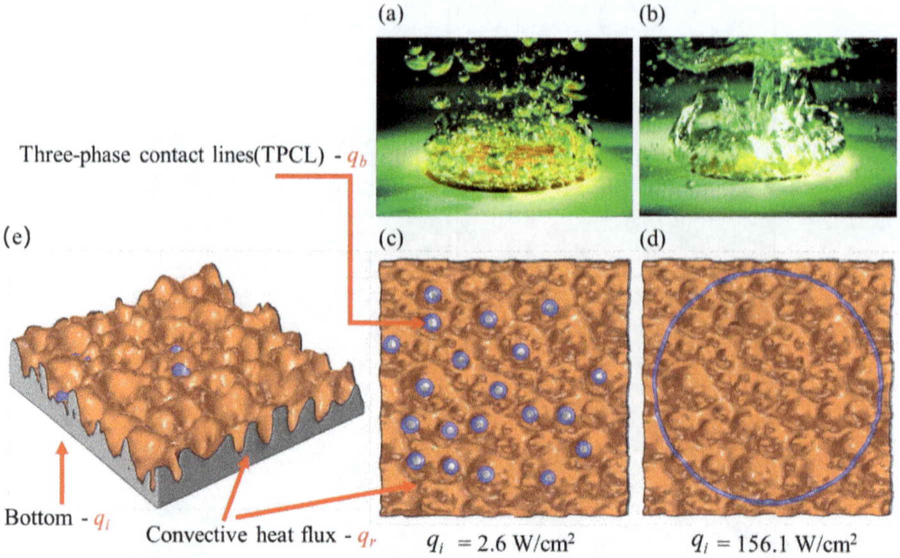

Figure 11.3. Two representative states of Sample#M2. Panels (a) and (c) represent the bubble dynamics and assumed TPCL heat-flux pattern at a low heat-flux level. Panels (b) and (d) represent the bubble dynamics and assumed TPCL heat-flux pattern at a high heat-flux level. (e) Side view of (c). (Reproduced with permission from [27]. Copyright 2021 Elsevier.)

micro-nano porous structures are provided in equations (11.4)–(11.8). In the boiling heat transfer process, a significant portion of the heat is transferred through evaporation within the micro region of the TPCL. Consequently, it is assumed that the heat dissipated by the phase change q_b flows exclusively from the TPCL. A constant heat-flux supply q_i on the bottom surface is utilized, in accordance with the experimental characteristic curves (figure 11.5). The q_r is the convective heat flux and the convective heat transfer coefficient h is set to 1000 $W{\cdot}m^{-2}{\cdot}K^{-1}$ in this work:

$$\rho c_p \frac{\partial \theta}{\partial_t} = \nabla \cdot (\lambda \nabla \theta) \quad \text{in} \quad \Omega \times \left(0,\, t_f\right) \tag{11.4}$$

$$\theta(\cdot,\, 0) = \theta_0(\cdot) \quad \text{in} \quad \Omega \tag{11.5}$$

$$\lambda \frac{\partial \theta}{\partial n} = q_i \qquad \text{on} \qquad \Gamma_H \times \left(0,\, t_f\right) \tag{11.6}$$

$$\lambda \frac{\partial \theta}{\partial n} = q_r = h(T_{\text{side}} - T_0) \qquad \text{on} \qquad \Gamma_R \times \left(0,\, t_f\right) \tag{11.7}$$

$$\lambda \frac{\partial \theta}{\partial n} = q_b \qquad \text{on} \qquad \Gamma_B \times \left(0,\, t_f\right). \tag{11.8}$$

Based on our previous analysis [36], with the gradual increase of heat flux q_i, local activities on the TPCL come to dominate the boiling heat transfer process. A ring-shaped local transient heat flux q_b (equations (11.9) and (11.10)) on the TPCL is suggested for the simulation. Here, C is regarded as a model parameter that can be determined by minimizing the disparity between the macroscopic averaged wall superheats obtained from experiments and numerical simulations. The different segments of q_b correspond to natural convection (before bubble nucleation), bubble growth time, and waiting time (after bubble departure). The entire simulation duration is set to 0.05 s to replicate the bubble growth period. Based on this mathematical modeling, we investigated three representative heat-flux supply q_i equal to 10, 50, and 150 $W{\cdot}cm^{-2}$, which correspond to two different boiling states:

$$q_b = [C \cdot \sin[100 \cdot \pi \cdot (t - 0.02)] + C]\ \text{W} \cdot \text{cm}^{-2} \quad t = (0.015,\, 0.035) \tag{11.9}$$

$$q_b = 0 \quad t = (0,\, 0.015) \cup (0.035,\, 0.05). \tag{11.10}$$

In addition, the porosity ϕ is considered in the numerical simulations of 3D heat-conduction processes. The effective thermal conductivity λ_e of the porous sintered copper material may undergo significant changes. As shown in equation (11.11), the effective thermal conductivity of porous structures is defined as follows (λ is the thermal conductivity of copper):

$$\lambda_e = (1 - \phi) \times \lambda. \tag{11.11}$$

11.2.3.2 The multi-bubble nucleate boiling model related to the fluid side

Parts of this section have been reproduced with permission from [21] and [27]. Copyright 2024 and 2021 Elsevier.

In this section, the plain copper surface and three reconstructed honeycomb porous structured surfaces with different pore structures (figure 11.2), namely M1, M3, and M4, will be included in the numerical simulation. A comprehensive general model has been developed to numerically simulate nucleate boiling processes. To the best of our knowledge, beyond our previous works [27], such a multi-bubble boiling model based on physically realistic honeycomb micro-nano porous structured surfaces has been established for the first time.

The nucleate boiling phenomenon is numerically investigated utilizing the volume of fluid (VOF) method [37]. User defined functions (UDFs) are employed to elucidate the Lee phase change model [38] and the mass transfer model [39] across the liquid–vapor interfaces. The integrated models effectively constrain phase change occurrences to predefined locations, eliminating the necessity for initializing bubble seeds. The VOF method, recognized as the most popular interface-capturing approach, has been extensively utilized in the simulation of boiling phenomena [40]. The interface between the vapor and liquid phases is tracked by solving the continuity equation for the volume fractions of each phase. In the VOF model, α_l and α_v are introduced as the liquid volume fraction and vapor volume fraction, respectively.

The continuity equations for the vapor and liquid phases are as follows:

$$\frac{\partial}{\partial t}(\alpha_v \rho_v) + \nabla \cdot (\alpha_v \rho_v \vec{u}) = S_{l\to v} \tag{11.12}$$

$$\frac{\partial}{\partial t}(\alpha_l \rho_l) + \nabla \cdot (\alpha_l \rho_l \vec{u}) = S_{v\to l}. \tag{11.13}$$

The source term of the evaporation process is $S_{l\to v} = -S_{v\to l} = c_l \alpha_l \rho_l (T - T_{\text{sat}})/T_{\text{sat}}(T > T_{\text{sat}})$. c_l is an adjustable evaporation coefficient, set to 50 in the current work.

The momentum equation is

$$\frac{\partial}{\partial t}(\rho \vec{u}) + \nabla \cdot (\rho \vec{u} \vec{u}) = -\nabla p + \nabla \cdot (\mu \cdot \nabla \vec{u}) + F_S + F_g, \tag{11.14}$$

where $\rho = \rho_l \alpha_l + \rho_v \alpha_v$ and $\mu = \mu_l \alpha_l + \mu_v \alpha_v$. The surface tension F_S including wall-adhesion accounted for by the continuum surface force model [41] is treated as volume force source term in the above equation:

$$F_S = \sigma \frac{\alpha_l \rho_l \kappa_v \nabla \alpha_v + \alpha_v \rho_v \kappa_l \nabla \alpha_l}{0.5(\rho_l + \rho_v)}, \tag{11.15}$$

where σ and κ, respectively, represent the surface tension coefficient for water and the vapor–liquid interface curvature, in which σ is set to $0.059\ \text{N} \cdot \text{m}^{-1}$. The curvature κ depends on the bending degree of the liquid–gas interface.

The energy equation is given as

$$\frac{\partial}{\partial t}(\rho c_p T) + \nabla \cdot (\vec{u}(\rho c_p T)) = \nabla \cdot (\lambda \nabla T) + E. \tag{11.16}$$

The energy source term is $E = h_{lv} S_{l \to v}$. In addition, this chapter incorporates the mass transfer model at the liquid–gas interface according to Tanasawa [39], who simplified the Schrage model by suggesting that the mass flux is linearly related to the temperature jump between the interface and vapor phase:

$$m = \frac{2\gamma}{2 - \gamma} \sqrt{\frac{M}{2\pi R}} \left[\frac{\rho_v h_{lv}(T_i - T_{\text{sat}})}{T_{\text{sat}}^{3/2}} \right], \tag{11.17}$$

where T_{sat} is determined by the Clausius–Clapeyron equation in the saturated state, R and M are the universal gas constant and the molecular weight of water, T_i is the interface temperature, and γ is the accommodation coefficient, set to 0.5 in this chapter.

In this study, nucleate boiling is numerically simulated using the ANSYS Fluent software [42]. According to Dhir *et al* [43], the development of a credible predictive model for nucleate boiling necessitates addressing several crucial aspects, including the density of nucleation sites, the thermal response of the heater, multi-bubble dynamics, and potential heat transfer mechanisms. As depicted in figure 11.4, the numerical simulation of nucleate boiling for P1, M1, M3, and M4 is conducted in a 3D square pool with a height of 18 mm (taking M3 as an example). The working

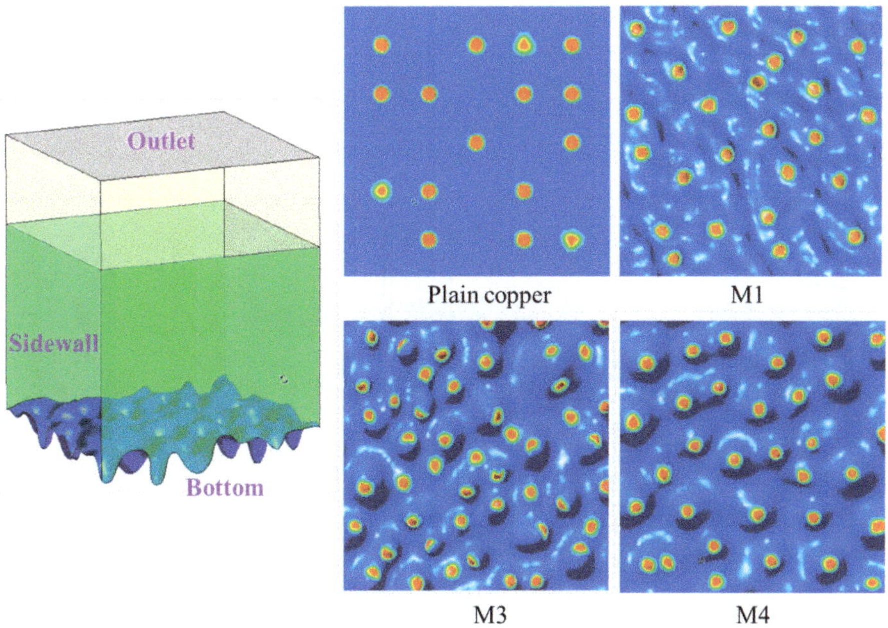

Figure 11.4. 3D view of the M3 computational domain (left) as well as the nucleate site (marked by red circles) distribution of P1, M1, M3, and M4. (Reproduced with permission from [21]. Copyright 2024 Elsevier.)

fluid in the pool is water. Due to computational constraints, a representative size of 900 μm × 900 μm (1/348 of the original surface) is utilized to conduct nucleate boiling within the square pool, measuring 16.8 mm in width and 16.8 mm in length, in accordance with the actual experiments. The boundary conditions of the entire computational domain consist of three components: the bottom, sidewalls, and outlet of the square pool. In this numerical simulation, the sidewalls are designated as no-slip stationary walls with adiabatic conditions. The bottom heated surface is treated as a no-slip wall with constant temperature, while the outlet of the computational domain is set as a pressure-outlet boundary. The backflow volume fraction of vapor is set to 1. Figure 11.4 illustrates the bottom of the computational domain with P1, M1, M3, and M4, respectively. Bubble nucleation is initiated directly by the superheat temperature. The method for determining nucleation sites in the current work is as follows. For structured surfaces M1, M3, and M4, the nucleation sites are determined based on the difference in boiling performance. The obtained nucleation sites for M1 and M4 are 49% and 68% of those for M3, respectively. The nucleation site densities and equilibrium contact angle θ of the four surfaces are listed in table 11.1.

Table 11.1. Some statistical information on various surfaces. (Reproduced with permission from [21]. Copyright 2024 Elsevier.)

	P1	M1	M2	M3
Bubble nucleation sites density N_{si}	16	24	48	32
Equilibrium contact angle θ (°)	45	35	0	0

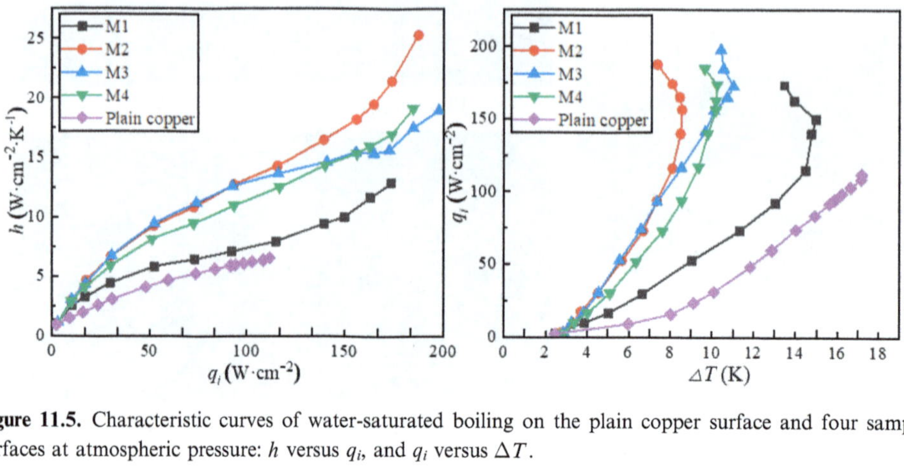

Figure 11.5. Characteristic curves of water-saturated boiling on the plain copper surface and four sample surfaces at atmospheric pressure: h versus q_i, and q_i versus ΔT.

11.3 Applications and analysis

11.3.1 Experimental analysis of boiling heat transfer

Parts of this section have been reproduced with permission from [21] and [27]. Copyright 2024 and 2021 Elsevier.

The characteristic curves illustrating water-saturated boiling on P1, M1, M2, M3, and M4 are depicted in figure 11.5, which indicates that all samples prepared in our experiment can enhance the CHF and HTC. Compared with the plain copper surface ($q_{\mathrm{CHF}} = 111.9\ \mathrm{W{\cdot}cm^{-2}}$), the CHF on the four samples is improved by over 1.5 times. This can be explained by the high bubble departure frequency (a statistical nucleation frequency increased over two-fold compared to plain copper surface) and abundant bubble nucleation on the honeycomb micro-nano porous structured surfaces, accelerating the rates of heat removal from the boiling surface.

The degree of performance enhancement varies among the samples. The bubble departure diameters of Sample#M2, Sample#M3, and Sample#M4 with complete honeycomb pores are smaller than those of the plain copper surface and Sample#M1. Referring to the literature [25], the bubble departure diameter decreases with increasing nucleation site density, which explains the reduced bubble coalescence. With the increase of electrodeposition current density, the average pore diameter of the prepared samples increased, ranging from 48 to 128 μm. Wall superheat depends on bubble nucleation that relates to the characteristics of cavities [44]. Sample#M2, with the appropriate average pore size of 80 μm (referring to Li *et al* [45], where the optimal active cavity size is between 10 and 100 μm) exhibits the best heat-removal capability, with a maximum CHF and HTC reaching 187.3 $\mathrm{W{\cdot}cm^{-2}}$ and 23.7 $\mathrm{W{\cdot}cm^{-2}{\cdot}K^{-1}}$, respectively. The cavities with suitable pore sizes are more conducive for bubble departure from vapor–liquid separation paths, while the cavities composed of nano-dendrites can facilitate the supply of liquid [46]. The wettability test results indicate that the tendency of wicking liquid capacity of the four samples is consistent with that of boiling performance. This phenomenon has been attributed to the ability of micro-nano structured surfaces to transport liquid effectively. Timely liquid supply can prevent wall drying, thereby delaying the onset of film boiling.

11.3.2 Numerical analysis of boiling heat transfer

11.3.2.1 The numerical analysis results associated with the solid side

Parts of this section have been reproduced with permission from [27]. Copyright 2021 from Elsevier.

The simulation results of P1, M1, M2, M3, and M4 for three representative macroscopic heat-flux levels, namely 10, 50, and 150 $\mathrm{W{\cdot}cm^{-2}}$ corresponding to the characteristic boiling curves, are discussed in this section. The first two heat-flux supply levels correspond to the low-heat-flux boiling state, while the third one corresponds to the high-heat-flux boiling state.

In the low-heat-flux boiling state, rings created by small bubbles with a width of 12 μm are assumed to be randomly generated on the boiling surface. For the five samples mentioned above, the corresponding model parameters C in equation (11.9) are determined by matching the macroscopic wall superheat ΔT measured by the experiment and a mean value of the distributed results from simulations. The percentage ϖ represents the degree of local heat transfer performance, defined as heat flowing through the TPCL divided by the applied heat supply at the bottom surface for all different surface samples, and is calculated according to

$$\varpi = \frac{\bar{q_b} \times S_1}{q_i \times S}, \tag{11.18}$$

where S stands for the total area of the bottom surface of the sample, and $\bar{q_b}$ is the average value of q_b (equations (11.9) and (11.10)) at 50 instants. S_1 is the TPCL area.

The evaluation value ϖ for the five samples at low heat-flux level (10 and 50 W·cm^{-2}) is listed in table 11.2. When the heat flux applies q_i equal to 50 W·cm^{-2}, the evaluation value ϖ for all samples is greater than 70%, indicating that heat removal by phase change accounts for most of the total heat input. This result is also consistent with previous research findings [47]. According to Van Strahlen's work [31], the overall heat transfer is the sum of heat transfer on the micro layer and the adjacent convective heat transfer. The ϖ value decreases with successively increasing thickness of the samples, which is mainly because the four samples obtained by electrodeposition have continuously lager convective heat dissipation areas, making the proportion of convective heat flux greatly enhanced. Nevertheless, such information cannot be obtained from a macroscopic experimental study. Due to space constraints, not all simulation results are presented. For illustrative purposes, considering the heat supply of q_i = 50 W·cm^{-2}, the predicted temperature distributions on the bottom surface of the five samples at five selected equidistant time instants (t = 0.01, 0.02, 0.03, 0.04, 0.05 s) are shown in figure 11.6. Due to the

Table 11.2. The ΔT for measured and simulated results, as well as calibrated model parameter C.

	Plain copper	M1	M2	M3	M4
q_i= 10 W·cm^{-2}					
Experiment ΔT (K)	6.3	3.9	3.3	3.3	3.4
Simulation ΔT (K)	6.0	3.9	3.4	3.2	3.3
The value of C	150	152	110	74	57
ϖ (%)	53.1	53.8	39.2	26.4	21.1
q_i = 50 W·cm^{-2}					
Experiment ΔT (K)	11.9	8.9	5.4	5.5	6.4
Simulation ΔT K)	11.9	8.6	5.6	5.6	6.4
The value of C	1190	1115	1110	1050	990
ϖ (%)	84.7	79.5	79.5	74.8	70.9
q_i = 150 W·cm^{-2}					
Experiment ΔT (K)	15.4	9.0	10.4	10.5	15.4
Simulation ΔT (K)	15.3	8.9	10.2	10.5	15.3
The value of C	6990	3545	3617	3312	6990
ϖ (%)		88.8	91.9	88.1	87.7

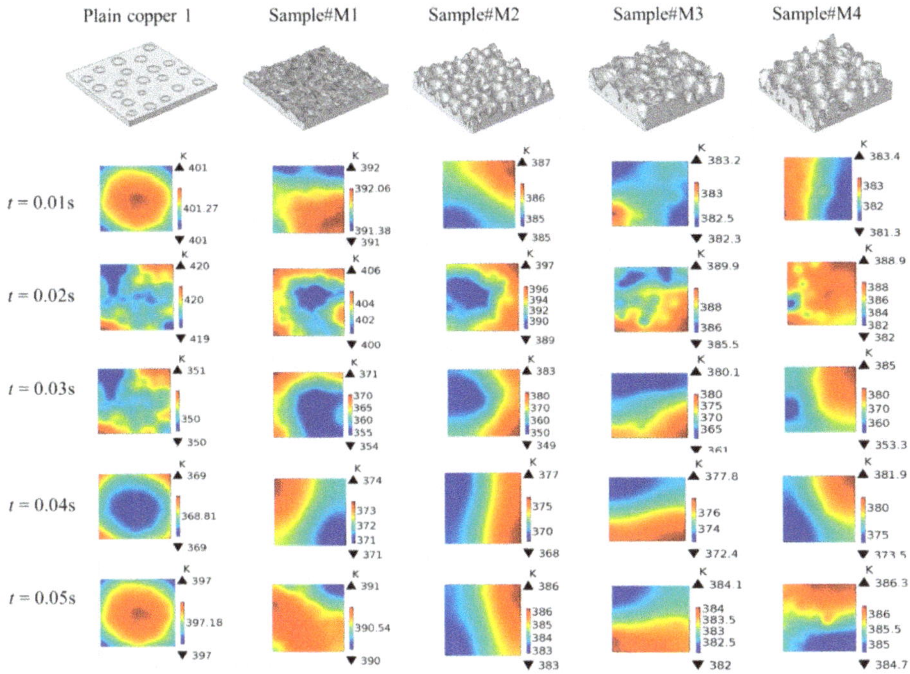

Figure 11.6. The predicted temperature distributions on the bottom of five samples at five different time instants $t = 0.01, 0.02, 0.03, 0.04, 0.05$ s ($q_i = 50$ W·cm^{-2}). (Reproduced with permission from [27]. Copyright 2021 Elsevier.)

structural disparities in sample geometries and randomly simulated TPCL locations, the temperature distributions exhibit varying local details, yet similar trends.

In the high-heat-flux boiling state, a central large annular heat-flux pattern with a width of 12 μm is assumed on the top surfaces for the same reconstructed micro-nano porous structures, with S2 representing the area of TPCL. The temperature distributions on the bottom surface of the four samples at five different time instants (t = 0.01, 0.02, 0.03, 0.04, 0.05 s) at the heat-flux level of $q_i = 150$ W·cm^{-2} are illustrated in figure 11.7. The temperature distribution on the bottom surface mirrors the heat-flux pattern of the top surface at 0.02 and 0.03 s, primarily due to the position of the TPCL ring during this period. However, as the simulation progresses, the irregularity of the ring's area becomes more pronounced, exerting a dominant influence on the temperature distribution on the bottom surface and leading to significant changes. Combined with the temperature distribution results at the heat-flux level of $q_i = 50$ W·cm^{-2} (figure 11.6), it is evident that there is no significant temperature difference observed at the bottom surface of the plain copper. However, experiments conducted with the micro-nano porous structures reveal relatively large temperature differences, primarily due to the assumed random distribution of TPCL on the boiling surface and significant structural differences. Both the micro-nano structure and nucleation distribution exert significant influence on the temperature at the bottom surface. The evaluation value ϖ for the five samples at a high heat-flux

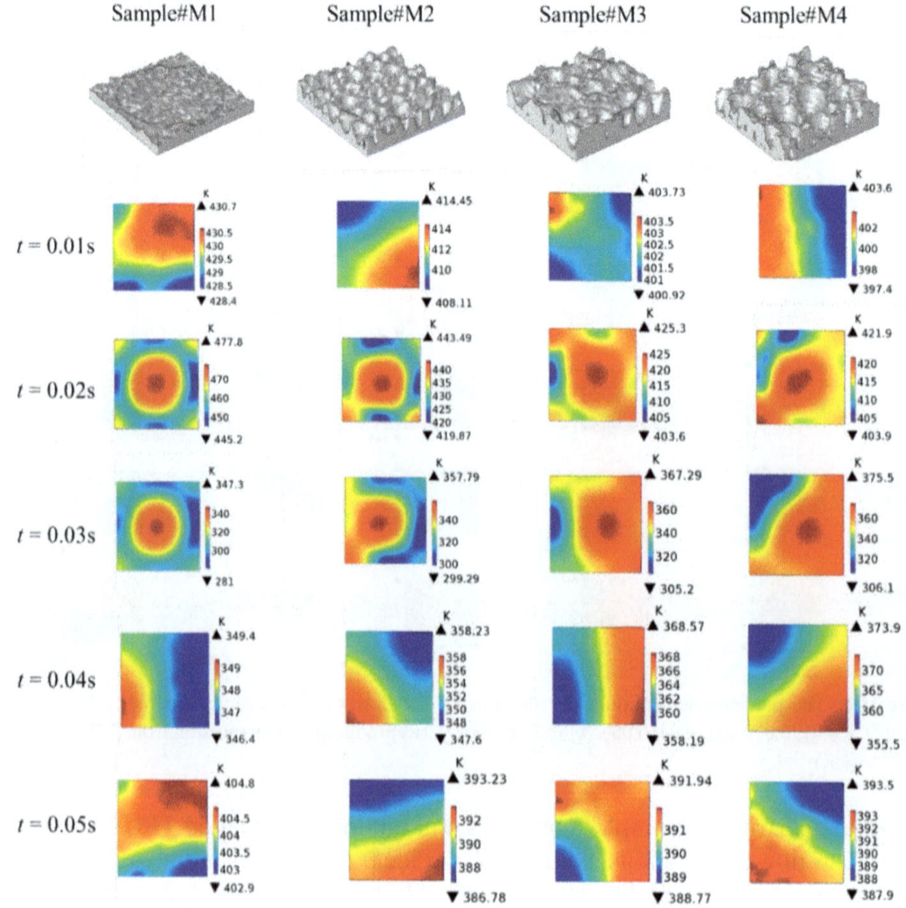

Figure 11.7. The predicted temperature distributions on the bottom of four samples at five different time instants $t = 0.01, 0.02, 0.03, 0.04, 0.05$ s ($q_i = 150$ W·cm^{-2}). (Reproduced with permission from [27]. Copyright 2021 Elsevier.)

level (150 W·cm^{-2}) is listed in table 11.2. In comparison with the low-heat-flux boiling state, the ϖ values of the four other samples (excluding the plain copper surface) in the high-heat-flux state exhibit significant improvements, reaching 88.83%, 91.92%, 88.07%, and 87.72% successively. This suggests that most heat dissipates from the TPCL, indicating that micro-nano porous structures can notably enhance heat dissipation. In other words, the effect of increasing heat flux surpasses the influence of area. Detailed temperature distribution simulation results at the high heat-flux level of $q_i = 150$ W·cm^{-2} are provided in figure 11.7.

Mathematically, the inverse problem investigated in this work is akin to reconstructing the Neumann boundary conditions of the 3D transient heat-conduction model. A multi-layer end-to-end convolutional neural network has been devised and trained to serve as a dynamic solver for inverse heat-conduction problems (IHCP), namely the estimation of unknown boiling heat fluxes highly variable in time and space from available temperature data on the heated surface. In this simulation-based inversion algorithm test, the temperature data utilized for the estimation procedure are generated by solving the forward heat-conduction problem (equations (11.4)–(11.8)) with the predefined C parameter value, along with the corresponding boiling heat flux (equations (11.9) and (11.10)). In the case of Sample#M2, serving as a representative example, a total of 100 sets of heat-conduction forward models with diverse C parameter values are solved using high-throughput computing techniques. These models are utilized to generate training data for the multi-layer end-to-end convolutional neural network. Empirically, the parameter C is varied within the range of 20 to 2000 for the testing purposes. The numerical simulations of the 100 heat-conduction forward models are efficiently executed on the Tianhe-2 supercomputer, employing a high-throughput computing strategy. The data obtained from these simulations are partitioned into training, validation, and test sets, comprising 80%, 10%, and 10% of the total data, respectively, for training the multi-layer end-to-end convolutional neural network. The multi-layer end-to-end convolutional neural network architecture comprises one multi-channel input layer, three convolutional layers, three deconvolution layers, and one single-channel output layer. For clarity, a portion of the results is depicted in figure 11.8. The structural similarity (SSIM) index is utilized to assess the accuracy of the estimation results, with an SSIM value of 1 indicating perfect alignment between the expected and estimated results. Given the complex geometric models and highly transient boiling heat-flux dynamics, the estimation results (with an average SSIM index of about 0.8556) are deemed satisfactory. However, there remains potential for improvement through further analysis of the network

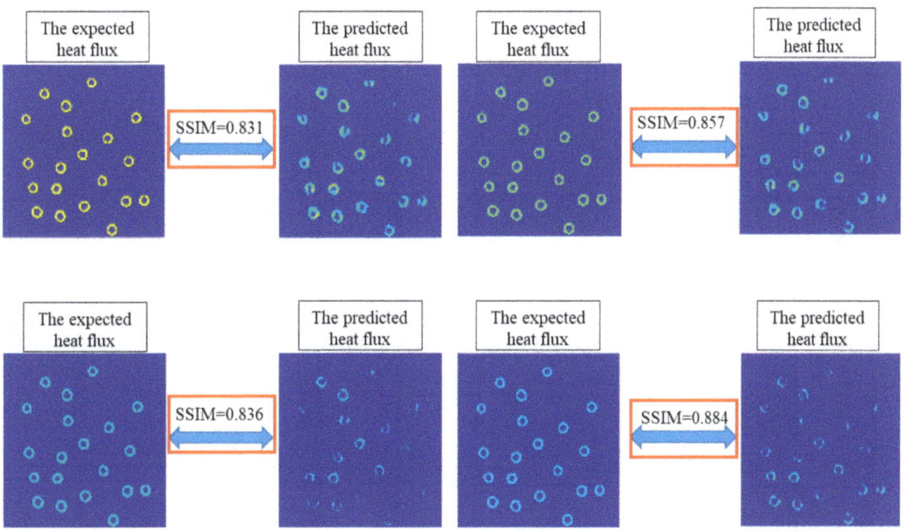

Figure 11.8. Comparison of expected and estimated local boiling heat fluxes by our multi-layer end-to-end convolutional neural network with respect to Sample#M2. (Reproduced with permission from [27]. Copyright 2021 Elsevier.)

structure, activation functions, and other related mathematical set-ups, which will be explored in our future research endeavors.

11.3.2.2 The numerical analysis results associated with the fluid side

Parts of this section have been reproduced with permission from [21] and [27]. Copyright 2024 and 2021 Elsevier.

Dynamic simulation of pool boiling serves as a valuable resource for gaining further insights into boiling behavior and local heat transfer details. To assess the accuracy of the model, the bottom surfaces of P1, M1, M3, and M4 (figure 11.4) are subjected to four types of wall superheats (figure 11.9) corresponding to different states along the boiling curve. The numerical simulation results reveal that the heat flux observed on P1 under different wall superheats aligns with the experimental measurements. However, the simulated heat flux observed on M1, M3, and M4 is lower than the experimental values at high heat-flux levels. Given the enhanced convective heat transfer resulting from the disturbance of abundant bubbles and the large extended surface area ratio within the porous structured surfaces, the built-in convective augmentation factor is appropriately calibrated to ensure alignment of the heat flux observed on the structured surface with experimental measurements. Figure 11.9 illustrates the characteristic curves of water-saturated boiling obtained from both experiment and simulation for the four surfaces. The developed models can be employed to capture intricate details of bubble evolution. In the following, for local bubble growth and heat transfer characteristics comprising multiple bubbles, the resultant heat flux distribution are analysed in detail.

The numerical simulation of multi-bubble dynamics is of great significance for further analysis of the enhancement mechanism of heat transfer performance induced by different pore structures. Figures 11.10(i)–(l) illustrate the multi-bubble dynamics of nucleate boiling on P1, M1, M3, and M4 from pool boiling experiments. Figures 11.10(a)–(d) and (e)–(h) show the bubbles on P1, M1, M3, and M4 during the preliminary growth and departure stages from simulations. The dynamic process of multi-bubble nucleation, growth, and departure is well simulated, and identical bubble behavior is observed in both the experiment and the simulation.

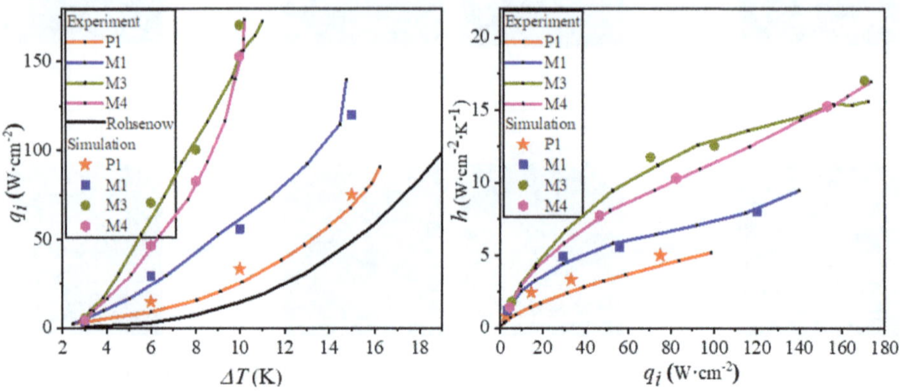

Figure 11.9. Validation of experimental data [27, 48] and numerical models of P1, M1, M3, and M4. (Reproduced with permission from [21]. Copyright 2024 Elsevier.)

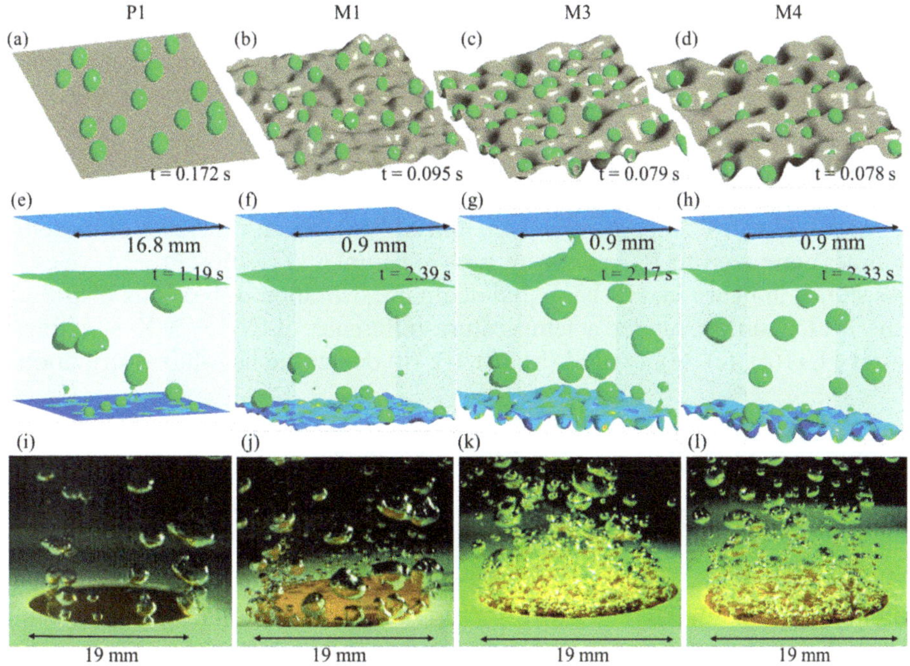

Figure 11.10. Panels (a)–(d) represent the preliminary stage of bubble growth. Panels (e)–(h) and (i)–(l) are the multi-bubble dynamics from simulation and experiment, respectively, under $\Delta T = 3$ K. (Reproduced with permission from [21]. Copyright 2024 Elsevier.)

As depicted in figure 11.10, the plain surface P1 is less favorable for nucleation due to the absence of defect cavities that could facilitate the nucleation process. Additionally, the growth and coalescence between bubbles have no barrier (figure 11.10 (e)), leading to the bubbles growing on P1 to coalesce and generate bubbles with larger diameters. In comparison to P1, the three structured surfaces M1, M3, and M4 feature more nucleation sites due to the presence of many defect cavities that facilitate bubble nucleation. As depicted in figures 11.10(f)–(h), abundant bubble dynamics are observed on M1, M3, and M4. The dynamics of bubble growth on the three structured surfaces with extensive porous structures vary, influenced by the structural differences in honeycomb pores and nucleation site density. As depicted in figure 11.10(b), M1 exhibits fewer nucleation sites and minimal restrictions on bubble growth due to incomplete honeycomb pores. Conversely, the bubbles emerging from M3 and M4 are encapsulated within the honeycomb pores, as illustrated in figures 11.10 (c) and (d). These honeycomb pores regulate bubble evolution, resulting in bubbles with reduced diameter. Moreover, the complete honeycomb pores in M3 and M4 act as effective barriers against premature bubble coalescence, which could reduce the CHF. Comparatively, M3 with an average pore size smaller than that of M4, possesses more honeycomb pores for nucleation under the same projected area (figure 11.2). This increased number of nucleation sites significantly reduces the bubble departure diameter and increases the number of detached bubbles, thereby facilitating the most abundant bubble dynamics in M3.

In addition, the bubble departure frequency f serves as a critical parameter influencing pool boiling performance. The collected f spectrum for 15 bubbles, monitored by tracking the heat-flux variation at 15 predefined nucleate sites within P1, M1, M3, and M4 has been statistically analysed. f is calculated by dividing the total number of bubbles detached from the 15 predefined nucleate sites within 3 s by the total time. The statistical f for P1, M1, M3, and M4 is, respectively, 1.02, 1.80, 3.29, and 2.49 s^{-1}. Comparatively, M3, and M4 exhibit an f increase of more than two times compared to P1, consistent with experimental results.

As shown in figure 11.11, the spatial–temporal evolution of heat-flux distribution at the heated surface under a temperature difference of $\Delta T = 3$ K is extracted. Figures 11.11(a), (c), (e), (g) and (b), (d), (f), (h) depict the heat-flux distribution on

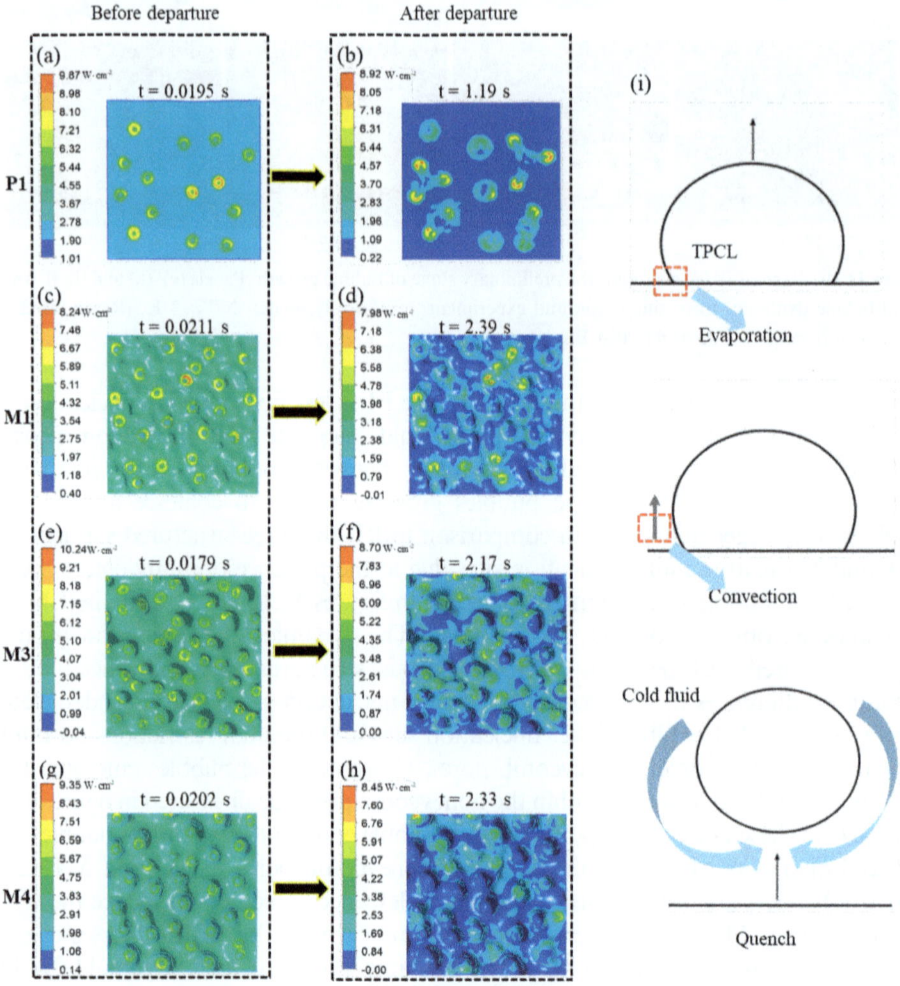

Figure 11.11. (a), (c), (e), (g) and (b), (d), (f), (h) illustrate the heat-flux distribution during the initial stage of bubble growth and partial bubble lift-off, respectively. (i) Quench heat-flux components. (Reproduced with permission from [21]. Copyright 2024 Elsevier.)

the heated surface during the initial bubble growth and following partial bubble lift-off. Before bubble departure, the bottom heat flux exhibits an annular distribution. The region covered by the bubble, also known as the dry patch, experiences low heat flux due to the low thermal conductivity of vapor. Conversely, the area near the TPCL experiences the maximum heat flux attributed to the strong evaporation effect. The overall heat dissipation effect across the surface is primarily governed by the evaporation occurring at the TPCL. After partial bubble departure, the peak heat flux, represented by the red/yellow ring-shaped area, is observed only in the region after bubble departure. The comparison results of heat-flux distribution before (figures 11.11(a), (c), (e), (g)) and after bubble departure (figures 11.11(b), (d), (f), (h)) demonstrate that the evaporation effect at the TPCL plays an essential role during the bubble growth. After the bubble departure, the overall heat transfer performance is dominated by transient heat conduction resulting from rapid liquid replenishment. The comparison results depicted in figures 11.11(b), (d), (f), and (h) indicate that the honeycomb micro-nano porous structured surface with abundant nucleation sites generates more escaping bubbles, resulting in the more peak heat-flux regions (the red/yellow ring-shaped areas) being observed in figure 11.11(f). This observation aligns with Rohsenow's correlation [48] $h = 2(\pi \lambda_l \rho_l c_{pl})^{0.5} N_s D_d{}^2 f^{0.5}$. The more nucleation sites, the better the heat transfer performance, where N_s are the nucleate sites. Combining the bubble f spectrum results, M3 and M4, characterized by complete honeycomb pores, exhibit an increase in f. This increase suggests that a high frequency of bubble departure facilitates the rapid backflow of bulk liquid to replenish the dry patch, thereby promoting the transient heat-conduction process by drawing cooler liquid towards the heated surface more massively and frequently.

11.4 Conclusions

Parts of this section have been reproduced with permission from [21]. Copyright 2024 Elsevier.

In this chapter, four honeycomb micro-nano porous structured surfaces (Sample#M1, Sample#M2, Sample#M3, and Sample#M4) with successively increasing pore sizes are first prepared. The Sample#M2 of average pore size 80 μm exhibits the maximum CHF and HTC of 187.3 $W \cdot cm^{-2}$ and 23.7 $W \cdot cm^{-2} \cdot K^{-1}$. To further investigate the underlying enhancement mechanisms of heat transfer performance associated with the honeycomb pore, two heuristic modeling approaches are proposed for model-based experimental analysis.

Specifically, based on the CT reconstructed honeycomb structured surfaces, a parametric 3D transient heat-conduction forward model has been developed to mimic the solid heater part of dynamic boiling process and a multi-layer end-to-end convolutional neural network is for the first time constructed and trained to solve the corresponding ill-posed inverse problems of 3D transient heat conduction efficiently. Numerical findings indicate that the percentage of heat dissipated from the TPCL can increase to 90% at high heat-flux supply levels.

Moreover, the phase change model and mass transfer model are employed to conduct CFD simulations of nucleate boiling based on three honeycomb porous structured surfaces with different pore structures and a plain surface. The simulation can correctly capture experimental trends and predict the local transient behavior of bubble growth and heat transfer. The overall heat transfer performance is primarily governed by the combined effects of evaporation at the TPCL, transient heat conduction from bulk liquid replenishment, and convective heat transfer. The simulation results of bubble dynamics indicate that surfaces M3 and M4, characterized by complete honeycomb pores, restrict bubble growth within the pores, resulting in a reduction in both the maximum TPCL and bubble departure diameter.

All the above-mentioned results can help to better understand the enhancement mechanisms of boiling heat transfer on honeycomb micro-nano porous structured surfaces and offer a promising alternative for the development of highly efficient temperature-to-heat-flux soft sensor techniques applicable to boiling applications and provide a guideline for the optimal fabrication of porous structures.

References

[1] Mehralizadeh A, Shabanian S R and Bakeri G 2020 Effect of modified surfaces on bubble dynamics and pool boiling heat transfer enhancement: a review *Therm. Sci. Eng. Prog.* **15** 100451

[2] Li W X *et al* 2020 Enhancement of nucleate boiling by combining the effects of surface structure and mixed wettability: a lattice Boltzmann study *Appl. Therm. Eng.* **180** 115849

[3] Ateş A *et al* 2023 Pool boiling heat transfer on superhydrophobic, superhydrophilic, and superbiphilic surfaces at atmospheric and sub-atmospheric pressures *Int. J. Heat Mass Transfer* **201** 123582

[4] Guo W, Zeng L and Liu Z 2023 Mechanism of surface wettability of nanostructure morphology enhancing boiling heat transfer: molecular dynamics simulation *Processes* **11** 857

[5] Prakash C G J and Prasanth R 2018 Enhanced boiling heat transfer by nano structured surfaces and nanofluids *Renew. Sustain. Energy Rev.* **82** 4028–43

[6] Fang X D *et al* 2016 Heat transfer and critical heat flux of nanofluid boiling: a comprehensive review *Renew. Sustain. Energy Rev.* **62** 924–40

[7] Li W *et al* 2020 Review of two types of surface modification on pool boiling enhancement: passive and active *Renew. Sustain. Energy Rev.* **130** 109926

[8] Stephan P and Hammer J 1994 A new model for nucleate boiling heat transfer *Heat Mass Transfer* **30** 119–25

[9] Moghaddam S and Kiger K 2009 Physical mechanisms of heat transfer during single bubble nucleate boiling of FC-72 under saturation conditions-II: theoretical analysis *Int. J. Heat Mass Transfer* **52** 1295–303

[10] Sato Y and Niceno B 2015 A depletable micro-layer model for nucleate pool boiling *J. Comput. Phys.* **300** 20–52

[11] Sajjad U, Hussain I and Wang C C 2021 A high-fidelity approach to correlate the nucleate pool boiling data of roughened surfaces *Int. J. Multiphase Flow* **142** 103719

[12] Li J Q *et al* 2019 Ultrascalable three-tier hierarchical nanoengineered surfaces for optimized boiling *ACS Nano* **13** 14080–93

[13] Wang Y Q *et al* 2020 PTFE-modified porous surface: eliminating boiling hysteresis *Int. Commun. Heat Mass Transfer* **111** 104441

[14] Yuan X *et al* 2023 Simulation on pool boiling heat transfer considering the integrated effect of engineered microchannels and mixed wettability *Chem. Eng. Sci.* **280** 119033

[15] Li Y *et al* 2021 Manipulating the heat transfer of pool boiling by tuning the bubble dynamics with mixed wettability surfaces *Int. J. Heat Mass Transfer* **170** 120996

[16] Gong S *et al* 2022 Mesoscopic approach for nanoscale liquid–vapor interfacial statics and dynamics *Int. J. Heat Mass Transfer* **194** 123104

[17] Zhang L N *et al* 2022 Boiling crisis due to bubble interactions *Int. J. Heat Mass Transfer* **182** 121904

[18] Song Y *et al* 2022 Unified descriptor for enhanced critical heat flux during pool boiling of hemi-wicking surfaces *Int. J. Heat Mass Transfer* **183** 122189

[19] Chen G *et al* 2020 Biomimetic structures by leaf vein growth mechanism for pool boiling heat transfer enhancements *Int. J. Heat Mass Transfer* **155** 119699

[20] Song Y *et al* 2022 Three-tier hierarchical structures for extreme pool boiling heat transfer performance *Adv. Mater.* **34** 2200899

[21] Hong M, Mo D C and Heng Y 2024 Bubble dynamics analysis of pool boiling heat transfer with honeycomb micro-nano porous structured surfaces *Int. Commun. Heat Mass Transfer* **152** 107256

[22] Yi T H, Lei Z S and Zhao J F 2019 Numerical investigation of bubble dynamics and heat transfer in subcooling pool boiling under low gravity *Int. J. Heat Mass Transfer* **132** 1176–86

[23] Chen H *et al* 2020 Bubble dynamics and heat transfer characteristics on a micropillar-structured surface with different nucleation site positions *J. Therm. Anal. Calorim.* **141** 447–64

[24] Wang J S, Diao M Z and Liu X L 2019 Numerical simulation of pool boiling with special heated surfaces *Int. J. Heat Mass Transfer* **130** 460–8

[25] Može M and Senegačnik M 2020 Laser-engineered microcavity surfaces with a nanoscale superhydrophobiccoating for extreme boiling performance *ACS Appl. Mater. Interfaces* **12** 24419–31

[26] Wang X *et al* 2023 Achieving robust and enhanced pool boiling heat transfer using micro–nano multiscale structures, *Appl. Therm. Eng.* **227** 120441

[27] Hong M *et al* 2021 Model-based experimental analysis of enhanced boiling heat transfer by micro-nano porous surfaces *Appl. Therm. Eng.* **192** 116809

[28] Wang Y Q *et al* 2020 PTFE modification to enhance boiling performance of porous surface *J. Heat Transf.-Trans. ASME* **142** 074501

[29] Mo D C *et al* 2020 Enhanced pool boiling performance of a porous honeycomb copper surface with radial diameter gradient *Int. J. Heat Mass Transfer* **157** 119867

[30] Robb R A and Ritman E L 1979 High speed synchronous volume computed tomography of the heart *Radiology* **133** 655–61

[31] Van Strahlen S J D 1966 The mechanism of nucleate boiling in pure liquids and in binary mixtures *Int. J. Heat Mass Transfer* **9** 995–1006

[32] Wayner P C, Kao Y K and Lacroix L V 1976 The interline heat transfer coefficient on an evaporating wetting film *Int. J. Heat Mass Transfer* **19** 487–92

[33] Lu S, Heng Y and Mhamdi A 2012 A robust and fast algorithm for three-dimensional transient inverse heat conduction problems *Int. J. Heat Mass Transfer* **55** 7865–72

[34] Luo J *et al* 2019 A novel formulation and sequential solution strategy with time-space adaptive mesh refinement for efficient reconstruction of local boundary heat flux *Int. J. Heat Mass Transfer* **141** 1288–300

[35] Dong L N, Quan X J and Cheng P 2014 An experimental investigation of enhanced pool boiling heat transfer from surfaces with micro/nano-structures *Int. J. Heat Mass Transfer* **71** 189–96

[36] Heng Y *et al* 2008 Reconstruction of local heat fluxes in pool boiling experiments along the entire boiling curve from high resolution transient temperature measurements *Int. J. Heat Mass Transfer* **51** 5072–87

[37] Hirt C W 1981 Volume of fluid (VOF) method for the dynamics of free boundaries *J. Comput. Phys.* **39** 201–25

[38] Lee W H A 1980 Pressure iteration scheme for two-phase flow modeling *Multiphase Transport: Fundamentals, Reactor Safety, Applications Hemisphere* (Washington, DC: Hemisphere) pp 407–31

[39] Tanasawa I 1991 Advances in condensation heat transfer *Advances in Heat Transfer* (New York: Academic)

[40] Tian Y S *et al* 2018 Numerical and experimental investigation of pool boiling on a vertical tube in a confined space *Int. J. Heat Mass Transfer* **122** 1239–54

[41] Brackbill J U, Kothe D B and Zemach C 1992 A continuum method for modeling surface tension *J. Comput. Phys.* **100** 335–54

[42] ANSYS 2019 *ANYSY Fluent Theory Guide and User's Guide* (Canonsburg, PA: ANSYS)

[43] Dhir V K, Warrier G R and Aktinol E 2013 Numerical simulation of pool boiling: a review *ASME J. Heat Transfer-Trans.* **135** 061502

[44] Kim J H *et al* 2015 Microporous coatings to maximize pool boiling heat transfer of saturated R-123 and water *ASME J. Heat Trans.-Trans.* **137** 081501

[45] Li C *et al* 2008 Nanostructured copper interfaces for enhanced boiling *Small* **4** 1084–8

[46] Li S H *et al* 2008 Nature-inspired boiling enhancement by novel nanostructured macroporous surfaces *Adv. Funct. Mater.* **18** 2215–20

[47] Liao J, Mei R W and Klausner J F 2004 The influence of the bulk liquid thermal boundary layer on saturated nucleate boiling *Int. J. Heat Fluid Flow* **25** 196–208

[48] Rohsenow W M and Mikic B B 1969 A new correlation of pool-boiling data including the effect of heating surface characteristics *J. Heat Transfer* **91** 245–50

www.ingramcontent.com/pod-product-compliance
Ingram Content Group UK Ltd.
Pitfield, Milton Keynes, MK11 3LW, UK
UKHW051042120926
8556IPUK00030B/728

9 780750 361729